TIN
STACKERS

TIN STACKERS

THE HISTORY OF THE PITTSBURGH STEAMSHIP COMPANY

Al Miller

WAYNE STATE UNIVERSITY PRESS   DETROIT

GREAT LAKES BOOKS

A complete listing of the books in this series can be found at the back of this volume.

PHILIP P. MASON
Editor
Department of History, Wayne State University

DR. CHARLES K. HYDE
Associate Editor
Department of History, Wayne State University

Copyright © 1999 by Wayne State University Press, Detroit, Michigan 48201.
All rights are reserved.
No part of this book may be reproduced without formal permission.

LIBRARY OF CONGRESS CATALOGING-IN-PUBLICATION DATA

Miller, Al, 1957–
    Tin Stackers : the history of the Pittsburgh Steamship Company / Al Miller.
        p. cm.—(Great Lakes books)
    Includes bibliographical references and index.
    ISBN 0-8143-2832-6 (alk. paper)
    1. Pittsburgh Steamship Company—History. 2. Steamboats—Great Lakes—History. 3. Steamboat lines—Great Lakes—History. 4. Shipping—Great Lakes—History. 5. Steel industry and trade—Great Lakes—History. I. Title. II. Series.
HE945.P79M55   1999
386'.22436—dc21   98-55159

Contents

Acknowledgments   7

List of Terms   9

Introduction   11

CHAPTER 1	Forged of Steel	15
CHAPTER 2	E Pluribus Unum	31
CHAPTER 3	Coulby Comes Aboard	51
CHAPTER 4	The Pace Quickens	85
CHAPTER 5	War and Peace	103
CHAPTER 6	The Roaring Twenties	125
CHAPTER 7	Tin Stackers Go to War	143
CHAPTER 8	A Billion Tons of Ore	163
CHAPTER 9	A New Leader, A New Look	191
CHAPTER 10	Into the Future	213
CHAPTER 11	Modern Times	243

APPENDIX 1	Vessels Built, Purchased, and Acquired	261
APPENDIX 2	Vessels Sold, Scrapped, and Lost	263
APPENDIX 3	Vessels of the Pittsburgh Steamship Company	267
APPENDIX 4	Original Vessels of the Pittsburgh Steamship Company	321

Notes   323

Bibliography   335

Index   339

ACKNOWLEDGMENTS

No one can write a book of this nature without the help of many people. It is time for me to offer my heartfelt thanks to all those who assisted me.

I do not exaggerate when I say this book would not have been possible without the generous assistance of the former Great Lakes Fleet sailors, shoreside staff, and managers who agreed to be interviewed. Bill Buhrmann, Ralph Bertz, and Bruce Liberty provided invaluable information and insights into the fleet's operations. Captain Guido Gulder spent an afternoon using his remarkable memory to recount the days of the Great Depression. Edmund Siegrist and Wendell Barrow described shipboard life for young men during World War II, and Christian Beukema helped me better understand the intricacies of the extended navigation season. Worthy of special mention is the late Captain Bill Wilson, who welcomed me aboard the *Edgar B. Speer* in 1986 and, a decade later, into his home while he so vividly described the life of a Great Lakes shipmaster.

I must also express my gratitude to the people who run the museums and libraries that hold so much information about the Great Lakes. Foremost among them is C. Patrick Labadie, director of the Lake Superior Maritime Visitors Center operated by the U.S. Army Corps of Engineers in Duluth. This book was his idea, and he has always been helpful in answering my questions, locating hopelessly obscure bits of information, and offering suggestions on how to find even more. Another person deserving special mention is Patricia Maus of the Northeast Minnesota Historical Center in Duluth, who helped me find many key pieces of information and photographs for this story and others. My thanks also go to the helpful professionals at the Center for Archival Collections at Bowling Green State University; the Great Lakes Historical Society in Vermilion, Ohio; University of Detroit Mercy

Archives; Library of Congress; National Archives of Canada; and the Duluth Public Library. In addition to providing information, many of these institutions permitted me to use photographs from their collections in this book.

Special thanks are due the people who have shared their love of ships with me. They include my friends Wes Harkins, an outstanding Great Lakes photographer, writer and historian; Bob Ashenmacher, who has talked our way aboard many ships over the years; and Dan Krummes, an author and librarian who always keeps an eye out for information and obscure publications to send my way.

Also deserving recognition are Doris Sampson, who allowed me to reproduce photos taken by the late Howard Weis, and the people at USS Great Lakes Fleet in Duluth and at Transtar Inc. in Monroeville, Pennsylvania, who took time to answer my questions and permitted me to quote from company documents and reproduce company photographs for this book.

Many other people—too numerous to mention here—over the past sixteen years have patiently shown me around their ships, taconite plants, and ore docks. They endured my questions, shared their stories, and added greatly to my knowledge. Although they probably never realized it, their help has proven invaluable time and again.

Finally, I want to thank my loving wife, Janie, who cheerfully endured the many hours I spent at the keyboard and offered her honest critiques of my work.

Al Miller
Duluth, Minnesota
June 1998

TERMS

Readers unfamiliar with Great Lakes shipping should know several terms to better understand the text.

First, much has been written about whether Great Lakes vessels are properly called *ships* or *boats*. While *boat* is a commonly used term to describe lake freighters, my interviews and research showed company employees and documents most often referred to their vessels as *ships*. For the sake of clarity, I have used *ships* throughout this book except where *boats* was used in a direct quote.

Second, nautical tradition provides distinct circumstances when a ship must be referred to as *she, he,* or *it*. Again, for clarity I refer to all ships as *it*.

Helpful terms to understand:

aft	toward the rear of a ship
amidships	middle portion of a ship
ballast	water pumped into tanks in the bottom or sides of a ship to add stability
beam	a ship's width as measured at its widest point
bow	a ship's forward area
draft	depth at which a ship sits in the water
gross ton	2,240 pounds
port	left side of a ship when viewed from the stern looking toward the bow
starboard	right side of a ship when viewed from the stern looking toward the bow
stern	rear portion of a ship
wheel	propeller

Introduction

The first rays of daylight streamed over the horizon as the freighter *Joshua A. Hatfield* slowly steamed down Lake Superior laden with iron ore wrested from the rich mines of Minnesota's Mesabi Range. Ahead, distant smudges of smoke marked the locations of the *Thomas F. Cole*, *Henry H. Rogers*, and *William Edenborn*—all fleetmates of the *Hatfield* and all pursuing the same course with the same cargo. A few hours earlier, the *Robert C. Stanley*, *Homer D. Williams*, and *D. G. Kerr* had quietly slipped past headed in the opposite direction, their cargo holds empty, their destinations one of the ore docks in the busy Minnesota ports of Duluth or Two Harbors. These seven ships—all bearing distinctive silver smokestacks—were among fifty-nine ore freighters plying the Great Lakes on June 28, 1955, belonging to the Pittsburgh Steamship Division of United States Steel Corporation. Scattered from Duluth to Conneaut, Ohio, these ships—known as "Tin Stackers," the "Pittsburgh fleet," or simply "Pittsburghers"—were busy carrying the thousands of tons of raw materials needed each day to fuel the nation's biggest producer of steel.

Also on Lake Superior that morning, other ships from the Pittsburgh fleet were already in port. The steamers *William G. Clyde* and *Francis E. House* lay alongside massive ore docks in Duluth as dockworkers lowered long, narrow chutes that quickly spewed tons of ore into the cargo holds of the waiting vessels. On the decks of both vessels, their first mates carefully directed the loading while a few other men stood by to run the winches and handle the lines that pulled the ships along the dock face to additional spouts ready to deliver more iron ore. Thirty miles to the north, the scene was repeated in Two Harbors, where the *Thomas W. Lamont*, *Eugene W. Pargny*, and *William P. Palmer* took on their loads of ore.

Down on Lake Michigan, ten Tin Stackers ranged from the Straits of Mackinac to Chicago. *August Ziesing* and *William J. Olcott* unloaded their ore cargoes for U.S. Steel's South Works in South Chicago while *Eugene P. Thomas* and *Peter A. B. Widener* did the same at the corporation's sprawling, smoking steel mill in Gary, Indiana. Under way on Lake Michigan that day were ships such as the *Arthur M. Anderson,* built just three years earlier, and the *William E. Corey,* built a half century earlier. On Lake Huron and in the swift waters of the Detroit River, eight vessels were downbound with ore for U.S. Steel's Ohio ports of Lorain and Conneaut. More ships were already unloading in Lorain or upbound on Lake Erie. Six more carefully followed the winding St. Marys River that links Lake Superior to the lower lakes. Throughout the day, these vessels would pass the twin cities of Sault Ste. Marie, Michigan, and Sault Ste. Marie, Ontario, known collectively as the Soo, and enter the U.S. government's Soo Locks to overcome the twenty-one-foot height difference between Lake Superior and Lake Huron. Of all these ships, a few carried coal to be made into coke or limestone to be used in making steel, but most were hurrying to their destinations assigned only to carry iron ore for the ravenous blast furnaces of U.S. Steel.

Great Lakes ships have always been special vessels. Most of those used during the 1950s were long and narrow, with their pilothouses far forward and their engines and cabins at the stern. In between was a deck free of obstructions that gave dockside unloading gear quick and easy access to the cargo. Called "straight-deckers," they were designed to efficiently carry iron ore, coal, limestone, and other bulk cargoes to the mills, factories, and construction yards that powered the industrialized hearts of the United States and Canada. In 1955, hundreds of ships of various sizes and descriptions sailed the lakes for U.S. and Canadian fleets. Standing out from all of them were the ships of the Pittsburgh Steamship Division.

The fleet was born in 1899 as the Pittsburgh Steamship Company. It was a small fleet, built to serve one of the steel companies that ambitious men were building in Middle America. From the ranks of these steelmakers emerged a few who would shape the nation's industrial future. Among them were little-known men like Isaac Ellwood and John Gates, who manufactured barbed wire, and a small-town lawyer named Elbert Gary. There also were rich and famous men, like steelmaker Andrew Carnegie, oil magnate John D. Rockefeller, and financier J. Pierpont Morgan. Their collective ambitions in 1901 produced the United States Steel Corporation. It was an industrial empire that spanned the nation, making everything from the nails used to build houses to the steel beams that shaped modern skyscrapers. From this "Steel Trust" emerged a new Pittsburgh Steamship Company, a fleet of ships with red hulls and silver smokestacks dedicated to carrying the million tons of iron ore that U.S. Steel's mills would require each year from the mines of Minnesota, Michigan, and Wisconsin.

The Pittsburgh Steamship Company was as grand as the steel company that spawned it. Assembled from several smaller fleets, the Pittsburgh fleet initially was made up of 112 ships and barges. Just the size of the new fleet made it noteworthy. Dwarfing other fleets on the Great Lakes and around the world, it boasted more ships than the U.S. Navy. Along with being the biggest fleet on the lakes, it was arguably the most progressive. By 1906 it had produced the first class of ships to reach 600 feet in length—a size that would remain the industry standard for thirty-five years. More innovations followed in vessel design, propulsion, and management. As just one link in the production chain for U.S. Steel, the fleet stressed cost controls and modern management techniques to be used aboard ships where captains were accustomed to the traditions of an earlier time. The fleet became an industry leader in stressing safety aboard ship. Following World War II, it was in the vanguard of vessel operators testing and installing radar aboard lake freighters.

To lead its armada, U.S. Steel chose men whose abilities were as broad as the fleet they guided. First to command was Augustus B. Wolvin, a former sailor and produce merchant who built fleets and revolutionized vessel design. In his wake came Harry Coulby, who ruled the Pittsburgh Steamship Company for twenty years and became the greatest fleet manager on the lakes. Becoming head of the Pittsburgh fleet meant more than just assuming a major role in U.S. Steel. The position automatically made a man one of the most important people in the Lake Carriers' Association, a powerful trade group that influenced Great Lakes shipping in everything from the placement of lighthouses to labor relations. Indeed, it was in resisting the efforts of labor unions to organize Great Lakes fleets that Pittsburgh Steamship Company took on one of its biggest roles in the Lake Carriers' Association.

Along with U.S. Steel, the Pittsburgh Steamship Company survived and prospered throughout the tumultuous twentieth century. It endured the economic downturn following World War I and struggled during the Great Depression of the 1930s. Its enormous capacity enabled the fleet to meet the industrial demands of armed conflicts from World War I to Vietnam. Just weeks after Allied forces invaded France in 1944, a hurricane struck the crowded Normandy beachhead, ravaging ships and supplies. Surveying the carnage, General Dwight Eisenhower marveled that America's industrial production allowed his army to shrug off a setback that would have been a disaster in other wars. The backbone of that industrial production was America's steel industry, and the steel industry depended on iron ore from the Great Lakes—much of it carried by the Pittsburgh Steamship Company.

America's steel industry prospered well into the second half of the century, but by the 1980s it faced a new kind of challenge as overcapacity, aging mills, declining demand, and aggressive foreign competitors sent the industry into a tailspin. Across the Midwest, mills closed and jobs disappeared. The fallout quickly reached the Great Lakes. Virtually every fleet in the United States and Canada scrapped its older, less efficient vessels, and converted most of those remaining to self-unloaders. Only by having the capability of discharging their own cargoes instead of using costly shoreside unloading rigs could Great Lakes vessels hope to remain competitive. The decade's economic tumult ended one of the Pittsburgh fleet's long-standing marks of distinction: after ruthlessly culling its older vessels, it no longer was the biggest American fleet on the Great Lakes. More significantly, the same decade brought to an end its ownership by U.S. Steel. Changing economics meant big companies no longer were enamored of the concept of vertical integration. U.S. Steel spun off its railroads and fleets into independent companies. The famed Steel Trust fleet became a common carrier dedicated to making a profit rather than solely serving U.S. Steel.

Today, a century after its formation, the Pittsburgh Steamship Company sails on as USS Great Lakes Fleet Inc. Its eleven vessels range in size from 1,000-foot giants dedicated to carrying taconite pellets to nimble 600-footers capable of squirming through narrow channels carrying a wide range of cargoes. The fleet maintains a special relationship with USX, as U.S. Steel is now known, but it also courts small customers that the old Pittsburgh fleet would have ignored. The fleet has evolved from a "link in the chain" of a steelmaking giant to an independent company in the efficiency-crazed 1990s. In many ways, it exemplifies the course American industry has taken in the twentieth century, and its story offers the clearest glimpse into the romance and hard-nosed reality of Great Lakes shipping.

1

Forged of Steel

The United States in the 1880s was slowly transforming itself from a rural, agricultural nation to an urban, industrial nation. As this new industrial nation emerged, its metal of choice increasingly was steel. Iron had been mankind's most useful metal for centuries. As the industrial age progressed, iron gradually replaced wood as the primary material for manufacturing tools, farm implements, household goods, railroad cars, bridges, and ships. Now steel was moving to the forefront. Stronger and more versatile than iron, steel was becoming the metal that would carry the United States into a new age of great cities, international power, and seemingly boundless wealth.

The key ingredient of steel is iron ore. From it, men can make iron. From iron, they can make steel. Iron ore was not difficult to find in the United States. Sir Walter Raleigh reported finding it in North Carolina as early as 1588. Other deposits had been mined around the country to fill the needs of small furnace companies that made iron. The greatest ore deposits, however, were discovered in upper Michigan in 1844. Knowing they had something of value, men traveled to the remote land to wrest the ore from the ground. It was a slow process. The land was harsh and surrendered its bounty reluctantly. Mining methods were crude. Transportation was sporadic at first and always difficult. The only thing that made mining in Michigan successful was the relatively cheap transportation available by loading iron ore onto boats in Marquette and shipping it down the Great Lakes to mills in Ohio.

Following the discovery of iron ore in Michigan, more people flowed into the region. The land around the upper Great Lakes—first Michigan, then Wisconsin and Minnesota—yielded more great deposits of iron ore. Some of it was hard, gray rock. Other ores were soft and red. More ore deposits were being discovered all the time, and the possibilities seemed almost endless.

Iron mining quickly developed into a robust industry in the vast northern lands. In 1855 locks were built at Sault Ste. Marie, Michigan, to enable ships to bypass the rapids in the St. Marys River, which links Lake Superior to the lower Great Lakes. Soon after the locks opened, the brig *Columbia* passed through downbound with 132 tons of iron ore piled on its deck. More vessels soon followed, and shipowners began building steamships designed specifically to carry iron ore. The first was *R.J. Hackett,* a 225-foot ship built in 1869 which departed from standard designs by placing the pilothouse far forward and the engine and cabins at the stern. Thirteen years later the *Onoko* used the same design but was built of iron. That vessel was eclipsed in 1886 by the *Spokane,* which translated the design into steel.

No less important than the evolution of ships was the development of dockside equipment that could efficiently unload the vessels. For years, dockworkers had struggled to unload iron ore by laboriously shoveling it into buckets that were lifted off the ship and swung ashore by mules or small steam hoists. It was not until 1882 that an enterprising inventor named Alexander E. Brown erected a mechanical unloading rig on his father's dock in Cleveland that could speedily hoist ore from a ship's hold and carry it to a storage facility ashore. Brown's innovation was timely, for the steel industry continued to expand in the 1880s. More mills were being built and more mines being developed to serve them. As demand grew so did the iron ore industry. In 1880 miners began working amid the swamps and mosquitoes of northern Minnesota to pull ore from the newly opened Vermilion Range.[1]

By the mid-1880s, fortunes were being made in steel. Mills were growing as manufacturers found more uses for steel and falling costs made the metal's use more economical. To control their costs and ensure themselves of needed materials and transportation, some steelmakers began buying their own mines and ships. The Mutual Steamship Company was a four-ship fleet established to carry iron ore mined in Michigan's Menominee Range and shipped out of Escanaba. In 1889 a wealthy Milwaukee man named Ferdinand Schlesinger and his business partners formed the Menominee Transit Company strictly to transport ore from the Chapin Mine on the Menominee Range.[2]

As the 1890s began, the steel industry was booming, carrying with it the mining and shipping industries. Around the country, ambitious men were setting up small mills to make steel in all its useful shapes: rails, nails, wire, pipes, plates, girders—the list was almost endless. Steelmaking was especially robust in the Midwest. In Illinois, Marshall Field and H. H. Porter of Chicago and Nathanial Thayer of Boston

Small steelmakers in the 1890s scrambled to assemble their own fleets of ore carriers. Among them were Marshall Field and H. H. Porter, who acquired the Mutual Steamship Company, including the steamer *Corona*, to carry ore to the mills of their Illinois Steel Company. Courtesy Northeast Minnesota Historical Center.

owned the Illinois Steel Company. To supply their mills they acquired the Mutual Steamship Company and its four vessels, *Cambria, Coralia, Corona,* and *Corsica.* Just west of Chicago, Isaac Ellwood, one of the inventors of barbed wire, decided to gamble and expand his stormy business relationship with John W. Gates, who had sold barbed wire for Ellwood before forming his own cutthroat wire company. Pooling their resources with another wire producer, Eugene Buffington, they opened the American Steel and Wire Company and made Gates its president. For legal services they retained Elbert H. Gary, a prosperous attorney from Warrenville, Illinois, who had served two terms as a county judge and was a bright star in Chicago legal circles.[3]

But the greatest player in the steel industry was, beyond a doubt, Andrew Carnegie. The son of poor Scottish immigrants, Carnegie had settled in Pittsburgh in 1848. His first job was as a bobbin boy in a Pennsylvania cotton mill earning $1.20 a week. After working for a time as a telegraph messenger, he learned the telegrapher's trade and went to work for the Pennsylvania Railroad, eventually becoming private secretary and telegrapher to a top company official. He advanced steadily until he was appointed superintendent of the railroad's Pittsburgh Division. His early investments in telegraph companies and sleeping car manufacturers proved fruitful, and later speculation in Pennsylvania oil lands made him wealthy.

After the Civil War, Carnegie left the Pennsylvania Railroad to devote himself to managing his investments. His financial interests varied widely, and included a foundry and a company that built iron railroad bridges. Carnegie initially thought his future lay in building bridges, but increasingly he was drawn to the foundry. In the early 1870s he began experimenting with the new Bessemer method of making steel. This process involved blowing cold air through the molten iron, producing heat so intense that it burned away carbon and silicon impurities. Using the Bessemer process, it suddenly was possible to make steel as cheaply as iron.

However, the Bessemer process had an Achilles' heel—phosphorus. No matter how steelmakers manipulated their furnaces, they could not burn away phosphorus, an impurity that rendered steel hopelessly brittle. European manufacturers were having trouble finding iron ore that lacked phosphorus. In the United States, the established ore fields of Pennsylvania produced ore laden with phosphorus, but the relatively new ore fields in upper Michigan—now the largest in the country—held mountains of ore almost entirely free of this impurity. In 1872 Carnegie traveled to England, where he first saw giant Bessemer plants producing steel on a grand scale. Discerning the profits that could be gained through mass production, he turned away from his other investments and devoted himself to making steel using the Bessemer process and Michigan ore. He organized a company to build the Edgar Thomson Steel Works in Braddock, Pennsylvania. In 1883 Carnegie purchased the rival Pittsburgh Bessemer Steel Company at nearby Homestead, Pennsylvania, and six years later added the Duquesne Steel Works just five miles up the Monongahela River from the Homestead Works. His four blast furnaces and three steelworks were all within a few miles of Pittsburgh, making it the center of the Carnegie Steel Company and the burgeoning steel industry. By 1890 Carnegie was among the wealthiest men in America.[4]

Assisting Carnegie in running his empire was Charles M. Schwab, who lived his own "up by the bootstraps" success story. As a boy he had worked in his father's stable near Carnegie's summer cottage and had frequently seen the tycoon traveling about the region. On one occasion he even held the reins of Carnegie's carriage horse. At seventeen he went to work in Carnegie's Edgar Thomson Works driving stakes into the ground for a dollar a day.[5] Ambition and intelligence won him promotions at an astounding pace. By age twenty-five,

Schwab was running the Homestead Works. By the time he reached thirty-five, he was running Carnegie's business interests.

Carnegie was a shrewd and relentless businessman, constantly fighting for every advantage he could get. He was a fanatic about controlling the costs of production, and had been ever since he first entered the iron business, maintaining that profits resulted from holding down costs, not from raising prices. As part of his drive to control costs, Carnegie began to take interest in the possibility of vertically integrating his company. During the 1880s he had dominated the business of converting raw materials into steel. Now, as the 1890s began, he wanted to also control the steps that came before and after the making of steel: acquisition of his raw materials, and conversion of his raw steel into finished steel products.

By now the nation's steel companies were locked in an ever-accelerating race to expand, consolidate, and dominate. Companies opened new mills and then slashed prices to undercut their competitors and gain a foothold in the market. Steelmakers fought to make sure they were not outmaneuvered by other companies and left with devalued or worthless stock. Adding to the frenzy, a giant new Minnesota iron range, the Mesabi, was discovered in 1890. It promised to be the biggest and richest ore deposit yet. Two years later the range's first mine, the Mountain Iron Mine, shipped 2,100 tons of ore to Superior, Wisconsin. There it was loaded aboard a barge prosaically named 102, one of the distinctive whaleback vessels built in Superior by Alexander McDougall and his American Steel Barge Company.[6] A year later, ten mines were operating on the Mesabi Range.

Then lightning struck. A financial panic wracked the nation in 1893. Hundreds of banks failed and fifteen thousand private lending houses went under.[7] Trade collapsed, credit dried up, consumer demand fell. Labor and social unrest spread through the cities. In the mining industry, small-time operators were hit hard as demand for ore plummeted and their bank credit evaporated. Out of this chaos strode John D. Rockefeller, the man who had amassed wealth of legendary proportions through his Standard Oil Company. In December 1892, Rockefeller had purchased $500,000 worth of railroad bonds from Duluth's Merritt brothers—the "Seven Iron Men" who were among the original developers of the Mesabi Range. In the Panic of 1893, the Merritts appealed to Rockefeller for direct financial support of their Lake Superior Consolidated Iron Mines. The oil magnate invested another $2 million in the Merritts' six mines, their railroad running from the Mesabi Range to Duluth, and their ore dock on the city's waterfront. It was not enough. In January 1894 several of the Merritts sold their ownership in the company to Rockefeller for ten dollars a share. For about $900,000, Rockefeller had snatched up a mining company that some later estimated was worth as much as $333 million. At about the same time he had invested in the Merritts' venture, Rockefeller also had put nearly $1 million into the American Steel Barge Company. Now these relatively minor investments had suddenly put him in position to play a significant role in the steel industry.[8]

No one really understands why Rockefeller moved into the iron ore business. In recent years he had devoted less time to Standard Oil, and the men with whom he formed the company had drifted away to pursue other interests. A compulsive worker and builder, he simply may have needed a new challenge. Regardless of the reason, Rockefeller made a decision that frightened all Midwestern steelmakers: "We had decided to mine, ship and market the ore," he said. Rockefeller began pumping millions of dollars into his new Lake Superior Consolidated Iron Mines Company. No stranger to the transportation business, he soon realized that he needed his own fleet to ensure that his ore was shipped profitably.[9]

The United States struggled for the next two years to emerge from the depression. As the economy picked up, so did the steel industry. The trend toward consolidation continued. Steelmakers wanted bigger mills so they could dominate their market and make more money through economies of scale. To control their costs and ensure a steady supply of materials, it became more important than ever for steel companies to possess their own mines and ships. Rockefeller understood this and took action to ensure that he remained a key player. He contacted Cleveland's Samuel Mather, ore merchant and ship operator, and asked him to arrange for construction of twelve ships. Concerned that shipbuilders around the lakes would raise their prices on such a large order, Mather conferred privately with each builder and left the impression that Rockefeller's order was only for one or two vessels. Eager to get work during such hard times, each shipyard carefully kept its bid as low as possible. In December 1895, Mather invited the shipbuilders to Cleveland for the bid opening. One by one, they were called into Mather's office and told they had won the contract. Each emerged triumphant, ready to console his competitors. Only after comparing notes did they discover that every shipyard had won contracts. The shipbuilders sheepishly realized that Rockefeller and Mather had manipulated them into keeping their bids low.

Despite their owners' chagrin, shipyards around the lakes were soon at work building Rockefeller a fleet of modern steel ore carriers

John D. Rockefeller entered the iron ore business in 1893 and soon began building the biggest and finest ships on the lakes for his Bessemer Steamship Company. Bessemer vessels were easily identified by the large letter "B" that adorned their smokestacks. Courtesy Northeast Minnesota Historical Center.

for his new Bessemer Steamship Company. Each was more than 400 feet long and bore the name of a famous inventor, such as *Sir Henry Bessemer, George Stephenson,* and *James Watt.* The same year, Ellwood, Gates, and Buffington established a lake fleet called American Steamship Company to serve their American Steel and Wire Company. To run the fleet they selected Augustus B. Wolvin, an ambitious vessel agent in Duluth. Through his separate Zenith Transit Company, which was closely allied with American Steel and Wire, Wolvin let contracts to build the steamers *Zenith City* and *Queen City.*[10]

Rockefeller vigorously pursued his new business. His agents were wheeling and dealing to find long-term buyers for ore from his Mesabi mines. The Bessemer fleet was rapidly becoming the biggest on the Great Lakes, and Frederick T. Gates, Rockefeller's chief adviser, was chartering more ships so that Rockefeller interests could control the price everyone had to pay to move their ore. Speculation was rampant in the press and in the boardrooms that Rockefeller would begin building steel mills in a bid to dominate the steel industry the way he dominated oil. Rockefeller said nothing to discourage such talk.

Carefully watching Rockefeller's actions was Andrew Carnegie. Following discovery of the Mesabi Range, Carnegie had hesitated to buy into any mining ventures despite the proddings of his younger assistants. But after visiting the region and seeing how cheaply ore could be scooped from the ground, he reluctantly agreed to join Henry W. Oliver in an effort to acquire ore properties. Oliver and Carnegie were old acquaintances, having once worked together as messenger boys in Pittsburgh's first telegraph office. In the intervening years, Oliver had built and squandered half a dozen small fortunes, a fact that made Carnegie regard him with suspicion. Now, Oliver's interest was focused on the Mesabi Range.

Oliver already held a financial interest in several mining ventures. As Rockefeller's company gained strength after the Panic of 1893, Oliver realized that an alliance between the two titans held the most promise for all parties involved. He approached Gates and began the delicate negotiations necessary for Carnegie's steel mills to buy ore from Rockefeller's mines. In December 1896, Carnegie agreed to lease and mine Rockefeller's ore deposits for the favorable rate of twenty-five cents a ton. In return, Rockefeller's ships would carry all of Carnegie's Mesabi ore to Cleveland. Rockefeller also agreed not to enter the steelmaking business—in fact, he had never intended to do so. Carnegie and Oliver promised that their Oliver Iron Mining Company would not buy any ships or any mines on the Mesabi Range.

The Rockefeller-Carnegie pact soon revealed its limitations. Within a year of signing the leases, Carnegie Steel was shipping more than 3 million tons of ore from the Mesabi Range. Rockefeller's ships could only carry half that amount, so Carnegie was forced to seek out other shipowners, all of whom charged far more than the generous terms set by Rockefeller. About the same time, Oliver finally persuaded Carnegie to purchase his own supply of iron ore—the Norrie mines in Michigan's Gogebic Range. Although he had dragged his feet for years, the Norrie purchase seemed to arouse a voracious appetite in Carnegie. He and Oliver began snapping up other mines in Michigan and on the Minnesota ranges. Among their purchases were several mines on the Mesabi Range—a move that violated the agreement with Rockefeller. Oliver was unabashed, and fell into wrangling with Gates and Rockefeller.

Once he began integrating his company's operations, Carnegie was more receptive to Oliver's advice that he acquire his own fleet of ore ships. It was another way he could control his production costs. By now, however, it was nearly too late. When Oliver began casting about for ships, few existing vessels were available for sale, and most shipyards were busy building Rockefeller ships.[11]

Oliver was undeterred. While Carnegie chartered vessels to carry his ore in 1899, Oliver scoured the lakes for ships. He made a pitch to buy the ten whaleback steamers and twenty whaleback barges owned by the American Steel Barge Company. Before Oliver could close the deal, Rockefeller's men got wind of it and paid $5 million to get the whalebacks for the Bessemer fleet.[12] Oliver then tried to buy a controlling interest in the venerable Mitchell fleet of Cleveland. Again, Rockefeller's men thwarted him by chartering the fleet's vessels. The Rockefeller interests were so eager to block out Carnegie that they agreed to pay Mitchell $1.25 a ton to carry ore when the previous year's rate had been just 60 cents.[13] These two moves by Rockefeller meant the Bessemer Steamship Company owned or had charters on sixty-seven ore carriers for the 1900 navigation season. With that many vessels under his control, Rockefeller could easily influence the rates others had to pay to ship their ore. "This move on the part of Rockefeller's representatives is prompted, of course, by the fear of ships not being found to care for all the ore that will be required in 1900," the authoritative *Marine Review* noted, adding, "but the immediate cause of it was a struggle between Rockefeller and Carnegie interests for supremacy in their dealings with each other."[14]

Eventually, Oliver's efforts began to bear fruit. In October 1899 he bought the Lake Superior Iron Company, a pioneering enterprise that controlled rich ore lands in Michigan's Upper Peninsula. A crucial part of the deal was the company's four relatively new ships: the steamers *Griffin, Joliet, LaSalle,* and *Wawatam.* Late in the year Oliver also picked up the new steamer *Clarence A. Black* from the Northern Lakes Steamship Company. From C. W. Elphicke and Company he purchased the steamer *William R. Linn* and barge *Carrington.* Oliver combined the Lake Superior Iron Company's ships with the *Black, Linn,* and *Carrington* and on November 6, 1899, chartered the Pittsburgh Steamship Company. The new company, which issued $4 million in bonds on January 1, 1900, was given a separate corporate identity but was controlled by the Oliver Iron Mining Company and the Carnegie Steel Company. Oliver took the role of president and selected Edwin S. Mills to be general manager in charge of daily operations in Cleveland.[15]

As the century neared its end, Carnegie finally had achieved the vertical integration that was the goal of so many steel producers. He consolidated his holdings and put it all under direction of Charles Schwab. His Pittsburgh Steamship Company began carrying ore from Carnegie mines to Conneaut, Ohio, which was known as "the Carnegie port."[16] In 1892 this little harbor east of Cleveland handled its first ton of iron ore. In the spring of 1898, Carnegie Steel purchased all the land around the harbor and spent $250,000 to modernize its docks and equipment. By the following year the port was handling 2.5 million tons of ore a year. From Conneaut the ore was carried to Carnegie's steel mills in Pittsburgh on his Pittsburgh, Bessemer and Lake Erie Railroad.

Through Oliver's persistence, the Pittsburgh fleet continued to grow. A year after forming the fleet, he finally was able to place construction orders for five large steamers and a barge to be rushed to completion using steel made in Carnegie mills. The new steamers were giants—more than 450 feet long and 50 feet abeam. Each was named for a famous university: *Cornell, Lafayette, Harvard, Princeton,* and *Rensselaer.* The barge, also large at 400 feet in length, was named *Bryn Mawr* for the prestigious women's college. Although these new ships were part of the Pittsburgh Steamship Company, sailors referred to them as "the college line." Like the other Pittsburgh vessels, their hulls were painted red and their cabins white. Each ship's smokestack bore a large P that clearly proclaimed the fleet's identity to passing vessels.

Now the final act began to play out in one of the nation's great industrial dramas. While Carnegie and Oliver had focused their attention on Rockefeller in 1898, their smaller competitors had moved to consolidate and expand their own operations. That year, Marshall Field, H. H. Porter, Nathanial Thayer, and other principals of the Illinois Steel Company asked Elbert Gary to attend their board meeting in New York. By all accounts, Gary was an able man who lived the pious Methodist lifestyle taught him by his parents. He never touched cards and preferred attending church events to other, more worldly forms of entertainment. He firmly believed in free enterprise and the healthy rigors of competition among businesses. Since becoming counsel to Illinois Steel nearly a decade earlier, Gary had carefully educated himself until he was an expert on every phase of the iron and steel industry. Like other knowledgeable steelmakers, he began to believe the nation and its steel industry were growing at such a rapid pace that only giant, integrated companies could operate efficiently. Anything less would result in a quilt-work of small, squabbling companies always building more mills in an attempt to undercut their rivals. To Gary, this was "wasteful competition."

The principals of Illinois Steel had asked Gary's advice on whether to buy a railroad running between Chicago and the company's mill in nearby Joliet. Gary advised them to look beyond a simple rail link. To truly be competitive, he said, they needed to make the company a "rounded proposition" by expanding its range of steelmaking operations. At the board meeting in New York, Gary was called upon to explain a suggestion he had made earlier that Illinois Steel merge with the Minnesota Iron Company and the National Tube Company. A combination of these companies made sense, Gary argued, because they complemented each other. Illinois Steel had mills in North Chicago, South Chicago, and Joliet, and National Tube produced tube and pipe at its plant in Lorain, Ohio. Key to the deal, however, was the Minnesota Iron Company. This firm traced its heritage to 1872, when woodsman George Stuntz found an outcropping of iron ore near Vermilion Lake in northeastern Minnesota. His discovery led to development of the Vermilion Range, and his efforts to exploit it eventually involved Charlemagne Tower, James Pickands, Samuel Mather, and Jay C. Morse—all pioneers in iron mining. They formed the Minnesota Iron Company in 1880 and five years later shipped their first ore through the newly opened port of Two Harbors, Minnesota. In another five years they had added the Minnesota Steamship Company, which grew to include twenty-two ships and barges operating under the astute management of Pickands and Mather. After listening to Gary's arguments concerning these three companies, the board members of Illinois Steel agreed that his proposal made sense. They adopted it on one condition: J. Pierpont Morgan had to finance the deal.[17]

In 1898, the name J. P. Morgan was synonymous with money. Just three years earlier he had almost single-handedly saved the United States from defaulting on its financial obligations when a run on gold threatened to empty the U.S. Treasury of its reserves. By selling gold to the Treasury and engineering a plan to discourage the purchase of U.S. gold by European buyers, Morgan enabled the government to save its reserves. As a financier, Morgan had profited on the bailout, and because of that many people reviled him. Like Gary and many others in the upper echelons of business, Morgan disliked price-cutting and destructive competition. He believed in "rationalizing" competition by combining smaller companies to form large, efficient producers.

Gary approached Morgan about the merger, and the financier agreed to underwrite it. The two worked closely together for three months before finalizing arrangements that combined Illinois Steel, National Tube, and the Minnesota Iron Company to form the Federal Steel Company. The new combination was capitalized at $200 million in stock. It was an integrated company that possessed a variety of mills, rich iron mines, a large Great Lakes fleet, and the Duluth and Iron Range Railroad for hauling ore from the mines to the five ore docks in Two Harbors. Before closing the deal, Morgan insisted that Gary be named president of the new company. Gary resisted because he did not want to give up his lucrative law practice, which brought in about $75,000 a year—a huge sum in those days—but Morgan dismissed his arguments and offered to let Gary name his own salary and his own executive board. After a day of careful deliberation, Gary signed a three-year contract to run Federal Steel from its headquarters in New York.

Creation of Federal Steel set off a new wave of mergers throughout the steel industry. In 1899 alone, smaller manufacturers merged to create the National Steel Company, American Bridge Company, American Steel Hoop Company, and the American Sheet Steel Company. One group of enterprising men snapped up 265 tin plate mills and put them under a single ownership that was quickly dubbed "the tin plate trust." During this time Federal Steel did well, but it was not achieving the results that Gary and the others had anticipated.

Andrew Carnegie regarded his growing competition with contempt. The creation of Federal Steel had caught him off guard, seeming to come out of nowhere to assume a position as a serious rival to the Carnegie Steel Company. But Carnegie's steel empire was poised on the brink of truly legendary achievements. He had recently ousted a partner he disliked, gaining a greater share of Carnegie Steel's ownership. He was contemplating an alliance with major railroads that in the next five to ten years could crush Federal Steel and his other major competitors. Yet at the same time, Carnegie was thinking about stepping away from his life's work. He and his wife had just spent eighteen months overseas, and Carnegie marveled at how much he enjoyed the respite. Also, he now was sixty-five years old. His advisers said Carnegie Steel could dominate the industry in another decade, but at his age Carnegie did not know how much more time he would have to spend pushing his company to ever greater profits. Some biographers have said that Carnegie was unwilling to risk his investment in an industry where he no longer was the unchallenged ruler. Others maintain a simpler view, saying he wanted to retire so he could pursue his growing interest in philanthropy. Whatever the reason, the greatest steelmaker of all time decided it was time to cash in his chips. He made it known to his top men that he would entertain offers to buy the Carnegie Steel Company.[18]

Carnegie's decision to retire was not entirely unexpected. Not long after Federal Steel was created, he had sent an emissary to Elbert Gary to propose a merger. After some discussion, they failed to reach an agreement and dropped the matter. Carnegie then approached John D. Rockefeller. His asking price for Carnegie Steel was $250 million. Rockefeller turned him down.

Carnegie then decided to take the offensive. He trumpeted plans for a series of new ventures in steel. He let contracts for a $12 million tube plant to be built on five thousand acres near Lorain. He announced that he would build a giant wire mill in Pittsburgh. Finally, he unveiled plans to build a railroad from Lake Erie to the Atlantic coast—a clear attack on J. P. Morgan, who held a considerable financial interest in the Pennsylvania Railroad. The news threw the steel industry into turmoil. Carnegie's projects, if carried out, would mean even more competition and more price-cutting. As the century drew to a close, the value of some very big investments was in jeopardy. Frightened men in the steel industry began clamoring for somebody—namely, J. P. Morgan—to stop Carnegie.[19]

Shortly after assuming his new job as head of Carnegie Steel Company, Charles Schwab approached Elbert Gary. He bluntly told Gary that Federal Steel was not the fully integrated producer it was intended to be. Furthermore, he suggested that Federal buy the Carnegie properties. This time Gary was receptive to the idea. He took it to Morgan, who would again be needed to finance the buyout. Morgan balked. "I would not think of it," he told Gary. "I don't believe I could raise the money."[20]

Morgan was still convinced that the giant merger was impossible when he happened to attend a dinner on December 12, 1900, given by two New York financiers in honor of Charles Schwab. After the meal, Schwab was called upon to speak. Carnegie had already left for home, but Schwab knew Morgan was still present. Now was his chance. For the next half hour Schwab proceeded to lay out his vision for the steel industry, putting into words what already was becoming apparent to all in the room. Instead of having one mill producing dozens of products, Schwab said, steelmakers should have dozens of mills, each dedicated to making a single product. The economies of scale in such a grand scheme would result in monumental savings in production costs. More rational location of plants would cut transportation costs, which then made up a third of the cost of steel. To accomplish this, Schwab continued, no existing manufacturer was big enough. The only way to do it would be to form a corporation larger than any in existence. This giant would be completely integrated, owning ore mines, coalfields, railroads, docks, ships, and all manner of steel mills and finishing plants.[21]

Morgan was intrigued. Afterward, he pulled Schwab aside and the two talked for more than an hour. A couple weeks later, Morgan invited Schwab to his home, where they met in the library for serious talks. Schwab again laid out his vision, this time with more details about costs and production figures. The meeting began at 9 P.M. and lasted until dawn. At this point, Morgan supposedly said, "Well, if Andy wants to sell, I'll buy. Go and find his price." While the quote may be apocryphal, the results of the meeting were not. Schwab, who had not informed Carnegie of his talks with Morgan, gingerly broached the subject with his boss. Carnegie listened quietly, then agreed to consider the offer overnight. When Schwab called at Carnegie's house the next day, Carnegie took out a pencil and a scrap of paper. He jotted down some figures showing his company's worth. They totaled $480 million. Schwab took the paper to Morgan's office on Wall Street. The financier examined it and said, "I accept this price."

The deal set off a frenzy of activity. Morgan laid out his plan to Gary and asked him to analyze it and determine whether it would work. Gary took the idea and expanded it even further. Using Federal Steel as a nucleus, he added Carnegie's properties along with those of half a dozen other merger combinations. Morgan approved Gary's plan. For the next month, Morgan's office throbbed with activity as arrangements were made to pull the deal together. Word leaked out that something big was happening. Newspapers picked up the story, and people across the country waited anxiously to see what the giant new trust would do.[22]

Finally, on February 25, 1901, incorporation papers were filed in Trenton, New Jersey, for the United States Steel Corporation. Charles Schwab was named as president and Elbert Gary as chairman of the board. The new industrial colossus dwarfed anything the world had ever seen. Using a complex web of stock buyouts and exchanges, it combined Federal Steel, Carnegie Steel Company, American Steel and Wire, National Tube, National Steel, American Tin Plate, American Steel Hoop, and American Sheet Steel. Each of these companies, in turn, brought with it dozens of mills and subsidiaries that made everything imaginable out of steel: beams, rails, rods, billets, pipe, nails, barbed wire, binder ties, galvanized sheet steel, slabs, and axles. Its mills were spread across the nation in Illinois, Ohio, Michigan, Indiana, Pennsylvania, Washington, Kansas, California, Massachusetts, New York, Maryland, and West Virginia. Also part of the deal were iron ore mines around Lake Superior, coalfields in the South, and limestone quarries in the Midwest. The corporation's sweeping charter said it would deal in raw materials, make iron and steel, and build buildings, machinery, and ships.[23] In March, U.S. Steel announced plans to sell $850 million in stock to finance its operations. More was sold a month later, making U.S. Steel the much-ballyhooed "Billion Dollar Corporation." To cynical newspapermen and the public, however, the new company became known simply as the "Steel Trust."

The biggest surprise emerging from the corporation's formation was that it did not include the mines and ships owned by John D. Rockefeller. As the new company came together, Gary became anxious. "We ought to have the Rockefeller ores," he told Morgan.[24] The financier resisted, largely because he detested Rockefeller. Nonetheless, Morgan soon agreed to personally approach the oil magnate to propose a deal. By mid-March, Lake Superior Consolidated Iron Mines belonged to U.S. Steel.[25] Rockefeller had invested $40 million in his mines and steamships, and he sold them to the corporation for $90 million.[26] Of that amount, $8.5 million went toward the stock of the Bessemer Steamship Company.[27]

Just as U.S. Steel now boasted a vast empire of mills and mines, it also lay claim to the country's largest collection of ships under one owner. The original merger brought together 56 steel ships and barges. This included the Pittsburgh Steamship Company, with 13 vessels; Minnesota Steamship Company, 22 vessels; American Steamship Company, 12 vessels; Menominee Transit Company, 5 vessels; and Mutual Steamship Company, 4 vessels. The purchase of Lake Superior Consolidated Iron Mines brought into the fold another 56 ships and barges from the Bessemer Steamship Company, which included the whalebacks of the American Steel Barge Company. All told, the corporation owned 69 ships and 43 barges, for a total of 112 vessels. Various tugs, fireboats, and other small craft added another 19 vessels to the total.[28] It was a daunting figure for U.S. Steel's competitors and the independent lake fleets. Virtually overnight a vast new rival had taken shape that many vesselmen feared would dominate ore traffic on the Great Lakes. "Greater than the number of vessels in commission in the combined fleets of the United States Navy, larger than the invincible armada sent against England by Philip of Spain, is the colossal force of vessels gathered together on the Great Lakes to carry iron ore for the United States Steel Corporation, generally known as the Steel Trust," a writer for Beeson's *Marine Directory* commented. "The uniting of the fleets has brought about a most alarming condition of affairs on the lakes. Vessel owners fear the trust is so preponderous as to regulate the rate of all freight traffic and that it will be impossible for them to compete against such an overwhelming foe. Their only hope at present is to rely on coal and grain and other commodities to furnish cargoes for their vessels."[29]

While vesselmen were wringing their hands over the new giant, the men organizing U.S. Steel had problems of their own. They had to take 112 disparate vessels from six fleets and mold them into a single, efficient organization. So far U.S. Steel had a lot of ships, but no fleet.

2

E Pluribus Unum

By the time the United States Steel Corporation was formed in 1901, the promise of spring was already teasing the Great Lakes. February's bitter cold had passed. The gradual warming of March was softening ice that held ships fast in ports from Duluth to Cleveland. With only two months left before the start of the Great Lakes navigation season, the men running U.S. Steel had to quickly find someone who could perform the labors of a nautical Hercules. The chosen man had to combine six fleets of vastly different sizes with vastly different ships into a single organization. Office staffs had to be merged and vessel masters and sailors brought under a single management. Purchases of everything from food to bunker coal had to be consolidated. Vessel dispatch needed to be centralized so the new company's 112 ships could load ore and coal and deliver it to hungry steel mills on time and at reasonable cost. It promised to be a formidable task, and it had to be done fast. The steel mills could not wait.

Then, as now, the fraternity of Great Lakes shipowners and operators was tightly knit. Most of these "vesselmen" knew each other personally. Anyone they had not met, they certainly knew by reputation. So the job of finding a man to run U.S. Steel's fleet focused on a relatively small pool of candidates who were known to all. Those in charge of the search soon set their sights on one man: Augustus B. Wolvin of Duluth.

Augustus Benjamin Wolvin was born in Cleveland, Ohio, on October 16, 1857, to Captain Benjamin S. Wolvin and Finetta Harrington Wolvin. The family moved to Chicago seven years later. Then, in 1871, they moved again to Pecatonica, Illinois. Contemporary biographies offer no reason for the move, but it is possible the family was among the 98,000 people left homeless by the Great Chicago Fire.

Benjamin Wolvin was a vessel master on the Great Lakes, which was a reputable middle-class profession. He worked for the Winslow fleet and over the years commanded the steamers *Dean Richmond* and

Augustus B. Wolvin of Duluth was chosen to manage the Pittsburgh Steamship Company after the formation of United States Steel Corporation. This photo likely was taken some years after he left the fleet at the end of 1903. Courtesy Northeast Minnesota Historical Center.

City of Rome. Reared in a maritime family, it was only natural that Gus Wolvin would be drawn to the Great Lakes. He went to work surprisingly young. Accounts differ on whether he was ten or twelve when he went to work as a cabin boy aboard his father's ship. He did well under his father's tutelage, working on the boats when they sailed from May to December, and studying during the winters. As he grew older, he began spending the winters studying in Chicago.[1]

Over the course of ten years, Wolvin steadily rose through the ranks of young men working for the Winslow fleet. In 1879, at age twenty-two, he was appointed master of the steamer *Annie Smith.* Over the next few years he commanded the *Raleigh,* the *V. Swain,* and his father's old ship the *City of Rome.* As an intelligent young man advancing in his chosen career, he undoubtedly made an attractive bachelor. That status did not last long. In January 1880, Wolvin married Carrie Kilgore, a young woman living in Pecatonica.

Wolvin's career as a vessel master was destined to be a short one. His mother and father died when he was twenty-six. Their passing left his younger brothers and sisters without anyone to care for them. Rather than abandon them to the orphanage, the promising young mariner decided to leave the lakes in 1883 and return to Pecatonica with his wife. For the next six years, he labored in the little farm town close to the Wisconsin border, working as a produce merchant during the day and overseeing the rearing of his younger siblings by night. By all accounts, he was as successful in business as he was at navigating ships.

When his siblings were old enough to care for themselves, Wolvin left his produce business to return to the Great Lakes. But rather than renewing his master's license so he could sail again, he and Carrie moved to Duluth, a rapidly growing city at the western tip of Lake Superior. The growth of the mining and shipping industries meant things were happening in Duluth. It was a place where an ambitious man who knew something about steamboats could do well for himself. Not long after arriving in town in 1888, Wolvin met Fred LaSalle, a coal merchant who was a partner in the small vessel agency of Horton and LaSalle. In January 1889, LaSalle dissolved his partnership and joined forces with Wolvin. The name of their business, LaSalle and Wolvin, became widely known around shipping circles.[2]

As a vessel agent, Wolvin's job was to act as representative for his clients' steamboats whenever they were in port. That meant ensuring that the ships called at the right docks and arranging delivery of food and other supplies. After a little more than five years, Wolvin dissolved his partnership with LaSalle to handle his own list of clients, which included the prominent Western Transit Company. Being a vessel agent was turning out to be good work for Wolvin, but he was learning that it was an even better way to meet important people.

As the 1890s brought about the formation of many new mining, shipping, and steelmaking companies, it also brought an opportunity that would change Gus Wolvin's life. In 1895, while the country was still recovering from the financial panic of two years earlier, the American Steel and Wire Company and its representative, Cleveland attorney James H. Hoyt, contacted Wolvin. The company was looking for someone who could make arrangements to get ships built and then manage them as a fleet. They wanted Wolvin to handle the job. He agreed to do it.

Wolvin set out to assemble the fleet, and began by purchasing the steamer *W. H. Gilbert* for the new Empire Transportation Company, of which he was president and Hoyt was secretary.[3] He then organized the

Zenith Transit Company in 1895 and immediately set about getting ships built for the new fleet. Here, for the first time, he exhibited a trait that would become characteristic of his work over the next decade. Wolvin wanted ships that were innovative. Zenith Transit's first ship, the steamer *Zenith City*, was built that year in South Chicago. At 387 feet in length, it was the longest on the lakes, along with its sister ship *Victory*, which was built for the Pickands and Mather partnership.[4] The steamer *Queen City*, built the following year, was even bigger, measuring 401 feet long. *Empire City* and *Crescent City* followed in 1897, each bigger than the one before it. The new fleet's biggest ship was *Superior City*, a 429-foot giant built in 1898 and the first vessel produced at the Cleveland Ship Building Company's new shipyard in Lorain. Wolvin served as president and general manager of Zenith Transit and ran the fleet from his office in Duluth. The fleet's financial backers were known in the trade press as "the Wolvin syndicate." Wolvin was paid for running the fleet and also received an ownership share in each of the new vessels. His financial worth began to grow, along with his reputation.[5]

Following formation of Zenith Transit Company, everything began going Wolvin's way—and at an increasingly rapid pace. He kept a relatively low profile in Duluth, but he was becoming one of its most prominent men in the maritime industry. He formed another important contact by becoming the Duluth agent for Pickands-Mather and Company and handling local matters for the American Steel Barge Company's whaleback fleet. One year after setting up Zenith Transit Company, Wolvin joined a group of well-to-do local men in rescuing the bankrupt Inman Towing Company. The Inman line was owned by the colorful Byron B. Inman, who operated a large fleet of tugboats in Duluth and neighboring Superior, Wisconsin. Inman was an astute tugman but a terrible businessman. All his tugs were mortgaged, he bought coal and other operating supplies on credit, and he lived beyond his means. The Panic of 1893, followed by two vicious rate wars with competing tug lines, torpedoed his shaky finances and sent him into receivership in 1896. That's when Wolvin and the others stepped in. First, they paid Inman's debts to the coal companies that supplied his tugboats with bunker fuel. Then they bought the best of his tugs at a federal marshal's auction, blocking out potential competitors from other ports. When the first ships arrived in Duluth for the 1897 navigation season, the tug line was running again under Inman's management.

Rescuing Inman ensured that vessels calling at Duluth would have reliable and cheap tug service, but Wolvin was not doing this as a

An icy *Queen City* and another Tin Stacker wait for the Poe Lock to raise them to the level of Lake Superior during a late-season trip through the Soo. Construction of this large lock in 1896 helped spur the movement of iron ore as well as the building of bigger ships. Library of Congress.

matter of civic duty. For many years vessel owners had grown increasingly frustrated with the delays and high costs their ships encountered at ports around the lakes because of underpowered tugs, strikes by tug crewmen, and hefty increases in towing fees. To solve these problems, the owners quietly organized their own "tug trust" and named it Great Lakes Towing Company. They wanted one company—under their control—to provide consistent, reliable towing services on all the lakes. Among those involved in the venture was Wolvin. Acting as an agent for this powerful group, he discreetly began buying tugs and docks and holding them as trustee for the new company. When Great Lakes Towing was formally organized in 1899, it became clear that one of Wolvin's first moves had been to purchase a controlling interest in the Inman tugs. Inman bitterly resented Wolvin's secret maneuvers. For his efforts, Wolvin earned a seat on Great Lakes Towing's board of directors and the title of second vice president.[6]

At the same time he was helping to form Great Lakes Towing, Wolvin was busy wheeling and dealing for more ships to feed the ore frenzy gripping the lakes. In March 1899 he sold the steamers *Texas* and *Pennsylvania,* still in the shipyard under construction, to Federal Steel for $275,000 each, which was well over the price he had paid to have them built.[7] Federal renamed the ships *Malietoa* and *Mataafa* and added them to its Minnesota Steamship Company. Three months later, Wolvin ordered four ships to be built—two in Lorain and two in West Bay City, Michigan. Wolvin handled the arrangements as trustee for the American Steel and Wire Company. Three of the new ships would be named for principals in the wire company: *William Edenborn, Isaac L. Ellwood,* and *John W. Gates.* The fourth was named *James J. Hill* in honor of the Minnesota railroad magnate who also was involved in Minnesota ore properties.[8] Again, these ships were giants: 478 feet long at the keel and nearly 500 feet overall, earning them the title of the first 500-footers. From their launching in 1900 until 1904, they would be the biggest ships on the Great Lakes.

Impressed by what they saw, others sought Wolvin's assistance. In the fall of 1899 he placed orders for construction of two "Welland Canal size steamers" on behalf of a syndicate of Cleveland and Buffalo men planning to build grain elevators in Montreal.[9] Wolvin's final move of this busy year was made on November 18, when he signed over the vessels of Zenith Transit Company to a new fleet named the American Steamship Company. The transfer put all the American Steel and Wire Company ships under single ownership. On the bills of sale for each ship, Wolvin was listed as president of Zenith Transit and managing owner of the vessels. Also added to the American Steamship fleet were the four ships under construction for the wire company, the *W. H. Gilbert,* and two other new ships, *William P. Palmer* and *A. B. Wolvin,* both of which were small vessels measuring 242 feet long.[10] In reporting the transaction, the authoritative *Marine Review* remarked prophetically that "the transportation of ore, alike to other branches of the iron industry, is to be eventually controlled by the combinations to a far greater extent than was expected a short time ago."[11]

By 1900 Wolvin was a prominent man in Great Lakes shipping. He was general manager and a vice president of American Steamship Company. He had been named president of Superior Ship Building Company, which operated the shipyard in Superior where Alexander McDougall had built his whalebacks and where conventional ships were now taking shape. His fellow vesselmen had selected him to serve on an important committee for the Lake Carriers' Association, the trade group representing Great Lakes vessel owners. The committee was charged with solving the vexing problem of getting reliable, economical labor to unload grain from ships calling at the forest of grain elevators in Buffalo, New York. On his own behalf, Wolvin was working with James C. Wallace, general manager of American Ship Building Company, and W. L. Brown, the company's president, to form the International Navigation Company. They planned to build half a dozen small ships, called "canallers," for service on the St. Lawrence River with its many canals and locks.[12] All this undoubtedly was lucrative to Wolvin, and he spread the largesse to his family, employing two young men who apparently were his nephews: John W. Wolvin and Roy M. Wolvin. It proved to be an instructive experience for the latter, who in years to come would play a key role in forming another giant fleet: Canada Steamship Lines.[13]

As more money came in and more business contacts were made, Gus Wolvin's stature among vesselmen continued to grow. In January 1901, members of the Lake Carriers' Association elected Wolvin president. His nomination was unanimous. "The election of A. B. Wolvin to the presidency of the Lake Carriers' Association places at its head one of the most successful men now on the lakes," the *Marine Record* noted. "Captain Wolvin is a son of a sailor man whose life was spent on the lakes. He lived on the water as a boy and became a sailor in early years. . . . In later years he has been known as a leading member of the lakes fraternity, has become wealthy and is popular with his associates on the lakes as a 'good fellow' at all times, as well as a shrewd business man."[14] A subsequent edition of the newspaper described Wolvin as "one of the most enterprising and resourceful men in the lake transportation business," and added that his reputation was one of "aggressiveness combined with fairness to all interests involved."[15] Writers at the *Marine Review* agreed with the glowing assessments of Wolvin. "He is a young man with a vast experience, a combination of energy and judgment, which is exactly what the far-reaching scope of the office needs."[16]

Experience, energy, and judgment: these were qualities that Wolvin had amply demonstrated in a relatively short time. They also happened to be qualities that made him an ideal choice for running the U.S. Steel fleet. Given his high profile and his role with the American Steel and Wire Company's fleet, it was only natural that his name would come up as a candidate for the job. The Steel Corporation was formed in mid-February, and by mid-March it had acquired John D. Rockefeller's vast fleet. Not long after that, the new corporation offered Wolvin the job of general manager of its fleet.

Surprisingly, it was not an easy decision for Wolvin to make. True, the U.S. Steel fleet represented a great challenge, and running it promised ample rewards. He would be well paid. The size of the fleet meant he likely would become the most influential man in Great Lakes shipping. Yet it was a troubling proposition. In all likelihood, the fleet would be headquartered in Cleveland, the heart of the shipping industry. That would drag Wolvin away from Duluth, where he had a fine new home, his vessel agency, his shipbuilding deals, and a new fleet he was starting for the Peavey grain company in Minneapolis. In the middle of April, while Wolvin was in Cleveland on business, he hinted publicly that he would take the new job "under certain circumstances"— meaning he would do it if he could run the fleet from Duluth and maintain his other businesses. U.S. Steel balked at the notion, so Wolvin turned down the offer. He had lucrative business commitments in Duluth, and that was where he and his wife wanted to stay.[17]

Just days after Wolvin declined the job, U.S. Steel appointed James Gayley of the Carnegie Steel Company to be the Steel Corporation's vice president in charge of mines, docks, and ships. Surprising everyone, Gayley immediately contacted Wolvin in Cleveland and told him he could run the Steel Trust fleet from Duluth. Wolvin accepted. He was now general manager of the biggest fleet under the U.S. flag.

Wolvin returned to Duluth on May 2. His first day back he was swamped with telephone calls and visits from well-wishers congratulating him on his new position and praising him for bringing the new fleet's office to Duluth. The new boss wasted no time basking in the limelight. He quickly set about moving his offices to the fourth floor of Duluth's new Board of Trade Building at First Street and Third Avenue West.[18] He also picked out the key men who would help him shape the new fleet. They were all experienced hands in the vessel business, and some had worked with Wolvin previously or for his rivals. W. W. Smith was named marine superintendent. He had held the same job with Pickands-Mather and Company when that organization ran Rockefeller's whaleback fleet. Joseph Hayes was selected as chief engineer, transferring over from the same role with American Steamship Company. Wolvin also brought along a young man from Cleveland named Allyn F. Harvey to be his principal assistant.

One of the first things Wolvin and his staff had to do was select a name for the fleet. Most newspapermen and people on the street were already calling it the "Steel Trust Fleet" because U.S. Steel was commonly known as the Steel Trust. While that name worked in newspapers and saloons, it struck a sensitive note in U.S. Steel headquarters during an era when trust-busting was very much in political vogue. Wolvin announced their choice to the Duluth press on May 1. "The name of the steamship company which will operate the boats is the Pittsburgh Steamship Company," he said. "We shall take the charter of the present company of that name, and all the boats taken over by the steel company will be operated under it."[19]

The Pittsburgh Steamship Company was set up as a subsidiary of the U.S. Steel Corporation. Its mission was clear and simple. It would serve as part of a conveyor belt to move iron ore and coal from shipping ports around the Great Lakes to receiving ports near the steel mills. It was a "cost company" rather than a profit-making venture. The fleet was not expected to make money; its operation was part of the cost of making steel. That cost would be figured into the final price of the steel products turned out by U.S. Steel mills. In the future, the Pittsburgh Steamship Company, along with the rest of the Steel Corporation's subsidiaries, would develop a strong culture of identifying and controlling costs. That culture, sometimes bordering on obsession, would pay handsome long-term dividends.[20]

To be president of the new subsidiary, Gayley named Daniel M. Clemson, another executive from the Carnegie Steel Company. Along with its headquarters in Duluth, the fleet would have another large office in Cleveland. Wolvin and his office crew in Duluth would set policy and oversee the fleet's work. The Cleveland office would handle day-to-day operations, including vessel dispatch, distribution of ore to mills, and allotment of ore to docks. To run the Cleveland office, the Steel Corporation chose Edwin S. Mills to be the fleet's assistant general manager.[21] Mills was another young man who had enjoyed a meteoric rise in the vessel business. He joined the Carnegie Steel Company in the late 1890s as a sales agent for ore. When the company began moving a lot of its own ore, he played a big role in buying and building the original Pittsburgh Steamship Company, eventually becoming its general manager. "There is probably not in all the United States a man of his age entrusted with anything like the important responsibilities that he takes up with the opening of lake navigation," the *Marine Review* observed.[22]

Mills was not alone in shouldering a lot of responsibility as a young man. U.S. Steel's top ranks were filled with relatively young men, starting with corporation chief Charles Schwab. The philosophy of youth extended down to subsidiaries like the Pittsburgh fleet. "The one thing noticeable in the giant organization is the young blood in it—that is young blood in its working force," the *Marine Review* continued. "Schwab, Gayley . . . Wolvin, Mills. There is method in it. Young blood means enthusiasm and energy. It means to dare. It does not mean as might be implied a lack of conservatism. It is going to move along carefully considered lines with well-matured plans, but it is going to move them along with irresistable force."[23]

Now the real work began. In both Duluth and Cleveland, a legion of clerks, dispatchers, accountants, office boys, and other men from the various fleets had to be brought into place to handle the new fleet's business. Legalities over ownership still had to be cleared up. Ships from the Bessemer fleet were officially sold to the Pittsburgh Steamship Company on May 17, followed six days later by the vessels of the Menominee and Mutual fleets. Enrollments for the ships from the American Steamship Company were not transferred until June 10. A dispatching system had to be set up to get the ships to the right docks for the right cargoes. Supplies had to be ordered and a distribution system set up. Forms and letterhead had to be printed. Captains, mates, and engineers had to be summoned and told when to report to their ships. A payroll system had to be arranged with forms and cash distributed through the fleet. Ships had to be located, their condition assessed, and preparations made to get them under way.

The ships and barges from each of the six fleets were painted in varying color schemes. Records are scarce on the matter, but as vessels came under formal ownership of the Pittsburgh Steamship Company their hulls were painted green, their cabins "straw yellow," and their smokestacks silver. The work was done as the ships unloaded at Lake Erie ports. Many had to get under way before the painting was finished, however, and for much of the summer Steel Trust ships sailed up and down the lakes with their paint jobs in varying stages of completion.

Wolvin was busy getting his arms around his vast fleet. He was familiar with the American Steamship Company and undoubtedly liked what he had there. The *Ellwood, Edenborn, Gates,* and *Hill* were the biggest ships on the lakes, and *Superior City, Empire City,* and the others were not far behind. The original Pittsburgh Steamship Company also contributed some fine ships, including those of the "college line." But it also was saddled with smaller vessels, such as the *Griffin,* only a decade old and, at 266 feet, only half the length of the newest ships. The Mutual Line was a mixed bag. *Cambria,* for instance, was built in 1887 and only 280 feet long, while *Coralia* was only a few years old

The *Robert Fulton* steams out of the Poe Lock at the Soo upbound for Lake Superior. This early photograph shows the steamer with the Pittsburgh fleet's original all-silver smokestack—and the sooty rim clearly shows why a black cap was added to the stacks in 1905. Library of Congress.

and more than 400 feet long. He could not be as charitable in assessing the Menominee Transit Company. All five ships were built in the early 1890s and none surpassed 300 feet in length. The Minnesota Steamship Company also offered a lot of variety. Its twenty-two vessels ranged from the *Manola*, a 282-footer built in 1890, to the new *Malietoa*, built in 1900 and 454 feet long. Unlike the other fleets, its ranks were heavily weighted with barges—ten to be exact. Steel barges had been built in great numbers between 1895 and 1901, a carryover from the old days when wooden schooners were shorn of their sails and rigging and turned into "tow barges" to be pulled behind steamers. At one time, shipowners thought building these steel "consorts" was an efficient way to increase the tonnage their small steamers could move during the season. The rapid growth in the size of ships, however, had rendered consorts obsolete by the turn of the century. The biggest and most diverse of the original fleets was the Bessemer Steamship Company. Of its fifty-six vessels, fully thirty-one were barges. Most of the barges were whalebacks, which already were considered small, awkward, and inefficient. The steamers were mostly whalebacks and, again, mostly small. The only bright spot there was a smattering of big, new steamers such as *Sir Henry Bessemer*, *Robert Fulton*, and *Sir William Fairbairn*, all more than 400 feet long.

Combined, the vessels of the Pittsburgh Steamship Company could carry about 10 million tons of ore in the eight-month shipping season. But U.S. Steel needed 13 million tons of ore each year. So before his first ship sailed, Wolvin already was behind in the tonnage race. To make up the deficit, he contracted with just about anyone who had available ships: Pickands-Mather, James and John Corrigan, Harvey H. Brown, C. W. Elphicke, W. C. Richardson, John Mitchell, James Davidson, Wilson Transit Company, International Steamship Company, Robert F. Rhodes, J. C. Gilchrist, and William Livingstone. The contracts quelled fears among independent fleet owners that the new giant would monopolize the ore trade. Suddenly, there seemed to be enough tonnage for everyone. The independent owners dropped plans to form their own trade group to battle the Pittsburgh fleet.[24]

Now, with the start of the 1901 navigation season imminent, another problem arose. The turbulent labor situation that had existed in Great Lakes shipping for the past decade came to a boil again with creation of the Pittsburgh fleet. Marine engineers, who were the officers in charge of the ships' engine rooms, threatened a strike against all fleets if the owners did not agree to hire more men for the engine room crews. After several days of threats, a strike was called against the Pittsburgh Steamship Company, with the engineers demanding that an additional assistant engineer be hired on all ships.

The walkout lasted about two weeks. During that time, Wolvin was able to find at least a few engineers willing to cross the picket lines and go to work. On May 9, 1901, the steamer *Charles R. Van Hise* pulled beneath the chutes at the Duluth, Missabe and Northern Railway ore dock in Duluth. The 458-foot ship, under command of Captain W. H. Campau, slipped out Duluth's ship canal without fanfare the next day, becoming the first vessel to get under way for the new Pittsburgh Steamship Company. A new era in Great Lakes shipping had begun.[25] Four days later, Wolvin worked out a compromise with the union, agreeing to add extra engineers on nine ships with water tube boilers and an engine room handyman on every steamer with four boilers. By May 16 the fleet's first upbound ships and barges reached Duluth, with *Masaba, John A. Roebling, LaSalle,* and *Bryn Mawr* going under the ore dock chutes to load.[26]

Even as the first vessels of the Pittsburgh fleet made their way down the lakes, Wolvin managed to find time to keep his own ventures going. On May 17 he was present for the launching of the *Frank Peavey,* the first of four 450-foot steamers he was assembling for the new Peavey Steamship Company. Frank Peavey was a wealthy Minneapolis grain merchant and chief stockholder in the fleet bearing his name, and Wolvin was a stockholder in the fleet as well as its manager.[27] As the *Peavey* tasted water for the first time, Wolvin and his business associates were pondering a deal to sell their steamers *Paraguay* and *Asuncion*, which had been built on the lakes for the saltwater trade. Wolvin, James Hoyt, and other men in the Wolvin syndicate also were putting together a proposal to present to Canada's minister of public works. The group planned to ask for a parcel of prime waterfront property in Port Colborne, Ontario, and government financial help in exchange for the syndicate's building a grain elevator and ten ships for service on the St. Lawrence River.[28] No one at U.S. Steel seemed to mind Wolvin's preoccupation with other affairs. Before May ended, he was named a vice president of the Steel Corporation.

Controlling 112 ships certainly gave the Pittsburgh fleet advantages over its competitors. For instance, there was flexibility in scheduling vessels. If one dock was crowded, a ship could be rerouted to another. If a vessel was unable to arrive in time for a cargo, another could be diverted to take its place. But having far more ships than anyone else also meant that the fleet's front office had far more headaches than anyone else. The new fleet quickly flooded the lower lake ore docks with cargo. "The capacity of the iron ore docks seems to be the only limit to the amount of iron ore now being sent forward," a Duluth newspaper observed on June 25. "It looks as if the steel trust managers are frightened over the prospect of an old-fashioned boom in lake freights next fall when wheat and corn begins to come down from the Northwest and West and they are now putting out every ton that can be handled over their docks."[29]

The busy pace and everyday perils of the Great Lakes began taking their toll. By late June, the steamer *Harvard* and whaleback barges 130 and 133 were dry-docked at the shipyard in Superior undergoing repairs. The *Isaac L. Ellwood* had smacked the barge *Constitution* at the Soo, and the master of the *Samuel F. B. Morse* had been fined $200 for passing another vessel illegally in the St. Marys River. July was no better. One of the fleet's big steamers was dry-docked for ten days after striking Ballard's Reef. No sooner was it repaired than it ran aground at Lake Erie's Pelee Passage. Then the barge *Maia* was just outside Two Harbors when it was brushed by a steamer from another fleet.[30]

And not only ships got into trouble. In September a dispute between the master and some sailors aboard the *Morse* nearly resulted in violence. The men had signed on for a full trip, but when the ship tied up at the ore dock in Ashland, Wisconsin, a number of them promptly quit. This was common practice in the fall. Drifters would sign on for unskilled jobs such as deckhand, fireman, or coal passer. When they reached a northern port, they would demand their wages and head into the forests to get jobs for the winter in the logging camps. The unfortunate shipmaster then had to scrounge for men before his vessel got under way again. In an attempt to curtail the practice, the *Morse*'s master refused to pay the quitters. The would-be lumberjacks gathered at the captain's office in the forward end of the ship to demand their pay. Harsh words were exchanged and mayhem threatened before they finally got their money.[31]

While it was difficult to keep unskilled crewmen on the ship, many vessel masters found it even harder to keep them sober. Captain Byron B. Inman, the Duluth tugman who had been bought out by Wolvin in 1897, was hired by him in 1901 to serve as a relief captain in the Pittsburgh fleet. If a ship's master became ill or was temporarily excused from duty, Inman filled the role until the master's return. That's how he came to be master of the steamer *German* as it was docked in Buffalo in October. "I arrived here last evening at 7 P.M., took a tug and went to the *German* . . . found her old Master had left and about half of the remainder of the crew drunk," he wrote his wife.[32]

There also were the long-smoldering labor problems. Seeing the engineers get satisfaction in May, firemen in many of the lake fleets demanded pay raises. Officials of Pittsburgh Steamship Company called together leaders of the other fleets and granted the men raises that paid them $45 a month before September 1 and $52.50 a month during the more hazardous autumn months. Not to be outdone, deckhands demanded and got the same pay. Then on November 1, they demanded $60 a month and got it.[33]

But just as sailors could cause trouble, they also could be heroes. On October 2 the Pittsburgh steamer *Crescent City* was plowing through rough seas off Vermilion Point in eastern Lake Superior when the pilothouse watch spotted the old steamer *M. M. Drake* and its barge *Michigan* wallowing in the waves. The barge had started to fill with water, and its crew signaled the *Drake* for help. When the *Drake* came alongside to rescue them, it was walloped by the tossing barge and disabled. A passing steamer came to the *Drake*'s aid and took off four men. Then the *Crescent City* came upon the scene and moved in to snatch the rest of the crew shortly before the *Drake* and the *Michigan* dove for the bottom. In recognition of the crew's bravery, Wolvin sent reward money to the *Crescent City*'s master to be divided among the men.[34]

Despite its woes, the Pittsburgh Steamship Company's first season seemed to be a success. Great Lakes ships moved 20 million tons of ore that season, with the Pittsburgh fleet accounting for about half that amount. Another 3.5 million tons moved to U.S. Steel mills on ships chartered from other fleets. The fleet continued to strive for efficiency by using its biggest ships to their full advantage in the ore trade. Late in the season, the *William Edenborn* set a record for the biggest ore cargo ever when it carried 8,348 tons from Two Harbors to Conneaut.[35]

The 1902 shipping season opened with promise. Ice melted quickly across the lakes, enabling ships to get an early start amid predictions of a huge leap in demand for ore. The Pittsburgh fleet got under way without the labor problems that had troubled it the previous year. As the vessels began moving, Wolvin's men again faced mishaps that were common in turn-of-the-century lake shipping. *Crescent City* and its consort, whaleback barge 130, rumbled over Lake Superior's Au Sable Reef in April but escaped with little damage. A compass error was blamed for a June 2 accident that left *Mataafa* briefly stranded on Knife Island reef, again on Lake Superior.

The fleet's fortunes took a turn for the worse when it suffered its first major vessel loss on June 7. The whaleback steamer *Thomas Wilson* was riding low in the water as it steamed out of Duluth's ship canal laden with ore. Less than half a mile away, the wooden steamer *George G. Hadley* was approaching the canal when it received orders from a passing tug to switch its destination to the Superior, Wisconsin, entry several miles away. Captain Mike Fitzgerald of the *Hadley*, apparently unable to see the rapidly approaching whaleback from his position on deck, immediately ordered a sharp turn to port. Aboard the *Wilson*, Captain M. C. Cameron was faced with instantly choosing between the lesser of two looming evils: he could turn left and run aground, possibly tearing out the bottom of his ship, or he could swing to the right and hope he turned tightly enough to avoid the *Hadley* as it turned in the same direction. He choose to turn right. Tragically, the *Wilson* could not make a tight enough turn. Within moments the *Hadley* knifed into the whaleback's steel hull amid the groan of timbers and the shriek of bending, ripping steel. Loaded whalebacks rode so low in the water that they resembled modern-day submarines running on the surface. With such little freeboard on the *Wilson*, it was inevitable the *Hadley*'s penetration would be deep and fatal.

The impact of the collision rolled the *Wilson* sharply to one side. It slid ahead a short ways, then stopped and began sinking by the bow.

Throughout the ship, startled men scrambled up ladders and bolted through hatchways as they frantically tried to escape the doomed ship. Within three minutes, the *Wilson*'s stern rose in the air and slid beneath the surface, less than a mile off the ship canal. There was no time to launch lifeboats or rafts. Men bobbed in the cold water, clutching at debris or pulling on life jackets tossed to them by sailors on the *Hadley*. Eleven men were rescued, but nine others perished in the cold water or drowned inside the ship.[36]

Wolvin had lost his first ship literally within sight of his office. It was in relatively shallow water, and he considered raising it or at least recovering some of its ore cargo. Divers hired to explore the wreck reported a deep V-shaped gash in the hull just forward of the deck turret that housed the boiler room. The decks were intact, but air trapped inside the sinking ship had blown off the wooden cabins. Their verdict: the ship could not be refloated.[37] They also were unable to find any of the missing men. After several weeks of effort, Wolvin gave up. "Divers report that the water is murky, even clear to the bottom, which would indicate that the undertow is deep, and has perhaps carried the bodies away from the vessel," he said. "It now looks as though the only thing to do is to abandon all. The cargo is hopeless, and so is the boat."[38]

The fleet's second vessel loss quickly followed the first, and again it was a whaleback. As the season neared its end, the 430-foot steamer *Maunaloa* was about thirty miles off Vermilion Point in eastern Lake Superior with barge 129 in tow. The ore-laden vessels were laboring through heavy seas when the towline parted. The steamer swung around in a circle and came in close to the barge to take a new towline—a task that was hazardous but routine. This time, however, it did not work. Waves slapped the vessels together, and the steamer's anchor ripped a gash in the barge's hull. Crewmen aboard the barge raced belowdecks and reported rapid flooding. The *Maunaloa* came alongside the barge again, this time to take off its crew of half a dozen men. The barge was soon gone. Its loss was put at $80,000.[39]

The 1902 navigation season had been tumultuous for Wolvin and the Pittsburgh Steamship Company, but the mishaps and sinkings were overshadowed by the stunning climb in ore tonnage carried on the Great Lakes. In the previous season, lake carriers had hauled a little more than 20 million tons of ore. This season the total was 27 million tons. The Pittsburgh fleet clearly needed more big ships, not just to handle additional tonnage but to replace smaller, less efficient vessels. Some relief came when Donora Iron Mines, a U.S. Steel subsidiary, bought the Union Steel Company and its two ships, *Howard L. Shaw*

Steam-powered Huletts dig out ore from the *Sir William Siemens*. From the Milton J. Brown Collection, Dowling Marine Historical Collection, courtesy University of Detroit Mercy.

and *Simon J. Murphy*, a pair of two-year-old vessels nearly 450 feet in length. These were the first ships purchased by U.S. Steel since its formation. Pittsburgh Steamship Company took over operation of both vessels even though they would not officially come under fleet ownership for a couple more years. They were a welcome addition, and everyone realized the fleet needed more like them. For the past two years, shipyards around the lakes had been busy building ships for the smaller fleets. Now that backlog was easing and new orders could soon be accepted. Wolvin approached his acquaintances at American Ship Building Company and asked them to consider building him another innovative ship. This time it would be 550 feet long—again, the biggest on the lakes.[40]

Unfortunately for Wolvin, the 1903 navigation season turned out to be as troublesome as those of the previous two years. U.S. Steel and its subsidiaries were still having trouble molding together the many mills and other assets brought together during the company's formation. On the Great Lakes, the Pittsburgh fleet was facing high costs along with its rash of accidents. Adding to the problems was growing dissent among the lake fleets and the men they employed. Various unions were fighting among themselves and with the fleets to represent the sailors, engineers, mates, and masters aboard the lake freighters.

As the season began, many of the mates employed by Pittsburgh Steamship Company heeded their union's urging and refused to sign season contracts offered them by the company. Wolvin responded by punishing the mates who were leading the resistance. Eventually, enough men agreed to sign contracts so the ships could get under way.

Next, fleet owners began to squabble among themselves. Angry that Pittsburgh Steamship Company had too much say in the Lake Carriers' Association, the smaller fleets engineered a reorganization that allowed the association to negotiate with labor unions on their behalf. Many members thought combining their strength in this manner would help them swing a better deal with the labor movement gaining strength among lake sailors and officers. After much haggling, the smaller fleets decided it was all right for the Steel Trust fleet to join the association. But then Wolvin decided he did not want Pittsburgh Steamship Company to be part of the group because he did not wish to surrender the right to deal directly with his workers. Finally, the reorganization plans were changed to satisfy Pittsburgh Steamship, and Wolvin agreed to join the new Lake Carriers' Association.[41]

Labor trouble continued throughout the season. The growing Lake Seamen's Union battled members of the Lake Carriers' Association until finally winning sizable concessions from the fleets. Although wages remained unchanged from the previous year, both sides agreed to limit the number of hours sailors had to work each day. Before, men could be required to remain on duty anywhere from eighteen to thirty-six hours while a ship was loading. The new contract set a regular workday in port at ten hours between 7 A.M. and 6 P.M. A rate was established for overtime pay, and mates were required to refrain from giving men unnecessary work to do on Sundays and holidays. On the heels of this contract, masters and mates won agreements with hefty pay raises.[42]

Pittsburgh Steamship Company agreed to the new labor contracts along with other members of the Lake Carriers' Association. But in September, trouble arose again with the mates. Angered that some Pittsburgh ships were sailing with mates who did not belong to their union, most mates went on strike. Wolvin ordered all his ships into port to lay up. He then hired ships from independent fleets to tow his barges so some cargo could keep moving. The Lake Carriers' Association disliked the way Pittsburgh Steamship Company was handling the dispute, and a rift developed between the two organizations. Finally, Wolvin gave in. He fired the nonunion mates to settle the strike and salvage the remainder of the season.[43]

As the navigation season neared its end that year, word began circulating that Wolvin would resign from Pittsburgh Steamship Company. Early in December he revealed he would, indeed, step down on January 1, 1904, along with fleet president Daniel M. Clemson. Wolvin did not offer a reason for his decision, so everyone rushed to supply his own. *Marine Review and Marine Record* (the two trade papers had recently merged) reported that the men were resigning as part of a U.S. Steel "retrenchment" aimed at getting costs under control. Others tried to link the resignations to the recent labor problems. "It is rumored in marine labor circles that the resignation of A. B. Wolvin from the trust steamship interests is for the purpose of giving marine labor organizations a battle next season," a Duluth labor newspaper said. "Mr. Wolvin was always regarded to be fair with organized labor in the settlement of disputes."[44]

On January 1, 1904, Gus Wolvin stepped down as vice president and general manager of the Pittsburgh fleet. It apparently was an amiable parting. Wolvin certainly was not facing any financial hardships. He was already busy forming another new fleet—the Acme Steamship Company—that was building another revolutionary new ship, this time named for himself. In the years to come he would continue to aid the Pittsburgh Steamship Company when called upon.

3

Coulby Comes Aboard

As Gus Wolvin's brief reign at Pittsburgh Steamship Company neared its end, U.S. Steel officials were already courting his replacement. No less than Elbert Gary himself was involved. He made it clear that he wanted thirty-nine-year-old Harry Coulby to command the Pittsburgh fleet.

Coulby's life sounded like something straight out of an inspirational dime novel. He was born on the first day of 1865 near the English village of Claypole. As a boy he labored on his father's farm, but the rustic life held no allure for him. He left home at fourteen and walked to Newark in Nottinghamshire to become a telegrapher for the London, Midland and Scottish Railway. He worked for free for three months as a "learner," then began earning twelve shillings a week. He spent four years in the trade before looking for advancement. His opportunity came in 1883, when he took a job with the West Indies and Panama Telegraph Company and boarded a ship in Liverpool to take him to Cuba and his new post. The adventure failed miserably. After two months he was sick with malaria and in trouble with the senior clerk. Desperate to escape, he stowed away on a steamship bound for New York.[1]

As soon as he arrived, the eighteen-year-old Coulby was sent to a Catholic hospital, where he spent two weeks as a charity case, eating and sleeping as he recovered from his bout with malaria. While he waited, he thought out his next move. He had once read about the faraway Great Lakes and the mysteries they held. Now they were close at hand—his personal promised land. As soon as he was released, Coulby set off on foot for Cleveland. The young pauper followed the New York Central Railroad north through Albany and then west through Buffalo. He did odd jobs wherever possible in exchange for food as he walked the six hundred miles to his destination. Once in Cleveland, Coulby tried to ship out as a deckhand on the steamer *Onoko*. Rejected because he had no experience, he instead got a job pushing a wheelbarrow at a construction site. After spending long days on the job, he joined other ambitious young men studying shorthand at night school. In 1884 his studies landed him a job as a stenographer for the Lake Shore and Michigan Southern Railway.

It was in the Lake Shore's front office that Coulby made the connection that determined his destiny. Amasa Stone had built the railroad, and one of his sons-in-law was John Hay, who had served as secretary to President Abraham Lincoln. Hay was writing a massive twelve-volume biography of Lincoln and needed a secretary to help him. Coulby got the job, and for the next two years he worked alongside Hay, transcribing the older man's writings and carefully checking facts against a mountain of correspondence and documents. When Hay finished the work in 1886, he decided to move back to Washington. Coulby was offered the opportunity to accompany him but turned it down. Impressed with his young assistant, Hay asked his brother-in-law, Samuel Mather, to give Coulby a job.

Coulby went to work for Pickands-Mather and Company for fifty dollars a month. In an office in Cleveland's Western Reserve Building, the stocky, square-jawed young man began learning the intricacies of running the Pickands-Mather fleets—freight rates, fuel consumption, navigation, weather, and the heartless competition among the lake carriers. He worked hard and learned fast. As Pickands-Mather grew, Coulby grew along with it. In the 1890s Pickands-Mather formed the Huron Barge Company. A few years later the firm took over management of the American Steel Barge Company's fleet of whalebacks. Coulby, the young man who began as a stenographer, by 1900 had become Coulby, the respected and energetic partner. When Pittsburgh Steamship Company was formed in 1901, Pickands-Mather lost its contract to manage the Minnesota Steamship Company. Not willing to let his company decline in prominence, Coulby formed a new fleet called the Mesaba Steamship Company. He also spent much of 1903 as president of Great Lakes Towing Company.

Now, Judge Gary wanted Coulby to jump ship and join the Pittsburgh fleet. As was the case with Wolvin three years earlier, Coulby seemed unable to decide whether he wanted to tackle such a monumental undertaking. "He has been extremely noncommittal and appears as one who is debating with himself," the trade papers said.[2] After a few days, Coulby made up his mind. He would take the job as president and general manager of Pittsburgh Steamship Company under certain conditions: he would resign as president of Great Lakes Towing but remain a partner in Pickands-Mather. U.S. Steel agreed.

Harry Coulby worked as a telegrapher and office secretary before rising to command the Pittsburgh Steamship Company when it was the world's largest fleet under single ownership. Historical Collections of the Great Lakes, Bowling Green State University.

In January 1904, Harry Coulby began his rule over the Pittsburgh Steamship Company. He was the new colossus of the Great Lakes, running the biggest fleet and wielding enormous influence over the Lake Carriers' Association. He was a complex figure of a man: big and bulky; practical yet reflective. On one hand, he dressed in fine English-tailored clothes and had a taste for good cigars. On the other, he liked nothing more than to talk of the farm folk who lived near his family's home in England. Although he had left at an early age, Coulby never forgot his little village, and in later years he often sent money to the church there to be used for charity. Always mindful of his two-week stay in a Catholic charity ward, Coulby reputedly never turned away a nun whenever one approached him for a donation. He liked to joke with the other partners of Pickands-Mather, and he frequently stopped business meetings to tell a story. As John Hay's assistant Coulby had developed a strong respect for Abraham Lincoln, and like the martyred president, he enjoyed telling his associates homespun stories that drove home a point.

Coulby was not reluctant to leave the comfort of his office to ride the ships in his fleet. As fleet manager for Pickands-Mather, he visited loading and unloading docks wherever his ships called. He knew the lakes and rivers well, and used that knowledge to make his fleet ever more efficient and to seek government action on removing hazards and improving navigation.

Characteristically, Coulby vigorously attacked his job at Pittsburgh Steamship Company. He quickly became known for two things: ruthlessly weeding out inefficiency, and relentlessly hammering labor unions that challenged him for control of his ships and men.[3]

Coulby began making changes right away. First, he moved the fleet's headquarters to its Cleveland office. He lived in Cleveland and wanted to run the fleet from there. L. W. Powell, an executive with U.S. Steel's Oliver Iron Mining Company, was named the fleet's vice president to oversee operations in Duluth. Allyn F. Harvey, who had served as Wolvin's assistant, was assigned to work with Powell and promoted to assistant general manager. Next, Coulby began reorganizing the Duluth office as part of U.S. Steel's financial "retrenchment." Some men were fired. Others suffered pay cuts ranging from 8 percent to 25 percent. In the auditor's office, several clerks earning $125 a month had their pay cut to $60 to induce them to quit.[4]

As the navigation season neared, Coulby took aim at the unions. His reputation preceded him. "Mr. Coulby has never been noted for tact, and it is said that the labor unions will find a hearty warrior when they come up against him," a Duluth newspaperman observed.[5] He was only on the job a few days when newspapers reported that Pittsburgh Steamship Company wanted to cut sailors' pay and dump the union rules in effect on its ships. A big fight was expected in the spring. Independent fleet owners began hoping the Steel Trust fleet would be paralyzed by a strike, enabling the smaller fleets to grab some of the lucrative ore-hauling contracts.

Organized labor to this point had a short and largely unimpressive history on the Great Lakes. The first organization to represent sailors was the Seamen's Benevolent Union of Chicago. Formed in 1863, it confined itself to "the moral, mental and mutual improvement of its members." Fifteen years later, Chicago sailors revived this organization, renamed it the Lake Seamen's Benevolent Association, and set out to achieve economic objectives. They aggressively enforced a closed shop. Union men refused to sail with nonunion men. They boycotted boardinghouses that sheltered nonunion sailors and, in one case, even boycotted an undertaker for burying a nonunion corpse. They set the tone for the unions that would follow.[6]

By 1880, vessel owners had decided to retaliate against the growing labor movement. They formed the Cleveland Vessel Owners' Association and set about trying to destroy the unions. They set up hiring halls to attract new employees and used thugs to keep union members away. Various unions contributed to their own downfall by fighting among themselves. Men who sailed the rapidly disappearing schooners refused to help the new steamboatmen, declaring they were not really sailors. The feuding weakened the unions and left the employers in control within a couple of years.[7]

The labor situation began to change in the 1890s. Great Lakes shipping was growing rapidly, and steamboats now dominated the trade. More ships meant more men coming to the lakes to work, and the advent of steam technology meant they did not need to spend a lifetime learning the sailing skills held by the old schooner hands. These unskilled sailors could come and go much like workers in a factory. However, they found that while steamboat jobs were more plentiful, they held little long-term promise for men in the lower ranks. Officers—masters, mates, and engineers—were skilled men with government licenses who were respected and paid relatively well, but unskilled men such as deckhands and firemen were treated poorly. They were hired for one trip and often discharged if the ship was going to spend more than a few days in port. Knowing the men would not be around long, vessel masters worked them hard and skimped on expenses such as food. Wages varied widely from year to year. If vessel owners expected a slow year, they slashed wages, knowing they would have plenty of men from which to choose. If a busy season loomed, they would boost pay to ensure they had enough labor. The unpleasant conditions scared off good, steady men and instead attracted "irresponsible adventurers"—drifters, drunks, and criminals.[8]

Everyone agreed that firemen were the worst of the lot. Reliable men found little allure in a low-paying job that forced them to labor in a stifling boiler room shoveling coal into the furnaces. Many who took these jobs were habitual deserters who jumped ship before completing the trip. Some vessels were "hard firing," requiring so much steam that men were physically incapable of making more than a couple trips. When firemen did return to their ship after a stay in port, they were most likely to be drunk and worthless for their next watch in the firehold.[9]

With more men working on the lakes and many of them unhappy with the treatment they received, the labor unions found fertile ground in which to sow the seeds of their comeback. By 1895 they had regained strength. They still fought among themselves, but they succeeded in organizing masters, mates, engineers, firemen, deckhands, coal passers, and cooks. At the same time, vessel owners tried to improve working and living conditions aboard their ships in an attempt to attract a more stable breed of sailor and to reduce the increasingly expensive employee turnover. The improvements were insufficient, however, and the unions quickly won the right to enforce shipboard work rules among their members. That only made matters worse. Union leaders were supposed to supply men to the ships, enforce work rules, and discipline members who left a ship without providing proper notice. But most of the union officials were professional officeholders. Many were former firemen or even bartenders, and they knew they could not stay in office by enforcing unpopular rules. The owners' attempts at reform failed, and it became even more difficult for the fleets to exert control over their crewmen and their own ships.

Men in the lower ratings were not the only ones unhappy. Officers who navigated the ships—the masters and mates—were increasingly disgruntled with the steady erosion of shipmasters' traditional rights and privileges. Changes had begun with the Bessemer Steamship Company and reached full force when Pittsburgh Steamship Company began imposing more management rules on its ships. In the past, the ship's master had controlled the vessel and conducted its business. But to combine six fleets into one and manage such a large number of vessels, Pittsburgh Steamship Company had to standardize its operations. Rules and procedures were set up, and masters who wanted to keep their jobs had to follow them. The fleet began shifting masters from ship to ship for more efficiency in manning its vessels. Management—not the master—hired men. Masters even needed permission from the office to visit their homes while in port. It was part of a broader U.S. Steel management philosophy, but it was a big change in tradition and caused a lot of discontent among shipmasters. To make matters worse, the masters and mates faced growing insolence from their crews. More and more union members began regarding their labor contracts as a set of shipboard rules that took precedence over any orders their captain might give.[10]

In an attempt to wrest control from the unions, the Lake Carriers' Association launched its own union in 1901 under the name Lake Carriers' Beneficial Federation. The organization made its debut with much fanfare, but withered under the pressure of growing union strength. Coulby and other members of the Lake Carriers' Association used the failure to oust the association's old leadership and reorganize the group in 1903. The new Lake Carriers' Association announced it would establish amicable relations with employees. But the announcement was a ruse. The fleets were biding their time until the right moment came for them to attack the unions head-on. In the meantime, a confusing hodgepodge of unions battled to represent the men and negotiate contracts.[11]

Faced with a contentious labor situation when he assumed command of the Pittsburgh fleet, Coulby wasted little time in picking his first fights. Joining with other fleets in the Lake Carriers' Association, Pittsburgh Steamship Company got the Marine Firemen, Oilers and Water Tenders' Union to agree to a small cut in wages and a clause that allowed shipmasters to lay off crewmen if their vessel was detained in port more than three days. The engineer position on barges also was eliminated. Next the fleets persuaded the Seamen's Union to agree to drop overtime pay and accept the clause that allowed layoffs if a ship was detained more than three days. The fleets succeeded in breaking the Marine Cooks and Stewards Union, and then offered chief cooks and stewards a flat rate of seventy dollars a month.[12]

The big challenge came against the American Association of Masters and Pilots. A few weeks before the ships were to sail, Coulby required captains in the Pittsburgh fleet to reapply for their jobs—a clear attempt to weed out any who had joined one of the three unions vying for their allegiance. The masters and mates persisted, however, and joined their comrades in all the major fleets in demanding the right to belong to a union and to hire their own crews. They also demanded a salary of $2,250 a year for masters commanding the biggest ships. That was a substantial increase over the $1,980 which the Pittsburgh fleet had paid its top captains the previous season. The masters also wanted to be paid for a full nine-month season regardless of how long their ships operated, and they wanted a clause permitting captains to be transferred only to bigger and better ships. The vessel owners refused these demands, so the masters threatened to strike and delay the start of the shipping season.

Some fleets immediately wanted to give in to the unions. Coulby cajoled them into standing firm, mainly by threatening not to charter ships from anyone who conceded the fight. He led the Lake Carriers' Association in maintaining that a captain was the owner's representative aboard ship and must answer only to the owner. If the captains belonged to a union and won the concessions they were demanding, the vessel owners' control over their own ships would be jeopardized.[13]

Neither side would back down, so the shipmasters went on strike. Few ships sailed in April or May, but eventually the fleets persuaded a few of the youngest and the oldest masters to go back to work. A handful of ships got under way. Other captains saw this, became fearful they would lose their ships, and broke ranks. On June 14 the union called off its strike. The American Association of Masters and Pilots was broken. There would never be another serious attempt to organize the shipmasters.[14]

Coulby undoubtedly was cheered by the victory over the masters' union. Surprisingly, by 1905 good feelings existed between Coulby and his vessel masters. He extended contracts to them that paid $1,980 a year for masters of the biggest ships, $1,760 for those commanding midsize vessels, and $1,540 for captains on the smallest ships. Coulby told the captains he would act reasonably in labor relations, and that he expected their utmost loyalty in return. He gave them permission to hire their own mates providing they got approval from the fleet office and made sure no "unionists" were among them.[15]

Other workers were more combative than the captains, but by now the fleets had the unions on the defensive. Coulby and the other owners and managers employed a divide-and-conquer strategy that was aided greatly by the unions themselves. The union representing firemen threatened to strike, but the Lake Seamen's Union and the Marine Engineers Beneficial Association agreed to supply strikebreakers to the Lake Carriers' Association if the walkout occurred. The fleets blacklisted mates who belonged to a union, figuring their jobs could be filled by unemployed captains or by promoting wheelsmen. The mates sought support from the International Longshoremen's Association, which called a strike for May 1. It quickly failed because other dockworkers, tugmen, and grain shovelers refused to join the walkout.[16]

A few unions, however, continued to cause problems for the fleets. Foremost among them was the one representing firemen. The union's contract entitled it to be the exclusive supplier of firemen for the lake fleets, including Pittsburgh Steamship Company. Union officials also were supposed to discipline members who broke work rules, such as the one requiring firemen to give twelve-hour notice before quitting a ship. But the union could not comply. It was supplying only about 60 percent of the firemen needed, and it rarely enforced work rules. This made life difficult for vessel masters who tried to follow the contract, and it disgusted Coulby. In a letter to one of his captains, he summed up the problem and added a note that he clearly hoped would prove prophetic: "The position of the firemen this year has become nearly unbearable. There are several reasons which make it worse than usual. They are having internal dissensions; they also have practically a prohibitive price or initiation fee which prevents their getting new members; they have not near enough men to furnish what is required; their officers seem to have lost control of them. It will certainly be a happy day when we are rid of the whole bunch. In my judgment, the sooner we reach that point, the better we'll be off."[17]

While he was battling the unions, Coulby also was beginning an aggressive and continuous program of building bigger ships and disposing of smaller ones. When formed, the Pittsburgh Steamship Company owned the sixteen largest steamers and seven largest barges on the Great Lakes. Its oldest ship had been in service only fifteen years, and half the fleet was less than five years old. By 1904 many of the fleet's 114 vessels were still considered large by the standards of the day, but a significant number were the small and inefficient whalebacks. The small steamers required just as many crewmen and cost about the same to operate as the fleet's biggest ships, yet carried half as much cargo. The fleet also was saddled with forty-three barges, which had to be continuously mated to steamers bound for the same destinations. The solution to this problem was to get bigger ships, and the only way to do that was to build them. So late in 1904 the Pittsburgh Steamship Company placed orders for its first new ships.

Chicago Ship Building Company won contracts to build the *William E. Corey* and *Elbert H. Gary.* West Bay City Ship Building Company would build the *Henry C. Frick,* while the Superior Ship Building Company got the job to build the *George W. Perkins.* The ships were named for important men in U.S. Steel Corporation, establishing a tradition that would continue throughout the history of the fleet. The ships' design was based on the revolutionary steamer *Augustus B. Wolvin,* launched in 1904. Championed by Gus Wolvin, his namesake was like no other vessel ever built on the lakes. Existing ships used vertical steel stanchions and horizontal beams inside the cargo hold to support the deck overhead. This cluttered the hold and slowed the big new Hulett unloaders and other dockside unloading machines. Slower unloading meant more time spent in port, and that cost money. Wolvin's new ship was built specifically to accommodate the Huletts. It had ballast tanks on both sides of the cargo hold. Steel arches stretching over the hold supported the deck, eliminating the need for stanchions inside the hold. This arrangement left the cargo hold wide open, making it a continuous steel hopper where unloading equipment could chew into the ore without dodging obstructions. Moreover, the hold's construction permitted the ore to be piled in steep, high mounds that were easy for the Huletts to attack. Instead of wooden hatch covers, the ship had labor-saving steel covers that opened by sliding back and telescoping into a neat pile at the edge of the deck. The ship featured a "flush deck" design that put all crews quarters inside the hull and left the deck free of cabins or other structures that might interfere with loading or unloading gear.[18]

After reviewing plans for the *Wolvin,* Coulby consulted several experts and concluded that the arch design was primarily intended to support weight from the top. In his new ships, he replaced the arches with straight beams that could support the deck and leave the cargo hold unobstructed while adding strength to the vessel's hull. He also did away with the flush deck design used by the *Wolvin* and instead placed cabins on the deck at the stern. At the bow he placed a cabin called a "texas house" atop the forecastle and then put the pilothouse on top of it. The *Gary* class, as the four ships were known, became the standard design for Great Lakes ore carriers. For the most part, this design would vary only in size for the next sixty-five years.[19]

Wolvin's 560-foot ship was the biggest on the lakes, and on its maiden voyage carried a record 10,694 tons of ore. Each of Coulby's four ships would be nine feet longer than Wolvin's, and each could carry more than 10,000 tons of ore on every trip. The leap in efficiency was monumental: one of the new steamers could carry more iron ore in one trip than most of the fleet's whalebacks could carry in three. Another measure of thrift could be found in the engine rooms of the new quartet. The *Corey,* for instance, had exactly the same size boilers and triple expansion steam engine as the *Manola,* which was built in 1890 and at that time was the biggest ship on the lakes. The *Manola* could carry 68,000 tons of ore in a season, while the *Corey* would be able to haul 225,000 tons with the same fuel consumption. "It is clear from these figures that the economical carrier is the single steamer of large carrying capacity and low power," the *Marine Review* observed, adding, "The day of the consort is past."[20]

First of the new ships to hit the water was the *Gary.* It went into commission June 1, 1905, and soon shattered records for the biggest ore cargoes carried from Escanaba to South Chicago and from Lake Superior to South Chicago. But the fanfare was reserved for the second vessel, the *Corey,* which was named for the new president of U.S.

Lined up like a row of pecking birds, five Hulett unloaders dip into the cargo hold of the *Elbert H. Gary* to scoop up iron ore. Construction of the *Gary* and three sister ships in 1905 signaled the start of the fleet's efforts to build large, innovative vessels. This photo shows the *Gary* on July 23, 1908, when it brought the first load of ore to U.S. Steel's new mill at Gary, Indiana. Since then, ships have carried more than 370 million tons of ore to the mill. Courtesy Dowling Marine Historical Collection, University of Detroit Mercy.

Steel and designated the fleet's flagship. The four ships were virtually identical, except the *Corey* had five luxurious staterooms in its forward end to accommodate company officials and guests of U.S. Steel. The rooms were decorated in birch and enameled white, and each had its own bath. A separate deckhouse held a private dining room, and an observation room on the texas deck was available for guests' private use. For the *Corey*'s launching on June 24, U.S. Steel chartered a train to carry three thousand people from Chicago's LaSalle Street Station to the Chicago Ship Building Company's shipyard. Men sporting straw boaters and corsetted ladies wearing flowered hats gathered in the dusty, littered shipyard to inspect the ship as it stood on the launchways. A few men and boys even ventured up to the side of the giant as the yard crew labored to ready it for launching. President Corey was in Europe and could not attend the ceremony, but Coulby and Vice President James Gayley were present to help Corey's sister Ada christen the vessel.

Following the ship's flawless launch, the guests boarded the train to return to downtown Chicago and a sumptuous Saturday night dinner. Ship launchings increasingly were accompanied by social ceremonies, and for the *Corey* the Pittsburgh fleet put on one of the best yet seen. The banquet room of the Chicago Auditorium was specially decorated for the occasion. A lighted fountain burbled gently in the middle of the room, the walls were decorated with American flags, and cut roses and carnations were strewn across the tables. Among the honored guests that evening was Augustus Wolvin. James Hoyt, serving as toastmaster, introduced him second, right after Coulby, and made a point of acclaiming Wolvin as "the man who had built the first leviathan on the lakes." Wolvin graciously demurred and credited William Corey with being the man who had persistently advocated construction of big ships as a means of reducing costs. He then proposed a toast to Corey as the prime mover of the trend toward bigger vessels.[21] After the celebration, the launching party joined Coulby in traveling to Superior, where the following day they witnessed the launching of the *Perkins*.

As the *Corey* and *Perkins* entered service in mid-August, they joined the rest of the fleet in sporting a new look. The hulls of the fleet's vessels were now being painted deep red and their cabins white. Traces of the original green were retained on trim and on anchors. The vessels' most distinctive feature, however, were their smokestacks, which were painted silver. This color was notoriously difficult to keep free of soot around the top of the stack, and crewmen spent a lot of time scrubbing and painting to keep them presentable. The chief engineer of the supply boat *Superior*, which served Pittsburgh vessels at the Soo Locks, solved the problem by painting a black band around the top of his boat's stack to hide the soot. A fleet official saw it on an inspection trip and ordered the same thing done on every vessel.

The distinctive silver smokestacks soon became the fleet's unofficial trademark. Around the lakes, dockworkers and sailors began referring to them as "tin stacks." The nickname stuck, and for many years to come the fleet's vessels would be known as Tin Stackers. Company officials gradually accepted the moniker, but for marketing purposes preferred the somewhat more prestigious nickname Silver Stackers.[22]

While the four new ships were taking shape, Coulby's men began seeking buyers for the fleet's smaller vessels. They found one in early 1905 for whaleback barges 117, 127, 201, and 202. In February they sold twelve more whalebacks to Benjamin Boutell, a Bay City, Michigan, vessel owner who was making the purchase for East Coast interests. The steamers *E. B. Bartlett, Joseph L. Colby, Colgate Hoyt,* and *A. D. Thompson* were sold along with barges *Sir Joseph Whitworth, John Scott Russell,* 105, 107, 109, 110, 116, and 126. All the vessels sold were under 275 feet in length and had the small, troublesome cargo hatches characteristic of whalebacks. Combined, the twelve vessels sold to Boutell could carry 639,000 tons of cargo a year, compared to the 800,000 tons for the Steel Corporation's four new ships.[23] "The financial returns clearly show that the ore trade belongs to the larger carrier and the problem [of] what to do with the smaller class of vessels is doubtless one which is concerning the Steel Corporation quite as much as any other interest," the *Marine Review* stated as it announced the sale to Boutell. "There is little money in the ore trade for the steel freighter of moderate dimensions."[24]

After the labor disputes, launching ceremonies, and vessel sales of winter, spring, and summer, Coulby and others at Pittsburgh Steamship Company were heartened by the relative peacefulness of autumn. The fleet had suffered comparatively little damage from collisions or storms throughout the tumultuous year. As the shipping season neared its end, Allyn Harvey, better known simply as A. F. Harvey, took a moment to reflect. "We shall be much pleased if November turns out to be milder than usual," he told a Duluth newspaperman. "The month will show up strong in the ore shipping column for the season if conditions are favorable. The impression prevails that we shall have a late fall, and that would be welcomed by the vessel interests."[25]

The benevolent autumn weather ended abruptly in the last week of November. A raging northeast storm roared across Lake Superior on November 23 and 24, forcing many ships to seek shelter. After the storm passed, ships pulled up their anchors or cast off from their docks and headed onto the lake in a hurry to carry the last few cargoes of the navigation season. Because large storms usually are followed by periods of calm, shipmasters were surprised when weather forecasters predicted another storm to follow on the heels of the first. Dozens of vessels already had gotten under way, eager to make up the time lost to the first storm. There was no way to warn them of the tremendous weather system building over the Dakotas and racing toward the Great Lakes.

The storm swept out of the prairies and slammed into Duluth at about 6 P.M. on November 27. Winds rose quickly, climbing to forty miles per hour by 7 P.M., then hitting fifty, then climbing to sixty by midnight. Wet snow, driven by the wind, reduced visibility to near zero. All over the city, utility poles snapped under the strain, spilling live electrical wires into the streets. Streetcars ground to a halt, unable to gain traction in the snow or cut off from electricity. Passengers had no choice but to abandon the cars and trudge through the snow to shelter, dodging store signs that broke free and cartwheeled down the streets. Passenger trains were delayed. Telephone and telegraph service failed as lines went down. Amazingly, the winds remained above sixty miles per hour for thirteen hours, with gusts up to sixty-eight miles per hour. Waves on Lake Superior grew to stunning proportions. A veteran Duluth tug dispatcher declared the water in the sheltered harbor to be the roughest he had ever seen it. Fifteen ships were due to leave port during the day, but the masters of several chose to stay securely tied up to their docks.[26]

Out on Lake Superior, dozens of ships and barges were under way. Because of the season's late start, many Pittsburgh Steamship Company vessels were still operating, and the number on Lake Superior was typical for any day during the season. Among them was the steamer *Mataafa*, laden with ore and towing the barge *James Nasmyth*. The pair left Duluth at 3:30 P.M. on November 27 in the face of freshening northeast winds. Flags signaling the approach of a northeast storm were snapping from the Weather Bureau's flagpole, but Captain Richard F. Humble, master of the *Mataafa*, did not deem the weather severe enough to remain in port. To the east, off the Apostle Islands, the steamer *Mariposa* was upbound for Duluth with a cargo of coal, while *Crescent City* was running light for Two Harbors. The steamers *Isaac L. Ellwood* and the new *William E. Corey* were steaming alone, while *Coralia* was towing the barge *Maia*, *William Edenborn* was paired with *Madeira*, and *Lafayette* had *Manila* under tow.

Judging from the lowering skies and falling barometers, the vessel masters on Lake Superior that day knew they were in for foul weather. Trouble hit as autumn's feeble light faded from the sky. *Mariposa* encountered freshening winds about 4 P.M. near Devil's Island. By early evening the steamer was plowing through heavy snow and strong winds. *Mataafa* was approaching Two Harbors about 7:30 P.M. when winds rose to storm force and unleashed heavy snow. As the winds rose to sixty miles per hour and held there, the seas grew to tremendous heights. The *Edenborn* was heaving so badly by 9:30 P.M. that its propeller kept coming out of the water. Down in the engine room, Chief Engineer Silas Hunter had to throttle back the engine to fifty revolutions a minute to reduce the strain on the machinery.[27]

Aboard the 406-foot *Crescent City*, Captain Frank Rice was using all his skills as a mariner to bring his ship safely through the rapidly intensifying storm. *Crescent City* was not carrying cargo, and even though it had taken on water ballast the wind and waves made it difficult to control. At 10 P.M., Rice calculated that his ship was off Two Harbors, but it was impossible to enter port. "It was blowing at that time, I should judge, from sixty to seventy miles per hour from the northeast, accompanied by heavy snow; almost impossible to see one end of the boat from the other," he recounted later. Rice decided to turn his ship into the wind and head back into the lake to put some room between his vessel and Minnesota's rocky shore. He sent orders down to the engine room to work the engine and boilers at full capacity. In the wind and high seas, *Crescent City* would need all the power it could muster.

At Rice's command, the *Crescent City*'s wheelsman wrestled the steering wheel around until the rudder was hard over. Nothing happened. The ship was not powerful enough turn into the raging wind. In the boiler room, firemen staggered across the heaving deck as they fought to shovel coal into the boilers to maintain maximum steam pressure. In the adjacent engine room, the engineers anxiously watched the quadruple expansion steam engine as it pounded at its highest speed. Rice tried turning the ship to starboard three times and to port three times. Even at maximum power, the steamer was unable to turn into the wind.[28]

With his first strategy defeated, Rice elected to steer toward Duluth. One of the mates continuously swung the sounding lead over the ship's side, calling out the depths so Rice would know when he was nearing shore. Shortly after midnight, when the ship reached ninety feet of water, he ordered both anchors dropped and paid out nearly all the chain. The anchors dug in and held, and the *Crescent City* remained in place with its engine slowly working ahead. All was well until 3 A.M. on November 28, when a tremendous gust of wind caught the steamer on its starboard bow, pulling the anchors free of the bottom. Captain Rice quickly rang the engine room for full speed ahead, but with the rudder hard to port and the engine working at full speed, *Crescent City* was unable to make headway. The ship dragged its anchors for about an hour. Then, through the snow, Rice could make out a pointed bluff that was alarmingly close. *Crescent City* was drifting straight for it. Rice knew the outcropping would impale his ship and wreck it, so he rang for the engine room to back the engine at full speed. *Crescent City* slid past the outcropping and rumbled ashore next to a low wind-blown bluff in a sparsely populated area several miles north of Duluth.[29]

"Finally, with a crash we struck, and the vessel lurched broadside on the rocks, so that I thought she would never right herself," wheelsman Charles Abram recalled. "A huge wave rolled clear over her, drenching every man to the skin and it was nothing short of a miracle that no one was carried over." Stranded in shallow water and pounded by waves, the *Crescent City*'s steel hull soon broke amidships. Nonetheless, fortune had smiled upon the steamer's crew. The ship's deck was level with the top of the bluff, and the gulf between ship and land was less than ten feet. Captain Rice ordered the men to rig a ladder from the deck to the bluff. The entire crew gathered around and stood by as Hattie Stevens, the cook's wife, was carried across the ladder by her husband and two other men. Rice and First Mate Thomas Thomson then called one man at a time to step forward and creep across the ladder to safety. Among the men was twenty-eight-year-old watchman Arthur Daggett, who was relieved to have survived his third shipwreck. He had been aboard a schooner that sank in Lake Michigan several years earlier, and the *Thomas Wilson* when it sank off Duluth in 1902. Finally, the mate and then Captain Rice crossed the ladder. Realizing that everyone on board had made it safely ashore, the bedraggled band of survivors cheered weakly.

Although they had successfully abandoned ship, the sailors found themselves in a new predicament. They were stranded in the woods in a blizzard. Nothing had been saved from the ship except the clothes they were wearing and the ship's papers, which the captain had secured before leaving. Again, fortune favored them. To the west they could see lights dimly reflected in the sky. They determined this was Duluth. While Rice and Thomson stayed near the ship, most of the shivering crew walked to the nearby Duluth municipal water-pumping station or found shelter in scattered homes. The second mate and the chief engineer walked into Duluth and eventually caught an early-morning streetcar carrying downtown workers to their jobs. Wet and disheveled, the pair drew curious stares as they rode to the Pittsburgh Steamship Company office to report the shipwreck.

Daggett later described how the sailors felt as their ship drifted toward shore. "I'll tell you, nobody realizes how a man feels when he is in a storm of that kind and the ship captain having lost his bearings. . . . We were positive Tuesday and all night that we would never get through. I had my mind made up that it was the last few hours I would be alive, and I think most of the rest of the boys felt the same," he recounted. "I want to say that even when we struck the shore and it looked pretty close at hand, too, that we did not see anything ahead of us except to be battered to pieces against it. The shore does not look friendly in these cases. It is a menace and we were afraid of it."[30]

While Captain Rice was battling to keep the *Crescent City* afloat, Captain Dell Wright and his men were locked in their own struggle aboard the *Lafayette*. The 454-foot *Lafayette*, with the 436-foot barge

Manila in tow, had followed its usual course from the Soo despite the storm. At 9:30 P.M. Captain Wright cautiously signaled the engine room for half speed, figuring he would hold that speed until abreast of Devil's Island. After several hours of battling the storm, he had only a rough idea of his ship's location. The navigational tools available to Wright and other mariners that night consisted mainly of a magnetic compass, a clock, and keen judgment crafted through years of experience. Outside the pilothouse, a combination of darkness, wind, and swirling snow robbed Wright of any hope of spotting the Devil's Island lighthouse or hearing its foghorn. The reliability of his compass would become increasingly questionable as the *Lafayette* neared the iron-laced outcroppings of the Minnesota shore. The clock could tell him how long he had been steaming, but it could not tell him how far the wind had blown him off course. As the western end of Lake Superior became ever narrower, Wright had precious little room to depart from his course, so he had to make a decision: he could continue on toward the Minnesota shore, or he could change course and hope that, if he did run aground, he would do so on the soft sand beaches near Superior.[31]

Wright chose to head toward the Wisconsin shore. The seas, however, determined otherwise. As the wheelsman wrestled the steering wheel, the wind and waves repeatedly lifted the steamer up and pushed it around. After each wave, the wheelsman struggled to bring the ship back on the proper heading, only to have the vessel blown off course again. Each lurch of the ship threw the compass off by seven or eight points, making it nearly impossible for the wheelsman or Wright to get a bearing. After several hours, they could only guess at their location. Adding to the menace was the *Manila*, wildly dancing about on the end of the towline. "At times I would see the barge's light abreast of me to the northward, and in a very few minutes abreast of me to the southward, it being impossible to keep the ship on her course," the captain said.

Twice during the night Captain Wright attempted to bring his ship about and point it into the wind, but both times the laboring steamer slid into the troughs of the waves and was unable to climb out. Wright had no choice but to stick to the course he thought was taking him toward Superior. About 5 A.M. the weary men in the *Lafayette*'s pilothouse heard the unmistakable roar of surf breaking in shallow water. Wright immediately ordered the steamer's helm put hard over, then signaled *Manila* to drop its anchors. He was too late. The *Lafayette* lurched into shallow water and ground to a steel-ripping halt about fifty feet from a heavily wooded, ice-covered shore. The *Manila*

Early victims of the great 1905 storm were the steamer *Lafayette* (stern visible in background) and barge *Manila*, which were blown ashore on Minnesota's wild north shore about six miles north of Two Harbors. Several of the *Manila*'s crewmen got to shore by climbing into the trees overhanging the barge's stern, shown at the far right edge of the photo. Courtesy Lake Superior Maritime Visitors Center.

followed suit, rumbling through the rocks and slamming into the steamer's stern before swinging broadside to shore.

Although Wright did not know it at the time, his command had come ashore near Encampment Island about six miles north of Two Harbors. The pounding surf made short work of the *Lafayette*. "The immense seas would pick her up like an egg shell, toss her off from twenty to thirty feet, throw her against the rocks with a crash that could be heard for miles," the captain reported. "I think about the third sea that struck her, broke her in two at the third after hatch, her stern swinging into a little cove and the forward part still pounding."[32]

Because it was lighter than the steamer, the *Manila* had ridden in closer to shore when it grounded. Captain George Balfour ordered his crew to scuttle the barge so it would fill with water and settle to the bottom rather than pound itself to pieces. After opening the sea cocks, fireman William Platt ran to his cabin to change into dry clothes. In the excitement he almost forgot the $324 he had stashed away.[33] He grabbed the cash, stuffed it into his pockets, and ran back on deck. As the barge settled, Balfour gathered his tiny crew and four *Lafayette* sailors who had jumped to the barge when the two vessels smacked together. The barge was so close to shore that a few men grabbed tree branches and crawled over them to shore. The rest pushed a ladder over the side and clambered up a cliff.[34]

Once safely on land, the men raced to the stern of the wrecked steamer. After several unsuccessful attempts to get a line ashore from the stern, the *Lafayette*'s second engineer went below and retrieved a large nut from a parts locker. He tied a light line to the nut and then hurled it into the trees. Captain Balfour was able to snatch the line from among the limbs and use it to pull ashore a heavy hawser and secure it to a tree trunk. The men trapped on the stern then began the most perilous journey of their lives. One at a time, they gingerly climbed astride the rope and began pulling themselves hand over hand toward shore. It was a harrowing experience. Wind stung their faces and the cold numbed their hands and feet. As the surf rocked the stern, the rope swayed and alternately pulled tighter, then went slack. Several men had successfully made the crossing when fireman Patrick Wade's turn came. Instead of crawling across headfirst like the others, he decided to slide down feetfirst. But when he reached the rope's lowest point, he found that his position prevented him from pulling himself up the remainder of the rope to land. Wade struggled to turn around, slipping his legs off the rope and hanging by his muscular arms. As he dangled there, the ship lurched and the rope snapped taut. Wade dropped into the darkness between the cliff and the smashing, grinding hull.

Once the remaining men on the stern had safely reached shore, the group moved to the steamer's broken bow to help Captain Wright and the men stranded there. After separating from the stern, the front half of the ship had swung against a wall of rock. Spying a route up the cliff, the men on the bow leaped one at a time onto a ledge, then scrambled seventy feet up the cliff as waves dashed against its face. All made it safely to join the other men already ashore.

The mariners now were safe from the seas, but they still had to face the elements. Bruised, tired, wet, and inadequately clad for the bitter temperatures, the men stumbled through the woods for several hours vainly searching for a cabin or barn. They finally returned to the wreck and built a bonfire for warmth. At midafternoon the tug *Edna G.* steamed past the wreck searching for survivors. Although the tug's crewmen spotted the shipwrecked sailors, the high seas prevented their launching a boat. The crewmen from *Lafayette* and *Manila* were forced to spend forty-eight bone-chilling hours in the woods before being rescued. Six were hospitalized in Two Harbors to be treated for frostbite.

The wind and waves that overwhelmed *Crescent City, Lafayette,* and *Manila* also caught the 478-foot steamer *William Edenborn* and its consort *Madeira* at about 6 P.M. as they were fifty miles east of the Apostle Islands. By 9:30 P.M., Chief Engineer Hunter was forced to check down the engine to fifty revolutions per minute—about half speed—because the propeller kept coming out of the water, causing the engine to race. Six hours later, Hunter heard a tremendous bang overhead. Supposing the towline had parted, he set out to investigate. On deck the wind and snow were so severe that he was forced to crawl on his hands and knees, pulling himself from one object to another. He finally reached the ship's fantail, where he discovered that only the eye of the wire towline remained in the towing chock. The *Madeira* was on its own.

In such heavy seas it would have been foolhardy for the *Edenborn* to try recovering the barge. The *Madeira* was equipped with heavy anchors and chains specifically for a situation such as this, and its crewmen were accustomed to looking after themselves. So the *Edenborn* slowly steamed off into the storm. An hour after losing the barge, Captain A. J. Talbot, master of the *Edenborn,* decided to turn his ship into the wind and drop anchor. "The snow was so thick and so blinding we could not see half the length of our ship in any direction," Talbot said, "and the sea [was] piling over her stern in volumes of solid water." Called back to duty after standing his watch, First Mate William Hormig found the blowing snow so fierce that he could not face into the wind while on deck. Throughout the night Talbot had set his course somewhat to the south of his usual route in an effort to keep away from the Minnesota shore. Now, at 4:40 A.M., he rang the engine room for maximum power in anticipation of making his turn into the wind. As the ship struggled to come about, it began sliding through gravel and rocks before planting its bow into the forested shoreline. Waves crashed over the deck, working the ship's hull so hard that the covers on three hatches dropped into the cargo hold. The writhing soon caused the hull to break amidships around the ninth hatch, opening up a crack nine inches wide that stretched across the spar deck and continued partway down both sides of the hull. Nonetheless, the twenty-five crewmen remained aboard the ship and went about their duties. Uncertain what to do, Chief Engineer Hunter stopped in the boiler room to check on the situation and was surprised to find the fireman still at his post as water cascaded down from above. "I will remain on watch here if you order me to, Mr. Hunter," the man said gamely.[35]

In the darkness, four men working on deck fell through open hatches into the hold. After shouting for several minutes to attract attention, they were discovered and pulled to safety. Another man, Second Assistant Engineer James Johnson, was running down the deck to help put a life raft over the side when he tumbled into an open hatch and drowned in the flooded hold.

The *Edenborn* had come ashore at the mouth of the Split Rock River, about forty-five miles northeast of Duluth. Captain Talbot did not know his precise location, but he could see it was heavily forested and not particularly suited for a nighttime expedition. With the front half of the ship firmly anchored onshore, he elected to gather the crew in the forepeak compartment, where the men would be reasonably dry and safe.

At daybreak, First Mate Hormig and two other men got ashore on a raft. They secured a line to a tree and jury-rigged a bosun's chair to be used as a breeches buoy in case the crew had to get off the ship. Then, guided by several local men who had shown up at the wreck, Hormig floundered off through the deep, wet snow to find help at a nearby lumber camp. He ended up paying a man fifty dollars to snowshoe through the night to carry word of the steamer's plight to Two Harbors. The tug *Edna G.* rescued the *Edenborn*'s crew the next day.[36]

The Minnesota shore was less hospitable to the *Madeira*. Once the towline parted, the barge drifted for more than two hours while Captain John Dissette and his crew of nine fought to keep the vessel afloat and prayed their anchors would stop them from drifting. Like many other sailors that night, they had only a vague estimate of their location, but they knew the wind and waves were pushing them toward shore. The barge continued to drift until about 5:30 A.M., when it began rolling in heavy surf, then smashed into a huge wall of rock that soared off into the darkness. *Madeira* had come ashore at Gold Rock, just north of present-day Split Rock Lighthouse and only a couple miles from the *Edenborn* wreck. Trapped against the base of the cliff, the *Madeira* slammed into it again and again. The vessel's hull buckled and split, trapping four men on the bow and the remainder aft.[37]

The men aboard *Madeira* knew their vessel was doomed. As each wave receded, the barge slid off the base of the cliff, only to slam back into the rocks on the crest of the next wave. The barge was rapidly breaking up beneath their feet, and they would die if they remained aboard. But in the maelstrom at the bottom of the cliff, there literally was no place for them to go but up. They could not even see the top of the cliff. The men on the bow tried casting a line into the darkness to see if it would catch on something, but that proved fruitless. Driven by their terrible desperation, Seaman Fred Benson decided to act. The big, powerful man grabbed a coil of rope and slung it over his shoulder. Clambering over the railing, he waited until the barge surged against the rocks, then hurled himself onto the face of the cliff. Landing on his feet, Benson scrambled up the cliff as far as he could before bracing as a wave blasted over him. When the water receded, he resumed climbing, gaining several more feet before the next wave struck. He kept scrambling until he reached the top of the cliff. Then he lowered his rope to the barge's reeling deck and helped the other three men make the climb. Those men then passed a line down to the crew on the stern and helped them scale the cliff. The only man they could not save was Mate John Morrow, who had climbed the after mast and then tumbled into the water when it snapped. The wet, miserable survivors built a fire in the woods and huddled around it until daybreak, when they set off for help. They found respite at a fisherman's shanty, where the occupants fed them and directed them to nearby a lumber camp. The next day they were picked up by the *Edna G.* at Split Rock. All the men suffered badly from frozen hands and feet.[38]

Not all the Pittsburgh ships on western Lake Superior succumbed to the storm. The *Mariposa* was off Two Harbors at about 10 P.M. on November 27, running against a heavy sea and blinded by snow. Concerned about running aground, the ship's master successfully turned his vessel into the northeasterly wind. The vessel labored through the night and well into the next day. When the snow lifted about at 2 P.M., *Mariposa* was still off Two Harbors, having made virtually no headway after steaming full ahead for sixteen hours. Although it was still afloat, the ship was taking a frightful pounding. Wave after wave battered the cabins on the stern, stoving in the steel bulkheads and threatening to carry them away. Down in the cramped, dim boiler room, fireman William Schelb and others fought to maintain steam pressure in the boilers. Two feet of icy water surged from one side of the compartment to the other as the men threw shovelful after shovelful of coal into the furnaces. Only power could save their ship, and power came from steam. Schelb ignored his fatigue and called for more coal from the coal passers.[39]

Some Pittsburgh ships managed to avoid the Minnesota shore but came to grief at other treacherous locations on Lake Superior. On the Keweenaw Peninsula, the 413-foot *Coralia* with barge *Maia* in tow grounded at Point Abbaye. In the Apostle Islands off northern Wisconsin, the *William E. Corey* ran aground at Gull Island, ending Captain Fred Bailey's long struggle to bring the flagship safely into port.

The *Corey* had rounded Whitefish Point at about 3 P.M. on November 27 when the wind began increasing in strength. By 11 P.M. it was near hurricane velocity and Bailey had no choice but to point his ship north and plunge into the storm. Shortly before 9 A.M. the next morning he spotted the low, menacing shores of Isle Royale four miles in the distance. He turned the *Corey* to the south for much of the day, battled back to the north during the night, and finally turned south again, hoping to get his bearings from one of the lighthouses in the Apostle Islands. With his engine checked down to slow ahead, Bailey was proceeding cautiously in blinding snow at about 6 A.M. on the twenty-ninth when the *Corey* rumbled across Gull Island Shoal. When visibility improved a couple hours later, he could easily determine his location. After waiting another day for the seas to subside, he ordered four men to row the ship's yawl twenty-three miles to Bayfield, Wisconsin, and send a telegram to the fleet's office advising it of the *Corey*'s plight.[40]

While many Pittsburgh sailors found themselves stranded in virtual wilderness, others encountered their greatest peril literally within shouting distance of Duluth. Shortly after midday on November 28, the *Isaac L. Ellwood* steamed toward the Duluth ship canal, which provides entry to the sheltered waters of the city's large harbor. Earlier that day, the small steamer *R. W. England* had failed to negotiate the tricky currents at the harbor entrance. The ship tried to turn away but was blown aground on the soft sand beach two and a half miles south of the canal piers. Now it was the *Ellwood*'s turn to try the entry—a maneuver other ship captains that day would dub "shooting the chutes."[41]

The 478-foot *Ellwood* had departed Duluth harbor loaded with ore at 5 P.M. on November 27. Weather reports warned that a storm was imminent, but Captain C. H. Cummings's barometer indicated that it would not be serious so he departed without apprehension. The *Ellwood* had orders to stop in Two Harbors for fuel and to pick up the barge *Bryn Mawr* before proceeding down the lake. Snow began falling as the *Ellwood* steamed past Knife Island. By the time it arrived off Two Harbors, visibility was practically nil and the wind was rapidly increasing in strength. Unwilling to risk entering port, Cummings elected to turn the ship into the wind and slowly steam away from shore.

For the next several hours, the *Ellwood* punched its way into growing seas. Cummings periodically had to ring the engine room for more power to keep the bow pointing into the wind. As the seas grew in height, the *Ellwood* began throwing its propeller out of the water, forcing the engineer on duty to stop the engine for a moment to keep it from racing. As the ship lost power, it fell off course and wallowed in the trough of the waves for fifteen or twenty minutes before the wheelsman could wrestle it back onto its proper heading. Finally, at about 1:30 A.M. on the twenty-eighth, Cummings decided he had no choice but to let the ship blow around and head back toward Duluth.

As the storm reached its furious climax, the *Ellwood* took a beating. Waves washed over the deck time and time again. Tarpaulins stretched over the hatches to keep out water were torn away; then hatch covers started coming loose. The situation was so serious that Cummings personally led a party of men onto the exposed deck to secure the covers. After conferring with his chief engineer, the master returned to the pilothouse and began steering the *Ellwood* toward the sound of Duluth's bellowing foghorn.

With 1,800 horsepower, the *Ellwood* was among the most powerful vessels on the Great Lakes. Shortly after 1 P.M., as the ship approached the Duluth ship canal, Cummings was counting on this power to ram his vessel through the huge waves and violent undertow surging through the canal's outer entrance. The *Ellwood* plowed toward the canal, now only a few hundred feet away. For a few moments all was well. Then the ship's stern struck bottom. With a shudder, the bow pitched downward and struck bottom, then the stern dropped and struck again. The ship lost headway and the waves seized control.

The *Ellwood* veered toward the right and slammed its starboard bow into the canal's north pier. The steamer recoiled from the blow, sheered across the canal, and struck the other pier amidships on the port side. The impact ripped a gaping hole in the hull, and water rushed in. The *Ellwood* staggered into the harbor, where waiting tugboats pushed the sinking vessel into shallow water.[42]

As the *Ellwood* settled to the bottom of the harbor, the steamer *Mataafa* was approaching Duluth with barge *James Nasmyth* in tow. After leaving Duluth the previous afternoon, the pair had gotten as far as Two Harbors before the storm struck. Blinded by snow and darkness, the steamer held its course until 2 A.M., when the wind and seas pushed it into the trough of waves, broadside to the wind. With helm hard over and engine straining at full power, *Mataafa* struggled for an hour to get back on course. At 5 A.M., it was blown around again. For the next hour the wheelsman held the rudder hard over as the steamer again tried to crawl out of the trough of the waves. Realizing he was beaten, Captain Humble let the wind blow his ship around, using it to help get the *Mataafa* and the *Nasmyth* headed back toward Duluth. The snow was so thick that only the tautness of the towline assured him that the barge was still there.

Like other shipmasters caught in the tempest, Captain Humble was using every tactic he knew to survive. Back his ship came past Two Harbors. He could not seek shelter there because the wind made it impossible to turn into the harbor. On they went along the Minnesota shore. Unable to see anything and fearing he was nearing Duluth, the captain attempted another turn into the wind. For a long, frightening hour, the ship wallowed in the trough of the waves, unable to complete the maneuver. Again, Humble let the wind blow the ship around, this time coming around full circle and using the momentum to get the ship's bow pointed into the wind. "The blizzard out on the lake was something terrible," he said later. "A person could see only a few yards in any direction. The waves were the biggest I ever have seen them, and were sweeping over the decks from both sides." As the day wore on, visibility improved enough for Humble to see that his ship was drifting sideways toward Duluth. He also could see that the seas regularly sweeping *Mataafa*'s deck were bending the steel bars holding tarpaulins over the wooden hatch covers. If the hatch covers were washed way, he knew the end would come very quickly.

Increasingly concerned, Humble ordered his ship brought back on course for Duluth. With the huge seas running that afternoon, he knew it would be risky to bring the *Mataafa* into Duluth's ship canal, but clearly impossible to get the barge safely through. The *Nasmyth* and its crew stood a better chance on their own. Two miles from Duluth, he blew the whistle signals that alerted the barge crewmen to prepare their anchors and to cast off the towline. As his ship steamed away, Humble could see that the barge's anchors had taken hold of the bottom.

As the *Nasmyth* slowly fell astern, Humble turned his attention to the ship canal. He was not the only one interested in the wild seas breaking against the piers and nearby shore. Hundreds of men and women had braved the storm-driven snow and downed electrical lines on St. Croix Avenue to gather at the canal piers. They watched in awe as huge waves exploded against the concrete piers or rolled ashore in the westernmost tip of Lake Superior. From his office in the new Wolvin Building on First Street, Gus Wolvin also was watching the spectacle. The Pittsburgh Steamship Company had recently moved its Duluth office into the building, so there were plenty of people to share his concerns for the approaching *Mataafa*.

As his ship neared the canal shortly after 2 P.M., Captain Humble rang the engine room for full power. He intended to punch through the roiling water at the canal's mouth and quickly get inside the piers before the current could pull the ship sideways. With agonizing slowness, the steamer rode the waves toward the piers. The wheelsman aimed the ship toward the center of the canal. All was going well. The pierhead abutments were just ahead. In a few moments, the ship would be safe.

Without warning, a massive sea rolled under the *Mataafa*'s stern, lifting it so high that the ship's bow plunged downward and struck the lake bottom. The long steel hull shuddered. As the ship recovered, another sea swung the bow to the right toward the north pierhead. Humble shouted for wheelsman George McClure to turn to the left. The ship was slowly answering the helm when its bow slammed into the pierhead. The captain ordered the wheel reversed, hoping to slide his ship inside the canal, but the vessel was out of control. It continued to swing around until broadside across the mouth of the canal with its bow pointed southeast. Waves battered the ship against the concrete piers as Humble ordered the helm reversed again. As McClure spun the wheel, the captain grabbed the brass Chadburn handle and rang for full speed ahead. With the ship's engine already at full power, this was a desperate, emergency signal. Below decks, the engineers knew it meant to use whatever reserve power was still available to force a few more revolutions from the propeller. Captain Humble now hoped to work the ship off the end of the canal and back into open water. As the ship lay broadside to the seas, huge waves rolled over the deck, smashing one of the lifeboats and carrying away the life raft. Down in the engine room, fireman Charles Byrne heard a loud thud and saw the engine race out of control. The propeller had struck bottom, shedding its blades at the hub. Chief Engineer William Most closed the throttle, then signaled the pilothouse that he had stopped the engine. "It's all up, boys," he shouted. The engineers, oilers, and firemen ran for the stairs leading to the upper decks.

Lake Superior now controlled the stricken *Mataafa*. Slowly, the steamer slid away from the ship canal, drifting toward shore and swinging 180 degrees until its bow pointed northwest toward downtown Duluth. Humble ordered both anchors dropped, but one had been jammed in place by the collision with the pier and the other failed to slow the ship. Within minutes the vessel rumbled aground about 100 feet beyond the ship canal's north pier and about 600 feet from shore. Huge waves, breaking in the shallow water, exploded over the stranded

Waves explode over the battered *Mataafa* about twenty minutes after the helpless ship blew aground near the Duluth ship canal in the great storm of 1905. Courtesy Lake Superior Maritime Visitors Center.

steamer in spectacular fountains of foam and spray. Humble tried to make his way aft to the engine room, but was driven back to the pilothouse by the seas.

From his window overlooking the lake, Gus Wolvin watched in alarm as the *Mataafa* battered itself against the piers. He quickly rounded up as many men as he could get from the Pittsburgh Steamship Company's office and then hurried to the scene of the wreck. Throughout the city, the clarion call of shipwreck spread through taverns, hotels, brothels, and boardinghouses. Thousands of the curious and the morbid gathered to watch the drama unfolding at their doorstep. As other ships successfully fought their way into the ship canal, the spectators cheered wildly. But mostly their attention was riveted on the *Mataafa*.

Aboard the steamer, the sailors anxiously assessed their situation. Most had never experienced such a frightful storm. As long as they were on the open lake, they had been confident their ship would bring them through. Now that it was aground and awash in the foaming seas, they were scared. Twelve of the men were trapped in the bow, most in the crews quarters in the forecastle. Captain Humble ordered them to put on life jackets and stay in shelter so they would not be washed overboard. The other twelve men were aft, biding their time in the galley or in their cabins. *Mataafa* had a "flush deck" on the stern, meaning the crews quarters were inside the hull and there were no cabins on deck. The only structures there were the smokestack, the mast, and some of the large ventilators that brought fresh air into the engine room. About an hour after the ship grounded, the pounding waves smashed in the engine room gangway—a hatchway through the side of the ship that allowed easy dockside access to the engine room. Icy water poured through the opening, flooding the engine room and crews quarters. The men were forced up onto the exposed deck, where death waited to greet them.

Clambering up from below decks were the *Mataafa*'s engineers, firemen, oilers, one of the deckhands, the second mate, and the galley crew—steward, cook, and porter. Huddling in the meager shelter of the smokestack, they sized up their options. Both lifeboats had been smashed by the seas. Swimming to shore through the maelstrom would be suicide. Onshore, there was no sign of the U.S. Life Saving Service crew stationed in Duluth. If the men stayed where they were, they would surely drown or freeze to death. They had only one choice: run 250 feet down the wave-swept deck to shelter at the bow.

The first to try it was Second Mate Herbert Emigh. He sprinted as far as he could, then ducked down and clung to the wire deck railing as a wave rolled over the deck. Then he was up again, running for his life. As he reached the bow, he waved for the others to follow. Next to try it was Fred Saunders, the porter. Saunders was black, along with cook Walter Bush and steward Henry Wright. They undoubtedly were a tightly knit group, holding the only jobs available to African Americans on most lake freighters. Now, in peril, they found themselves equal to all the others. Timing the waves, Saunders began his dash. Four times waves washed over him, pulling his feet out from under him as he clung desperately to the railing. Each time he thought he was doomed, but each time he was able to hang on. Finally, he reached the forward end and was pulled into shelter by the men there.

Encouraged by the success of the first two men, fireman Thomas Woodgate started down the deck. A wave caught him midstride and carried him toward the edge of the deck. At the last possible instant, he caught himself in the wire railing, where he hung by his hands. Demonstrating remarkable strength, he pulled himself back onto the deck and resumed his run. Another wave caught him and flung him over the side. Again, he caught himself, waited for the deluge to pass, and then hauled himself over the railing. He started to run again, and another sea caught him and carried him away. One more time, he caught himself, hung from the railing, and grimly pulled himself aboard. Battered and possibly injured, he stumbled back to the smokestack. As he retreated, Woodgate was passed by fireman Charles Byrne, who succeeded in dashing to the bow. Whether they were scared by the beating

Woodgate suffered or simply biding their time, no more men tried to reach the bow. Then, with a report clearly heard onshore, the *Mataafa*'s hull broke amidships and the stern began to settle even lower in the water. The nine men huddled by the smokestack had run out of options.

Onshore, Gus Wolvin and his gang of office men were helpless. Although no longer employed by the Pittsburgh fleet, Wolvin took charge of the situation nonetheless. He ordered several employees to go to the Life Saving Station just inside the ship canal to summon its crew. When the men arrived at the station, however, they discovered the lifesavers were two and one-half miles down the beach, rescuing the men aboard *R. W. England,* the freighter that had grounded after failing to make the ship canal. The lifesavers finally arrived at the *Mataafa* as the afternoon light was fading. Under the circumstances, their preferred tactic was to use a small cannon to fire a line to the stricken ship so the sailors could rig a breeches buoy. The lifesavers first fired a line over the *Mataafa* amidships, but no one on board dared try to retrieve it. The second shot could not be found in the dark. The third shot fell across the bow. Although several sailors managed to make the line fast to the ship's forward mast, it tangled and quickly became covered in ice. The men abandoned the useless line and retreated inside. Shouting through his megaphone, Captain Humble implored the lifesavers to take his men off by boat. His words were lost in the wind as darkness fell.

It was clear now that the shipwrecked sailors would not be rescued that night. For Captain Humble, his job as a captain for the Pittsburgh fleet no longer was about joining a union, or fighting Harry Coulby, or wrangling over pay. It now was simply about duty. His ship might be aground and broken in two, but he was still its master. As such, he had one remaining duty to perform: keep his men alive until morning.

Humble ordered the sailors to scrounge all the unbroken oil lamps and dry blankets they could find and then gather in his cabin. They lit the lamps and draped the blankets over their shoulders. The captain then ordered them to begin bouncing up and down. Wet, cold, hungry, tired—they were all likely to freeze to death. No one could be allowed to stop moving, to sit down, to sleep. That would mean death. For the next twelve hours, Captain Humble, First Mate Walter Brown, Second Mate Emigh, and one of the wheelsmen exhorted, cajoled, and begged the men to keep moving, to keep swinging their arms, to keep pummeling each other. Relentless waves plucked the glass from the cabin windows and stove in the door. Water sluiced across the floor and spray froze on the walls. For what must have seemed like eternity, amid the acrid smoke and dim light in the captain's modest cabin, the fifteen men shuffled about in a macabre dance.

About 3 A.M. on the twenty-ninth, the lamps began flickering out and the cold grew even more intense. Captain Humble feared the end was near. Desperate, he left the group and waded down a hallway flooded with three feet of icy water to the windlass room, a large compartment housing the anchor windlass and chain. The seas had gone down by now, and the windlass room was no longer flooded. The captain found some dry rags, a can of kerosene, and a box of matches. Wielding a fire axe, he battered apart a wooden bathroom. Using the kerosene and wood as fuel, he coaxed a fire to life, then called for the crewmen to join him. They huddled around the fire until the lifesavers returned by boat at 7:30 A.M. to begin taking them ashore.

Captain Humble and his mates helped load seven men into the lifesavers' boat. As the boat pushed off from the *Mataafa*, the survivors who remained aboard ate food and drank brandy that the lifesavers had left with them. Then Captain Humble and Mate Brown steeled themselves for what they knew would be a grim task. They crawled the length of the ice-covered ship, carefully climbing over the break in the hull. At the stern they discovered what they had expected. "We found four bodies frozen in the ice on deck near the smokestack, and could see that no one could possibly be alive around there," Humble reported. "We then went back forward and the life boat came out a second time. We helped the balance of the crew in the boat, then the mate slid down a line and then I slid down, the last one into the boat." Captain Humble had performed his duty.[43]

Across Lake Superior, the great 1905 storm sank or damaged thirty vessels. Of those, ten were steamers and barges of the Pittsburgh Steamship Company. Considering the number of wrecks it suffered, the fleet lost surprisingly few people. Twelve Pittsburgh men died, including all nine who remained on the stern of the *Mataafa*. The death toll could have been much higher were it not for the bravery and ingenuity of so many sailors. In nearly every instance, they rescued themselves.

As the reports of mounting vessel losses began coming into the Cleveland office, Harry Coulby chartered a train to take him to Duluth. The locomotive steamed into town late on the twenty-ninth, pulling a coach, a business car, and two freight cars filled with wrecking gear and pumps. Accompanying Coulby was W. W. Smith, marine superintendent of the Pittsburgh fleet, who was to personally take charge of salvage operations. The pair immediately went to a hotel near the train

The broken, ice-covered *Mataafa* lies abandoned in early December 1905. Buildings in downtown Duluth are visible in the background. Courtesy Northeast Minnesota Historical Center.

station to meet with H. W. Brown, head of the fleet's Duluth office; marine surveyor Joseph Kidd; and the masters of the *Lafayette, Mataafa, Crescent City, Edenborn, Mariposa,* and other Pittsburgh ships to assess the damage and begin making plans to recover the lost vessels. Early the next morning, Coulby and Smith went to the fleet's office on the seventh floor of the Wolvin Building. Smith later went on to Two Harbors, where he boarded the tug *Edna G.* to steam up the shore to inspect the *Edenborn, Lafayette,* and *Manila.*

Although he was busy mapping a strategy to save his ships, Coulby took time on December 1 to address two issues that had set tongues to wagging in Duluth. "While I have been in Duluth I have heard there was some criticism on the street of the tardiness of the lifesaving crew in rendering assistance to the *Mataafa*," Coulby said in a statement to the newspapers. "I do not think this criticism is warranted, from the fact that they had no knowledge that the *Mataafa* would need their services, and, as the steamer *England* was in distress, it was, of course, their first duty to go to her."

"I also understand there have been rumors that the boats of this company are expected to leave port immediately after they are loaded, regardless of weather conditions," he continued. "I am quite unable to understand how such a rumor as this should get abroad for the reason that it is in direct contradiction to the positive orders the masters of our steamers have had repeatedly. This company spares no expense to keep its ships absolutely seaworthy and our captains have positive orders that no official of the company will be permitted to give them any orders as to when they shall leave port or seek shelter."[44]

Captain Humble issued his own statement the same day to support Coulby's claim. "The weather was rough, but the boat was entirely seaworthy and there did not appear to be any hazard in setting out. No one forced my hand in the smallest regard and never in my experience with this company have I, or any other captain, been urged to put out when he felt that the hazard of the weather was too great. In fact, our instructions are positive that we must use our own judgment in this matter and there has even been a word of caution added. I want this said so that the people may know that if there was any error in judgment it was my own."[45]

Other shipmasters concurred. "I received no special instructions from the management of the Pittsburgh Steamship Company relative to my leaving port at that time," said Captain Cummings of the *Ellwood*. "In fact I never receive orders to leave port at any special time, that being a matter left to my own judgment, and I always take into consideration the weather conditions."[46]

Coulby and Smith hired every piece of wrecking equipment belonging to Union Towing and Wrecking Company, the Duluth subsidiary of Great Lakes Towing Company. Salvage operations began immediately. Two vessels clearly were total losses: the *Lafayette* and *Madeira* had been torn apart on the rocks. Oddly enough, the *Lafayette*'s stern remained intact and eventually would be refloated and brought to Duluth. Some vessels would be left in place during the winter and refloated the following spring. Coulby and his captains agreed that the *Mataafa* was so badly mauled that leaving it aground during the winter would not cause any more significant damage. Salvage crews would focus their immediate efforts on the remaining vessels. Men were already pumping out the *Ellwood* to prepare it for refloating and repair, and the barge *Manila* needed immediate attention. The *Edenborn*, badly damaged at Split Rock, would be brought off. Workers would patch the break in the *Crescent City*'s hull, then jack up the vessel and pull into it deeper water. The top priorities, however, were freeing the grounded and exposed *Corey* and the steamer *Coralia* with its consort *Maia*. The latter pair came off with relative ease, but the *Corey* proved to be a much tougher problem. At first a Pittsburgh steamer was dispatched to simply pull the *Corey* into deep water. When that failed, more ships were sent to the scene. A week after the *Corey* grounded, the job of salvaging the ship had become a major operation. Four Tin Stackers and three tugs were gathered around the stranded ship and 158 men were working the wreck. Coulby and Wolvin arrived on a Pittsburgh ship carrying dynamite to be used in blasting away part of the rock formation holding the *Corey* in place. After considerable effort, the flagship finally came free at 7:20 A.M. on December 10. After waiting a day for the weather to improve, it was taken to Superior to begin $100,000 in repairs.

In Duluth, survivors of the wrecked Pittsburgh ships were put up in hotel rooms, where they lay in bed for a few days, eating, smoking cigars, and drinking brandy. After recovering with the help of these stimulants, they returned to work on salvaging their ships or caught trains for home. Virtually all agreed that the storm was the most terrible they had ever seen. "What do I think about it?" mused F. D. Seeley, first mate of the *Lafayette*, as he nursed a sprained knee. "Why, I have sailed this lake for fifteen years and never dreamed that the wind could play such havoc. I never imagined that the sea could roll to the height of the wicked waves which swept the deck of the *Lafayette*. . . . It was awful, and no language of mine can make you understand the relentless force, the power of that storm."[47]

4

The Pace Quickens

As spring came to the Great Lakes in 1906, salvors began working on refloating the *Mataafa,* aground and broken off Duluth. The ship eventually was salvaged, rebuilt, and put back into service for the Pittsburgh fleet along with the other survivors of Minnesota's North Shore—*Crescent City, Edenborn,* and *Manila.* Salvage crews also saved the stern of the *Lafayette,* towing it back to Superior, where the engines were removed and eventually installed in another steamer.

Although it was an expensive setback, the devastating storm of 1905 did not slow Harry Coulby's drive to expand the Pittsburgh Steamship Company's tonnage capacity. The company already was building a new class of ships that would eclipse the giants of the *Gary* class. The Chicago Ship Building Company launched four of these ships in 1906. They were dubbed the *Morgan* class and included *J. Pierpont Morgan, Norman B. Ream, Henry H. Rogers,* and *Peter A. B. Widener.* With its extra guest accommodations, the *Morgan* became the fleet's new flagship.

Again, these were landmark vessels. The previous year Coulby had told a meeting of the Lake Carriers' Association that shipbuilding on the Great Lakes was in the midst of a revolution. Not only were shipyards producing longer ships, they were building wider ships. He called for navigation channels, loading docks, and other facilities to be improved to accommodate ships as wide as 75 feet.[1] Now Coulby was taking the first step toward these wider ships. The *Morgan* was 58 feet wide—two feet wider than previous ships—and 601 feet in overall length. This made the *Morgan*-class vessels the first true 600-footers on the Great Lakes and the new standard by which all other ships would be measured. The increase in size gave the ships the capability of carrying about 12,000 tons of iron ore.

The *Morgan*-class vessels were intended to become the backbone of the fleet. Four more were built in 1907: *Thomas Lynch, George*

F. Baker, Thomas F. Cole, and *Henry Phipps.* Like their predecessors, these ships quickly proved themselves to be the height of efficiency and profitability. In its first year of operation, *Thomas F. Cole* steamed 36,000 miles and carried more than 248,000 tons of iron ore loaded in Duluth and Two Harbors. After expenses were deducted, it earned the fleet $51,677, compared to the $4,347 profit eked out that year by tiny barge 118.²

No new ships entered service in 1908, probably because of a national economic recession and a sharp downturn in lake shipping. The following year, however, shipyards turned out *Eugene J. Buffington, Alva C. Dinkey,* and *J. P. Morgan Jr.* They were followed in 1910 by *William B. Schiller, William B. Dickson, William J. Olcott,* and *William P. Palmer.* Outwardly, these ships appeared to be essentially the same, although minor modifications were made. The *Schiller* and *Morgan Jr.,* for instance, were the first ships in the fleet to have anchor wells—recessed pockets that allowed their anchors to be pulled in flush with the hull to avoid striking docks or other ships when maneuvering in tight quarters. Starting with the *Dinkey,* an additional keelson—a large beam running the length of the ship's bottom parallel to the keel—was used to stiffen the hull. Construction of the *Palmer* marked the beginning of the Pittsburgh fleet's experiment in building ships using the Isherwood style of construction, in which hull frames were longitudinal rather than transverse. This type of ship was expected to be stronger and more flexible—important factors as ships grew longer. Isherwood construction would be used for a total of nine Pittsburgh ships before being discontinued after 1917.³

Pittsburgh Steamship Company did not build any new ships in 1911—again, it was a poor year for shipping. But the company did purchase three more big ships from the Weston Transit Company. Each boasted an overall length of 605 feet and a beam of 60 feet. These vessels, *LeGrand S. DeGraff, William B. Kerr,* and *William M. Mills,* became the fleet's biggest ships and held that status for several years. They were renamed *George G. Crawford, Francis E. House,* and *William J. Filbert,* respectively.

The final move in this wave of expansion came in 1913 with the launching of the final three vessels in the *Morgan* class. *Percival Roberts Jr.* and *Richard Trimble* followed the usual design. *James A. Farrell* became the new flagship. It had an entire extra deck built into the forward cabin to provide luxurious bedrooms and an observation room for guests. The extra deck gave the ship a distinctive three-story forward superstructure. A galley and dining room for guests occupied a

A pair of whaleback barges leave the Poe Lock on a sunny evening under tow of a Tin Stack steamer, which has slowed to receive supplies from the fleet's supply boat, *Superior*. In the fleet's early years, Pittsburgh steamers often towed one or two of the company's many barges. Library of Congress.

deckhouse that sat on the spar deck between the first and third hatches. All told, Pittsburgh Steamship Company had built eighteen stout steel ships and purchased three more that had an overall length over 600 feet.

With the addition of twenty-one ships capable of carrying 12,000 tons or more of iron ore on each trip, the fleet eagerly unloaded many of its oldest and smallest vessels. Seven whaleback barges and two conventional steamers were sold in 1909, followed by another steamer in 1911 and five more steamers and a barge the following year. All the vessels sold were under 400 feet in length, and many had been built in the 1890s.

While the fleet's expansion program was proceeding smoothly, its relationship with organized labor was not. Members of the Lake Carriers' Association, led by Coulby and the Pittsburgh Steamship Company, signed one-year labor contracts with several unions for the 1906 and 1907 navigation seasons. Under the 1907 contract, the Lake Seamen's Union was given exclusive right to supply men for the jobs of wheelsman, watchman, lookout, and ordinary seaman. The union agreed that if it could not supply enough men to a specific ship, the master could hire anyone he chose for one round-trip. After that, the union would again have the right to fill the jobs. Vessel owners were required to provide the men with clean sleeping quarters that had good mattresses and clean linens once a week. They were exempted from painting, cleaning brass, or scrubbing decks on Sundays and legal holidays. For their work, wheelsmen, watchmen, and lookouts were to receive $50 a month from the opening of navigation until October 1. After that they were to get $65 a month for sailing during more the hazardous autumn months. The monthly pay for mates on barges was $70, a respectable 8 percent increase over the previous year.[4] Neither the fleets nor the unions saw their labor agreements as anything resembling a lasting peace. Each pact was merely a one-year truce in a long war. Members of the Lake Carriers' Association disliked the labor contracts and felt they ended up with a poor bargain each time. Their only hopes were that they could stabilize wages from year to year and persuade the unions to curb the continuing problem of desertion among crewmen.[5]

As a temporary peace reigned during 1907, the steel companies and the bigger fleets were quietly setting a trap for the unions. Throughout the year, every available steamer and barge was pressed into service to carry as much iron ore as possible. Steel mill superintendents piled ore on every piece of available property. When the mill yards were full, docks began stockpiling ore. By the end of the season, the fleets had hauled 41.2 million gross tons of ore, nearly 4 million tons more than the previous year. Virtually every mill on the lower lakes had a comfortable backlog of the vital resource.

During the winter of 1908, Coulby sprang the trap. On March 26 he led owners of the bulk cargo fleets in declaring an open shop aboard their ships. From now on, no labor union would have exclusive right to represent men of the Pittsburgh Steamship Company or the other steel fleets. Next, Coulby sent each of his chief engineers individual contracts to sign. The contracts stated that the engineers would carry out the company's rules and duties, and that they could be dismissed for "satisfactory reasons." They had to agree to bar from their ship any member of the Marine Engineers Beneficial Association. Further, they had to agree that if their ship was laid up before the end of the season, they would serve as an assistant engineer on another vessel at their regular chief engineer's pay.

The engineers pondered their contracts, then considered the huge stockpiles of ore at every lower lake port. They understood that if they went on strike, the steel company fleets would simply leave their ships tied up at their docks. While the engineers went without work and pay, the steel mills would be comfortably operating on the backlog of ore. Realizing they had been outmaneuvered, nearly all the engineers signed their contracts and returned them to the Pittsburgh fleet office in Cleveland's Rockefeller Building. There would be no strike in 1908.[6]

The fleet owners, however, were not content with their gains. On the heels of Coulby's victory, the Lake Carriers' Association launched its own attack. It declared that no union representatives would be allowed aboard its members' ships. Then it began blacklisting union men to ensure they would not work or cause trouble. Next came an announcement that all men working aboard ships must sign oaths stating they were not union members. "I hereby renounce all allegiance to any and all labor unions, particularly the stewards, seamen, firemen, and oilers, and I declare it to be my intentions not to join either as long as I follow sailing for a living," the association's oath stated. "I am therefore a nonunion man and if I can get a position on a lake vessel in the event of any strike of any description involving the _____ _____ union, I will stay by the ship and faithfully perform my duties as such employee despite any strike or orders by any union."[7]

As the navigation season got under way, masters of Pittsburgh vessels took it upon themselves to rigorously enforce the fleet's open shop policy. On June 10 the barge *Maia* was docked in Toledo when its master and the local representative of the Lake Carriers' Association assembled the entire crew on the barge's stern. They told the men to throw away their membership books for the Lake Seamen's Union, adding that any man who continued to pay union dues would not be allowed to sail for the Pittsburgh Steamship Company. The captain of the barge *Manda* waited until he docked in Conneaut to fire three men for belonging to the union. Seymour Higgs, another union member, quit in protest. At first the captain refused to pay Higgs, but then, after considerable argument, offered to pay him a dollar a day. Higgs started uptown to get a lawyer before the barge master reluctantly agreed to pay him in full, then gruffly warned him to stay away from the union or he would lose all his money. Rudolf Schreiber complained that he was fired from a job as watchman on the *George F. Baker* because he spoke in favor of the union while talking to another sailor. Some men, mostly the union activists, swore out statements before local notaries public. The tactic provided a record of their complaints but accomplished nothing else.[8]

Speaking to a gathering of the Ship Masters' Association, Coulby made it clear that he was not going to share control of his fleet with any union or be hampered by union rules and restrictions. "What we are trying to do is simply to get back to the old conditions aboard ship," he said. "For my company, I can say that we are going to win if it takes one day, one month, one year or five years. If any man pulls a book of rules on you, he is not an open shop man. Put him on the dock. If any engineer, first, second or third, wheelsman, watchman, mate declines to obey orders, put him on the dock. We will help you fill their place."[9]

Even without a strike, 1908 was a difficult year for Great Lakes shipping. The combination of a nationwide economic slump and the huge backlog of ore prompted Pittsburgh Steamship Company to operate its ships only from June to November. The smallest whaleback steamers and barges remained idle the entire season. The slowdown cut deeply into the fleet's earnings. In 1907 the Pittsburgh fleet had earned $10.7 million and posted a net profit of $1.6 million, or .117 cents per ton of cargo carried. During the slump of 1908, Pittsburgh vessels earned $8.1 million and showed a profit of $520,000, a slim .046 cents per ton.[10]

The anti-union offensive continued during the winter of 1908-9 when the Lake Carriers' Association unveiled its "welfare plan." To be eligible for employment by any of the association's fleets, a man now had to belong to the welfare plan. To belong to the plan, he had to pledge to follow company rules and pay a small monthly insurance premium. Each man would then be issued a "continuous record discharge book" that listed his physical description and provided a history of his prior service. For each job a man held, that ship's master would mark whether his work had been good or fair. If a man's work was deemed unsatisfactory, the shipmaster could confiscate the discharge book and mail it to the Lake Carriers' Association. If a sailor died, the welfare plan would pay his designated beneficiary anywhere from $75 to $500, depending on the man's rank. In addition, the Lake Carriers' Association would set up assembly rooms in major ports where plan members could safely read, sleep, or study for advancement while waiting to be called for a job.[11]

Ostensibly, the welfare plan was a means to increase efficiency by attracting stable, qualified men and providing them with the security of life insurance. It undoubtedly did exactly that, but it also provided members of the Lake Carriers' Association with a handy way to screen out union members and exert leverage on men who were hired. After all, a man's behavior could be controlled by the threat of confiscating his discharge book and leaving him unable to get a job on another ship. Any man who quit sailing would lose the money he had paid into the welfare plan. The plan drew immediate and long-lasting criticism from union activists. "By this system the seamen are to be held in constant fear of an adverse opinion being rendered against him by any captain he serves under," union officials later told a congressional panel. "Always he must dread that the master will refuse to return his book when the term of employment is ended. If he desires to quit when the captain does not want him to he must hesitate—terror of being deprived of the book is relied upon to bind him to the ship. He must not dare resent ill treatment; he must not complain; under constant espionage, he must live in fear, always in dread of the fatal mark."[12]

Implementation of the welfare plan was the last straw for the unions. Despite an economic downturn in the spring of 1909, the Marine Engineers Beneficial Association called a strike by its members. By May 1 several other unions had followed suit. One of their grievances was the welfare plan and its continuous-record discharge books. The unions charged these books essentially were a blacklist. A sailor whose book was confiscated would have no appeal; he would not even be able to find out why the book was taken. Shortly after the strike began, a state arbitration board in Ohio met in an attempt to settle the dispute. Union officials appeared to state their case, but no one showed up from the Lake Carriers' Association. Over the next several months, the arbitration board appealed to the association to attend the sessions. Coulby responded by saying the open shop principle could not be arbitrated.[13]

The strike was a dismal failure. The fleets went to the East Coast to hire saltwater sailors, who gladly accepted the better pay, better working conditions, and better food they found aboard the lake freighters. Members of the International Longshoremen's Association refused to honor picket lines, enabling ships to continue loading and unloading. Only a few weeks passed before the fleets were operating at nearly full strength. Their position became even stronger as the economy warmed up and demand rose for iron ore. Many union members feared losing their jobs for good, so they defected and went back to work. Those who did not go back put up picket lines at ports around the lakes. As the season wore on, the strikers grew increasingly frustrated. Shouted insults escalated into fistfights. Men leaving their ship to find a tavern or to visit their homes often were accosted by pickets. "In many cases where argument failed of effect, violence was resorted to, and throughout the season assaults were committed upon engineers in the service of their vessels," the Lake Carriers' Association reported. "Instances of slugging and assault upon men employed upon our vessels were of frequent occurence in all ports." Of the 120 Great Lakes sailors killed in 1909, 6 were shot in strike-related disputes.[14]

The strike dragged on for three years, but its effect was negligible after the first few weeks of the 1909 navigation season. The lake fleets were affected little except for the costs incurred from accidents caused by inexperienced men. The strike's failure left the unions in disarray and the Lake Carriers' Association in firm control of deciding wages and hours for men aboard ship. Nonetheless, members of the association instituted pay raises for the 1910 season ranging from $1.50 to $10 a month. The lowest ratings were porters, earning $30 a month, and deckhands, making $31.50. At the upper end of the pay scale were mates on the largest ships, making $130, and chief engineers, making up to $175.[15]

Although the unions failed to win their strike, they were credited with drawing attention to poor working conditions, long hours, and low pay. The Lake Carriers' Association, led by Pittsburgh Steamship Company, began an sustained effort to increase stability in the workforce. Some of these efforts reflected the anti-union attitudes and social prejudices of the day. Better discipline was enforced by recruiting men to report on any shipboard agitators. The fleets sought to replace irresponsible men with those considered to be culturally "suited" for the work. For instance, the Lake Carriers' Association recommended filling jobs such as bosun or wheelsman with Scandinavians, who came from countries with nautical traditions. For the job of fireman, it urged members to hire from among the ranks of recently immigrated Greeks, Poles, Italians, Austrians, and Slavs. It noted that the latter generally were frugal, eager to earn and save money, and ignorant of unions. Many masters and mates hailed from small towns along the shores of the Great Lakes. These men traditionally were expected to hire the "forward end" crew of mates, bosuns, wheelsmen, watchmen, and deckhands. The association recommended recruiting men from small farming communities a few miles inland from the lakes. "These farmer youths are unacquainted with unionism, have no prejudices against their employers or the employing class, are regular in their habits and for a time at least, are glad to accept without questions the change of work and the relatively high wages offered them."[16]

Much to its credit, the Pittsburgh Steamship Company also focused on improving working conditions aboard its ships. In 1905 Coulby began holding annual winter meetings with his shipmasters, and four years later began similar meetings with his chief engineers. It was hardly a new idea. The Minnesota Steamship Company had con-

A Tin Stack steamer tows the yawing barge *Martha* through the Duluth ship canal. The fleet had sold most of its whaleback barges by 1909, but bigger barges like the *Martha* remained part of the company for several more decades. Hugh McKenzie photo, Ken Thro Collection, Lake Superior Maritime Visitors Center.

ducted such meetings in the 1890s. But the meetings held by the Pittsburgh fleet received prominent attention and established the trend among the other Great Lakes fleets. Coulby used the sessions to discuss problems and seek solutions from the masters and engineers. Proposed changes in policy were discussed by committees and voted on by the entire group. Coulby said he would not issue orders for a change in policy or procedure unless at least half the men voted in favor of it. New ideas often were tried aboard a few ships and, if successful, implemented throughout the fleet. These meetings led to significant changes for Pittsburgh sailors. Considerable emphasis was placed on providing better and more plentiful food; sanitary toilets and bathing facilities; and payment by monthly check rather than cash, with an opportunity for men to have their money deposited directly with the Cleveland Trust Company. As a result of the meetings, the fleet also agreed to push for improvements to the Lake Carriers' Association assembly rooms and for classes to be held over the winter to help men studying to earn higher licenses as engineers, mates, or masters.[17]

The Duluth, Missabe and Northern Railway ore docks in Duluth, shown here about 1910, were one of two primary loading ports for the Pittsburgh Steamship Company. Trainloads of iron ore were dumped into the docks' elevated pockets, which could then be emptied by gravity into waiting vessels. Hugh McKenzie photo, courtesy Northeast Minnesota Historical Center.

The annual meetings usually lasted two or three days and were held in a Cleveland hotel. They covered dozens of topics ranging from the proper means of handling laundry to establishing standard routes between ports. Technical sessions usually were conducted by the fleet engineer or Coulby's assistants, such as A. F. Harvey. But during much of the session, Coulby presided, at times leading discussions or seeking consensus among the men, at other times lecturing, joking, inspiring, or scolding. Above all, he used the meetings to preach his gospel of efficiency, of improving the fleet by improving its men and ships. "The Pittsburgh Steamship Company, as you know, has come on these lakes to stay," Coulby told his engineers and masters at the start of the 1911 meeting. "It is a pretty important link in the chain of the Steel Corporation from its raw material to the finished product; they cannot get along without it and we have not only to keep our fleet up and keep adding to it and modernizing it but we have to do something equally important and that is the developing of our organization, and I want the thought to be in the minds of all the men who work for us and more particularly of our officers and the men holding the higher positions as Captains and Chief Engineers, that just so long as you want to continue sailing, that so long as you perform your duties satisfactorily to the Company, you will have a definite, permanent position with the Company."[18]

Coulby used the meetings to introduce a mandatory retirement age of seventy for masters and to unveil U.S. Steel's pension plan. He urged masters and engineers to hold shipboard meetings with their officers to ensure that everyone understood critical rules and procedures. He also encouraged them to groom their mates and assistant engineers for promotion. "I have always been willing to extend to the men employed under me as much consideration as I have ever received myself, and that is to give them a show. I think every man is entitled to a show to make good and my plan has been if he could not make good, he should step aside and let some other fellow have a chance; that holds true all the way down the line," he said. "I want you to take a personal interest in your subordinates, to see that they are getting a fair show and see that they are in line for promotion as opportunity offers and as the men demonstrate their qualifications for the better positions."[19]

Ore docks employed hundreds of men, including these "ore punchers," whose job was to prod wet, sticky ore with steel rods to ensure it all ran out of the railcars and into the dock pockets. The number of cars shown atop this ore dock offers a hint of the structure's size. Hugh McKenzie photo, courtesy Northeast Minnesota Historical Center.

The meetings also provided Coulby with a platform to express his interest in improving shipboard safety. Working aboard Great Lakes ships in these years was a hazardous occupation. While shipwrecks received much of the attention for causing injury or death, each year many sailors died by falling into cargo holds, tumbling overboard, or becoming entangled in machinery. One of Coulby's favorite topics was the need to avoid accidents by educating the crew and by altering machinery to make it safer. "There is nothing in connection with the Steamship Company today that I am as much interested in as the protection of life and the protecting of our employees from injury," he said in 1911. "Now, gentlemen, that is not a duty that you owe to the Pittsburgh Steamship Company so much as it is a duty you owe to the community, to your fellow man and to your fellow worker."[20]

As part of his drive for safety, Coulby took a dim view of drinking by his officers. Again, he held forth on the topic at the 1911 meeting. "When anybody is dropped out of the employ for drinking they will not be reinstated. . . . There was a good deal of pressure brought upon me to reinstate McLaughlin, captain of the *Wawatam*. I gave it a good deal of thought and it was with some misgivings that I dropped him out. He was a man whom I had thought was one of our coming men. He made good in every position he had ever been in until he got to be master of the *Wawatam*. Now it does seem pretty hard, I appreciate, to drop a man out of this line for any cause and I don't know of any other thing we are likely to do it for except for drinking. I have been giving the question a great deal of thought for some time and I have made up my mind that we will not reinstate anybody who is dropped from the employ for drinking. I look at it from this standpoint: If it were only a question of property I might be more lenient, but when we send a ship out there are from twenty to twenty-five men on board. The most of them have families on shore depending upon them and I feel I should insist that the officers on that ship should not try to do their work while under the influence of liquor. As I have said, when you get through and your boat is laid up and you are on your own time, I am no custodian of your habits but when you have reported for duty and until you are through in the Fall, if we catch anybody under the influence of liquor they are going on the dock and so far as our Company is concerned they will stay there."[21]

Even during serious discussions, however, the meetings sometimes allowed a rare glimpse of Coulby's sense of humor. At one meeting he was chairing a long discussion among the engineers on the best brand of "boiler compound" to use for keeping boiler tubes free of rust and lime deposits. At one point, Coulby asked engineer Lon Arnold for his opinion on the matter.

"We haven't used any compound for some years; we have cut it out. I am using the Learmonth Purifier," Arnold replied.

"Well, the compound must be a good thing, Lon," Coulby countered, "for there are so many fellows selling it that used to be in the insurance business."[22]

The annual meetings also gave Coulby and his vessel masters an opportunity to carefully review the numerous collisions, groundings, and other mishaps involving ships from the Pittsburgh fleet. While the period from 1905 to 1913 was marked by the fleet's acquisition of giant new ships and the sale of older ones, it also was marred by the loss of four ships and near disaster for many others.

Coincidentally, the four lost ships were all built in the 1890s, and three were under 300 feet in length. The *Grecian* grounded on June 7, 1906, near Detour, Michigan. It was released but sank three days later off Thunder Bay Island while under tow to a shipyard. The 308-foot whaleback *John B. Trevor* stranded at Rainbow Cove on Isle Royale on October 11, 1909. The Pittsburgh fleet hired a salvage crew to free the ore-laden ship, but after a month of fruitless effort the fleet abandoned it to the insurance underwriters. The string of accidents continued in 1911 when *Joliet* sank after colliding on September 22 with fleetmate *Henry Phipps* opposite Port Huron, Michigan. Because of their age and relatively small size, none of these vessels was considered a major loss for the fleet.

Just eight years after the 1905 tempest ravaged Pittsburgh ships on Lake Superior, an even greater disaster struck the Great Lakes. Known in later years as the Great Storm of 1913, it developed as three massive weather systems collided and swirled over the lakes from November 7 to 10. The fury wrecked seventeen ships and killed an estimated 244 sailors, making it the deadliest storm in Great Lakes history.

As was typical for that time of year, many ships were still sailing, including a full contingent of Pittsburgh vessels. On Lake Superior, the *Alva C. Dinkey, George Stephenson,* and *Cornell* were plying their trade along with the whalebacks *Alexander McDougall* and *Henry Cort.* Farther down the lakes, the big *George G. Crawford* was upbound light for the upper lakes. Not far behind was *Matoa,* carrying coal from Toledo to Hancock, Michigan. Some masters had seen the flags hoisted by the U.S. Weather Bureau that warned a storm was imminent. Others knew from their plummeting barometers that trouble was brewing.

Aboard the *Robert W. E. Bunsen,* Captain George Holdridge was in the pilothouse, anxiously tapping the weatherglass and making a somber prediction to Second Mate Arthur Dana. "Boy, you are going to see a storm such as you never saw before," he said.[23]

The storm first rolled across Lake Superior on November 7, bringing high winds, heavy seas, and thick snow. Visibility fell to a few hundred yards, then to just a few yards. Ships were sailing blind, their masters praying they would not encounter another vessel. Loaded and with little freeboard, the *McDougall* plowed along as Captain F. D. Selee anxiously sought shelter behind Whitefish Point. The first mate tied himself to the bridge wing and kept swinging the sounding lead over the side, calling out the depth as the captain used a combination of skill and intuition to aim toward the buoy that marked the turn into the bay. Finally picking up the sound of the buoy's underwater bell, Selee swung the *McDougall* around the marker and into the sheltered waters without ever having sighted land. Behind them, the *Cort* was having a tougher time of it. Downbound and heavily laden with iron ore, the *Cort* was overtaken by every following sea. Waves rolled down the whaleback's deck, stripping away its running lights, deck railings, and even the light mounted on the foremast. As each wave receded, it left behind a coating of ice, which quickly accumulated and added its weight to the forces trying to sink the ship. The *Cort* looked like an oddly shaped confection when it finally arrived at the Soo.[24] The *Dinkey*'s master, who commanded a ship much larger than the whalebacks, chose to take his chances on the open lake. He pointed the big steamer's bow into the wind and steamed ahead for twenty-three hours until the weather began to moderate and he could resume his course to Duluth. "We had our whistle going continuously for twenty-five hours," he said, "but because of the roaring gale and blizzard I don't think I heard it once."[25]

With an overall length of 474 feet, the steamer *Cornell* seemed better able to handle the storm than the whalebacks. The ship was ninety miles above Whitefish Point when the storm struck at 2 A.M. on November 8. All was going well at first, but then the first mate fell violently ill while on duty in the pilothouse. Captain John Noble ordered the ship turned before the wind to help steady it while men carried the mate down to his cabin. When Noble tried to bring his ship back on course, he was unable to get it out of the trough of the waves. The *Cornell* wallowed in the troughs for the next thirteen hours, broadside to wind and sea, drifting out of control. The ship was within a mile and a half of shore when Noble saw his chance. He ordered both anchors dropped and storm oil poured over the bow. With the engine working at full ahead, he was able to turn the ship's bow into the wind. For another twenty-four hours the ship lay at anchor in just forty feet of water, its engine pounding at full speed. Late on the ninth the storm seemed to moderate, and the *Cornell* weighed anchor and fought its way into deeper water. But as the gale worsened later that night, the steamer once again slid into the trough of the waves. Again using his anchor, Noble got the ship pointed into the wind. About 5 A.M. on the tenth, a tremendous wave broke over the stern, smashing in doors on the after cabin and flooding the galley, dining room, and crews quarters. Again, the ship slid into the troughs, only to escape once more by dropping anchor and running the engine at full power. Twice more the ship fell into the troughs, twice more it managed to escape. Down in the engine room, Chief Engineer Charles Lawrence had his crew on duty seventy-eight hours straight, and during much of that time the men were convinced they were doomed. Although normally the warmest place in the ship, the engine room grew so cold that the men donned scarves and mittens as they worked. When the wind finally subsided on the evening of November 10, the *Cornell*'s crewmen could scarcely believe they had survived.[26]

As the storm swept on, it continued to develop and grow stronger as it struck Lake Huron and Lake Erie. Inside the breakwall off Cleveland's waterfront, the Pittsburgh Steamship Company had moored eleven barges—the *John Smeaton, John A. Fritz, John A. Roebling, Manila, Maida, W. LeBaron Jenney, Sydney G. Thomas, George H. Corliss, Alexander Holley, Martha,* and *Marcia*. They were lined up broadside to each other and linked by two-inch wire cables. By the evening of November 10, the vessels were banging together and ripping the steel mooring bitts right out of their decks. The *Thomas, Holley,* and *Jenney* eventually broke free and blew aground. All told, the barges suffered $100,000 in damage. It could have been worse. They were nearly joined by the new steamer *Richard Trimble*, which had ducked into shelter behind the breakwall. The ship's captain dropped both anchors, took on 8,000 tons of water ballast, and kept his engine working at full ahead to keep from blowing ashore in these "sheltered" waters. When the winds finally eased, the *Trimble* had dragged its anchors until it was within three hundred feet of shore.

All the trouble on Lake Superior, Lake Michigan, and Lake Erie, as severe as it was, seemed secondary to the terror on Lake Huron. It was there that the storm seemed to peak, catching numerous ships whose masters had ignored the Weather Bureau's warning flags, signal lights, and verbal messages. The maelstrom centered on southern Lake Huron, where eight ships went to the bottom during the dreadful hours of November 9 and 10.

Among those cursed to sail Lake Huron that day was *George G. Crawford* under command of Captain Walter Iler. Upbound in ballast, the ship struggled through the opening hours of the storm. Early Sunday afternoon, November 9, the ship was blown around and lay in the troughs of the waves. Unable to head into the wind, Iler turned around and retreated down the lake. "When we were running before the sea the waves were so large they would fill up the after part of the ship and go over the very top of the cabin, down through the skylight, which was smashed through, and into the engine room," he said. "The waves broke all the windows in the cabin and filled the dining room and kitchen with water. None of the men could stay in their rooms."

As he neared the southern end of the lake, Iler was forced to turn the *Crawford* again to avoid running aground. Despite losing both anchors in the maneuver, he managed to get the steamer sailing north, into the wind, once again. As his ship battled its way back up the lake late Sunday afternoon, the intermittent curtains of snow parted long enough for Iler and the other men in the *Crawford*'s pilothouse to spot another ship. Iler thought it was the steamer *Argus* of the Interlake Steamship Company. The ship was obviously in distress. It was caught in the trough of the waves, wallowing from side to side and unable to escape. As Iler and his men watched in stunned disbelief, the *Argus*'s hull simply broke apart. "The *Argus* just appeared to crumple like an egg-shell and then disappeared," he said. In the raging seas, the *Crawford*'s crew could do nothing but watch and pray as their ship was swept away from the wreck.[27]

"I never would have believed that any storm could seize a ship like the *Crawford* and hurl her about into the trough of the sea," Iler said later. "It was the wind from the north that did all the dirty business. I never saw a gale on the lakes to equal it, and others with whom I've talked since say they can't recall ever hearing of such a storm in the past. I was in the storm of 1905, but that was a summer zephyr compared with the one that raged over Lake Huron that Sunday."[28]

Less fortunate than the *Crawford* but far luckier than the hapless *Argus* was the Pittsburgh fleet's *Matoa*. Only 290 feet long, the steamer was hauling coal from Toledo to Hancock when caught in the storm on southern Lake Huron at about 6:30 A.M. on November 9. "The seas broke in a smother of foam over the decks," *Matoa*'s master reported, "and before we knew what had happened they stove in part of the after cabin, flooded the messroom and kitchen and let a mountain of water into the engine room." After battling the storm for another sixteen hours, the ship was showing signs of breaking up when it grounded near the Pointe Aux Barques Lifesaving Station early Monday morning, November 10. Unlike the *Mataafa* disaster in 1905, the men aboard the *Matoa* were relatively secure as they waited out the storm. All the crewmen made it to the forward cabins, where they used oil heaters to stay warm. After sunrise, a few even made their way aft to get a small coal stove to provide more heat. Hampered by damage to their station and equipment, the lifesavers did not reach *Matoa* until the morning of November 11. They found the steamer's crew comfortable and safe. So safe, in fact, that they refused to leave the *Matoa* to go ashore in the lifesavers' surfboat, which was tossing on the still-turbulent lake. "The whole damn crew, to a man, refused to take that little joyride," one of the engineers said. Although *Matoa* could be salvaged, the Pittsburgh Steamship Company did not consider it worth the expense and abandoned the ship to the insurance underwriters.[29]

As the 1913 navigation season ended, the Lake Carriers' Association and fleet owners around the lakes were worriedly trying to determine how a storm could have destroyed so many big, new ships. The Pittsburgh Steamship Company had taken its licks in the storm, but had come out far better than it had in 1905. The end of the season also marked Harry Coulby's first decade as head of the Steel Trust fleet. He had withstood assaults by storms, unions, and the sheer size of his own armada. Now, storm clouds of a different sort gathered over Europe, promising a new type of challenge to Coulby and his Tin Stackers.

5

War and Peace

The events of 1914 were troubling yet distant to most Americans. Mexico was locked in political turmoil, and the unrest was bubbling across the border into Texas and the new state of Arizona. At home, people fortunate enough to have jobs were grappling with a new federal income tax while thousands of the less fortunate were joining Jacob Coxey as he led his second "army" of unemployed men in a march on Washington, D.C. In the midst of these troubling times came news on June 28 that a Serbian assassin had shot the archduke and duchess of Austria. Over the summer, the tangled web of Europe's military alliances inexorably drew the continent into war. As the situation worsened, President Woodrow Wilson resolutely declared the United States would remain neutral in the conflict.

As the European powers slaughtered a generation of their young men, the United States remained largely unaffected, although the nation's sympathy increasingly was on the side of Britain and France after German submarines began attacking ships on the high seas. By 1916, however, American industries began experiencing strong demand for their goods because of the war. As industrial production grew, increasing demands were made for cargo hauled by Great Lakes ships.

Pittsburgh Steamship Company's *Henry Phipps* inaugurated the busy 1916 navigation season on April 17 on Lake Michigan. Other ships quickly joined in, and the combined lake fleets carried 8.4 million tons of iron ore in May—the most ever in a single month. They promptly topped that record by carrying 9 million tons in June. The rapid pace continued until September 1, when a nationwide railcar shortage began causing delays in moving ore from the mines to the loading docks and from the lower lake docks to the steel mills. By the end of the season, lake fleets had carried 64.7 million tons of ore, well above the previous seasonal record of 49 million tons set three years earlier.[1]

The *A. F. Harvey*, the first of two Tin Stackers named for the third president of Pittsburgh Steamship Company, and its consort barge lock through the Soo bound for the lower lakes. The *Harvey* was one of several ships purchased by the fleet in 1916 to meet the growing cargo demand caused by World War I. Young's Marine Studio/National Archives of Canada/PA-151612.

Led by Harry Coulby, the Pittsburgh Steamship Company continued to increase its tonnage capacity by building ships that were not only longer and wider, but also easier to unload. The fleet sold half a dozen small steamers over 1914 and 1915. Their tonnage was replaced in 1916 when the fleet made a major purchase of vessels from the famous Hawgood fleet, which was being broken up and sold off. The Pittsburgh fleet snapped up six ships, which were renamed *E. C. Collins, A. H. Ferbert, John H. McLean, Pentecost Mitchell, R. R. Richardson,* and *MacGilvray Shiras*. The vessels were relatively short—all were 430 to 440 feet long—but they were fast, reasonably new, and, considering the sudden industrial demand prompted by the war, available at the right time.[2]

The fleet's most notable additions in 1916, however, were the new *D. G. Kerr* and *William A. McGonagle*. These were big, brawny ships: 600 feet long overall and 60 feet abeam. From bow to stern, they were the Great Lakes freighter at its zenith. The *McGonagle* was built by Great Lake Engineering Works in Ecorse, Michigan. Its hull was made of hundreds of the finest inch-thick steel plates, held together by pneumatically driven steel rivets. On the ship's forward end, most of the deckhands, deckwatch, and wheelsmen lived in the forecastle, while the mates shared cabins in the stout steel deckhouse. Many earlier vessels had large, rectangular windows on the deckhouse, which proved susceptible to storm damage. New ships like the *McGonagle* had portholes, which were stronger than windows and could be secured from inside with watertight steel covers. Atop the deckhouse, the pilothouse was paneled in oak, with windows on all sides offering a 360-degree view. No detail was too small for the shipbuilders' attention: the threshold between the single owner's cabin and the captain's bathroom was covered with a sheet of brass, and tiny tacks hammered into the brass formed a diamond surrounding the letter P—symbolic of the Pittsburgh Steamship Company.

From outward appearances, the ship's stern was dominated by the white, steel deckhouse that sat on the spar deck aft of the cargo hatches. Inside were the galley, an oak-paneled dining room for the officers, a smaller messroom for the other men, and cabins for the chief engineer, assistant engineers, firemen, oilers, steward, and porters. But the stern was really dominated by the boiler room and the adjoining engine room, which housed the 2,200 horsepower triple expansion steam engine—the most powerful of any engine in the fleet.

Engine "room" and boiler "room" were really misnomers, for these spaces were really cathedrals to the latest technology in marine engineering. The engine room was an open compartment extending from the very bottom of the ship up to the spar deck and on through the center of the deckhouse to a set of skylights on the structure's roof. A person standing on the bottom of the hull—in what was called the crank room, where the propeller shaft whirled around seventy times a minute—looked up through the equivalent of a four-story building to the skylights.

To enter, the engine room crew—men from the forward end were generally not welcome unless invited by the chief engineer—passed through a watertight door in the rear of the deckhouse. They descended a steel companionway to a deck housing the black, oily steering engine that translated the turns of the ship's steering wheel six hundred feet away into movements of the rudder. On this deck, on both the port and starboard sides of the ship, were the engine room gangways—steel Dutch doors in the side of the hull that could be opened for loading parts or for ventilation while under way and bolted shut during foul weather. Descending another companionway brought men to the main engine room level. At this point they were below the ship's waterline, in a purely mechanical world permeated by the soft, pungent smell of warm oil. The room was kept meticulously clean, and all parts and tools were carefully stowed. A small desk provided a place for the chief to keep his log and attend to paperwork. Nearby was a brass engine room telegraph or Chadburn that relayed commands from the pilothouse to the engine room. Behind the desk were a rack for tools, a telephone station for communicating with the pilothouse, and the switchboard that provided direct current electrical power throughout the ship. Small auxiliary pumps and motors were positioned around the engine room.

Dominating the engine room was the towering triple expansion engine—so named because steam entering the engine passed through three cylinders, driving the massive pistons up and down and expanding in each cylinder as it cooled and lost strength. The engine stood on steel legs, which supported the cylinder chamber at the top. Extending down from each cylinder was a piston rod and then a connecting rod that attached to a crank, which in turn connected to the propeller shaft one level below the main engine room. As the pistons pumped up and down, they worked the piston rods and connecting rods, which spun the crank and turned the shaft. At full speed, the engine was a noisy but smooth-running symphony of moving parts. While an assistant engineer attended to the gauges and throttle on the side of the engine, an oiler walked on a steel grate catwalk among the flashing rods, flicking his hand onto moving parts to feel for heat, and deftly squirting oil from a long-necked oilcan. Engine rooms were noisy and hot, and men on watch took every opportunity to stand for a minute or two beneath the ventilators that scooped in air above the stern cabin and funneled it below deck.

Passing through another bulkhead brought the engine room crew into the gloomy recesses of the boiler room. When launched, the *McGonagle* had three Scotch boilers. On each watch, two or three firemen shoveled coal from the bunker at the rear of the boiler room into the roaring furnaces beneath each boiler. It was dirty, exhausting work in a part of the ship where temperatures commonly exceeded one hundred degrees. Firemen had to carefully tend their fires to ensure steady boiler pressure at all times. On each watch they used long steel rakes and slice bars to wing the live coals over to one side of the furnace and

The powerful *Zenith City,* flying the Pittsburgh Steamship Company house flag from its foremast, pushes out of a massive ice jam off Duluth on May 22, 1917. Ships moving urgently needed iron ore early in the navigation season that year were hampered by unusually heavy ice which remained a menace on Lake Superior well into June. Hugh McKenzie photo, Ken Thro Collection, Lake Superior Maritime Visitors Center.

pull the burned-out cinders or "clinkers" out the furnace door and onto the boiler room floor. There they were doused with water and washed overboard through a special water-powered chute. For thousands of young men, their first job aboard a Pittsburgh ship was as a fireman. For most, the backbreaking work was powerful incentive to move up in rank or to seek employment ashore.

As the *McGonagle, Kerr,* and other ships of the Pittsburgh fleet prepared for the 1917 season, the United States declared war on Germany on April 6. Demand for iron ore continued at unprecedented levels, and Great Lakes shipping responded with unprecedented effort. Spring was slow in coming to the lakes that year. Ice on Lake Superior was so thick that no ore could be shipped in April and the ore ports were closed for days at a time in May. Nearly one hundred ships were icebound in Whitefish Bay on May 4, and floating ice remained a menace on the lake until the end of June. To make matters worse, numerous small lakers had been "sold saltwater" over the previous two years and sent to the Atlantic Coast through the St. Lawrence River canals to meet the heavy demand for oceangoing ships. To ensure that all vital cargoes were carried—and to stave off the possibility of government intervention—the Lake Carriers' Association assumed the role of overseeing vessel movement among its members. When necessary, the association ordered individual ships to deviate from their normal duties to carry cargoes deemed vital to the war effort.[3]

To meet their contract commitments for iron ore, lake fleets pressed every ship and barge into service, carrying 10 million tons in July. The Pittsburgh fleet received timely help in the form of four new ships: *D. M. Clemson, Eugene W. Pargny, Homer D. Williams,* and *August Ziesing*—all nearly identical to the *Kerr* and *McGonagle*. The fleet used its array of big ships to maximum advantage, setting numerous cargo records in the process. *Kerr* carried the biggest cargo ever on the Great Lakes when it loaded 13,732 tons of ore on September 26 at Escanaba for delivery to South Chicago. On its next trip it carried 13,476 tons. *Pargny* hauled a massive 13,485 tons from Duluth to Gary, but was beaten by *Ferbert,* which managed to carry 13,503 tons on the same route. With help from U.S. Steel's railroads, Pittsburgh ships also worked at reducing the amount of time they spent in port—one of the most important factors in determining how much tonnage a ship could carry in a season. The Duluth and Iron Range Railway dock in Two Harbors loaded 12,032 tons of ore into the *Thomas Lynch* in just two hours and five minutes. The Duluth, Missabe and Northern in Duluth put a cargo of similar size aboard *George F. Baker* in two hours and fifteen minutes. Big cargoes and prompt attention at the docks enabled the *William B. Schiller* to carry 414,683 tons over the course of the season. It was Great Lakes shipping at a level unimaginable just a decade earlier.[4]

Communicating with the hundreds of ships nosing about the lakes was not easy in these days when wireless sets aboard freighters were cumbersome rarities. Shipmasters did most of their talking using their vessels' steam whistles. When ships met in a narrow waterway or on the open lakes, their masters let each other know how they intended to maneuver by blowing government-prescribed passing signals. When a tug was helping a ship in port, the master of each vessel used whistle signals to order changes in speed and direction. Ships also used whistle signals to send and receive messages from docks, communicate with barges, and identify themselves when passing shore-based reporting stations.

The Lake Carriers' Association published a lengthy manual listing the long and short whistle blasts that identified each fleet and every individual ship. The signals were essential because so many ships were nearly identical in appearance and indistinguishable from a distance. The Pittsburgh Steamship Company used these signals extensively to speed the movement of its vessels as they arrived and departed ports around the lakes. The fleet's identification signal was a long blast, followed by two short blasts and a long blast. There also were signals to identify each of the four classes the fleet used to rate ships by size. Then, each ship had an individual signal. *James A. Farrell,* for instance, identified itself with a long blast, a short and a long; *McGonagle*'s signal was a short and three long. To identify itself, a Pittsburgh ship blew its fleet signal, waited five seconds, blew its class signal, waited another five seconds, then blew its individual signal.

The fleet's regulations required extensive use of whistle signals. Each ship had to signal as it passed Mackinaw City, Michigan. A ship reporter stationed midway between the lighthouse and the car ferry dock noted the time each vessel passed and telegraphed the information to the fleet's dispatch office. It was the best way to time the arrival of ships at various docks and to keep track of vessels so none went unnoticed if wrecked or disabled. As a Tin Stacker approached Two Harbors, it sounded its identification signal. If the ore dock whistle responded with a single long blast, the ship entered port to load. If the dock responded with one short and one long, the ship proceeded to Duluth. Two long blasts meant the ship should set course for Superior. When an approaching steamer sounded its identification signal at South Chicago, the steel mill dock responded by repeating the signal, then added one blast if the ship was to unload in the north slip or two if it was scheduled for the south slip. Three blasts required the ship to wait for orders, while four meant it should proceed to the U.S. Steel mill in Gary. If a steamer approaching a port was scheduled to pick up a barge, it would identify itself. If a barge was ready, it answered with its own name. If not ready, a tug or the dock blew the steamer's fleet, class, and name signals and added three long blasts if the steamer was to wait or four short blasts if it was to proceed without the barge. It was an effective if complex—and noisy—system of communication.[5]

The incredible demand for iron ore and other cargoes continued into 1918, and the *Kerr* continued its record-breaking pace. It carried 14,084 tons of limestone out of Calcite, Michigan, on September 3— again setting the mark for the largest cargo ever carried. Six weeks later it loaded 478,000 bushels of wheat in Duluth—the largest wheat cargo ever carried on the lakes.

With unprecedented demand for cargo came unprecedented demand for sailors. That brought a return of labor unrest to the lakes. To stave off trouble, lake fleets gave sailors several sizable wage increases in 1917 and 1918. In 1917, at government direction, the Lake

Carriers' Association also eliminated its disputed continuous record discharge books. It issued new ones the following year, but these lacked the "discharge and opinion" section that so many sailors found objectionable.

After an armistice was declared in Europe on November 11, 1918, the Lake Carriers' Association reflected on the role Great Lakes shipping played in the Great War: "It is known now of what vital importance to the almost famished people of England and France were the successful efforts in bringing down the wheat during those bitterly cold and stormy days of December one year ago; likewise it is known now that the preponderance of American steel was of telling effect in bringing about the preliminary peace."[6]

Although more than 61 million tons of iron ore were carried in 1918, the navigation season was the shortest on record, lasting just seven months and six days. It was a harbinger of the trouble that lay ahead. As the nation adjusted to peace, industrial production fell sharply. Strikes by workers at ore and coal docks slowed Great Lakes shipping. Many ships did not sail until midsummer in 1919, while others ended their seasons early. The Pittsburgh fleet handled much of the 47 million tons of ore shipped that season. Many ships from other fleets hauled cargo on their upbound journeys and came back down light—a curious reversal from the usual practice.[7]

By 1920 the nation was in the throes of a recession. The Pittsburgh fleet did not have all seventy-eight of its steamers under way until late May, and ore shipments were finished before the end of November. Unneeded ore piled up on Lake Erie docks, and a strike by railroad yardmen at the season's height delayed seventy to one hundred ships a day. The following season was even worse. Some fleets cut their rates by up to 30 percent in an attempt to attract customers. Even so, about half the ships enrolled in the Lake Carriers' Association did not sail. Jobs were scarce. Longtime masters sailed as mates, while veteran mates took jobs as bosuns, wheelsmen, even deckhands. It was common to see fifteen licensed officers in a crew of twenty-eight men. A few ships sailed with twenty licensed men aboard.[8]

As if to add a dismal punctuation mark to the season, the Pittsburgh Steamship Company's *Superior City* was lost in a collision with the *Willis L. King* above Whitefish Bay in Lake Superior. The two vessels were steaming toward each other just after dark on August 20 when confusion arose over whether the ships would pass each other on the port or starboard side. The 429-foot *Superior City* turned across the path of the oncoming ship, and the *King* rammed it at full speed.

The ore-laden *Superior City* plunged to the bottom within two minutes. As the crew struggled to release the lifeboats from their davits atop the after deckhouse, the cold lake water rushed into the engine room and hit the hot boilers, causing a massive explosion that killed virtually everyone on the ship's stern. Only four men survived of the crew of thirty-three.[9]

The Lake Seamen's Union seized upon the troubled times as an opportunity to renew its efforts to organize the lake fleets. In 1920 union officials signed an agreement with many of the small, independent fleets that did away with twelve-hour workdays aboard ship in favor of an eight-hour day. They also managed to squeeze a 25 percent pay raise out of the owners. Ever eager to thwart the unions, Pittsburgh Steamship Company and other fleets in the Lake Carriers' Association countered by raising the pay of their men by 30 percent.[10]

Amid the dismal economic news and renewed labor disputes occurred a feat that astonished men in the lakes trade. It was done almost on the spur of the moment and, not surprisingly, involved the *Kerr*. Captain William P. McElroy had brought the *Kerr* into Two Harbors on September 6, 1921, with a cargo of coal. He expected to quickly unload that cargo, then cross the tiny harbor to the ore docks of the Duluth and Iron Range Railway. But other ships were ahead of the *Kerr* for the single berth at the coal dock. By the next morning, the dock cranes were just beginning to divest the steamer of its load.

At this point, George Watt, dock agent for the Duluth and Iron Range, sought out Captain McElroy. Watt had a simple proposition. To make up for the delay in unloading the coal, Watt and his men would try to load the *Kerr* in record time. If it worked, Watt reasoned, his men could break the record of twenty-five minutes set in 1911 at Superior by the steamer *William E. Corey*. McElroy agreed to give it a try.

To prepare for the job, Watt's office men pulled out their records of the *Kerr*'s previous loads. Poring over the papers, they carefully calculated the number of tons they needed to put in each of the ship's thirty-five hatches. It was vital work. If they failed to figure the tonnage just right, the rapid loading could bend or twist the ship's hull and cause serious damage.

While the unloading buckets chewed away at the *Kerr*'s coal cargo, workers on the ore dock began dumping iron ore into the dock pockets. The *Kerr* was slated to receive a load of low-phosphorous ore from the Pioneer Mine on Minnesota's Vermilion Range. Switch engines pushed strings of ore cars out onto the dock, spotting them over the loading pockets designated for the *Kerr*. Dock men quickly opened the

The *D. G. Kerr* set many cargo records in its early years of operation. Among them was a mark never equaled: loading 12,507 tons of iron ore in sixteen and one-half minutes on September 7, 1921, in Two Harbors, Minnesota. These photos show the *Kerr* at 4 P.M. after unloading its coal cargo. At 4:45 P.M., the vessel had just begun loading, while the tug *Edna G.* stood by. By 5 P.M., the ore dock chutes are up, the *Kerr* is loaded, and the *Edna G.* has stepped in to hold the ship against the dock face after a mooring line snapped near the bow. William Roleff photos, courtesy of Northeast Minnesota Historical Center.

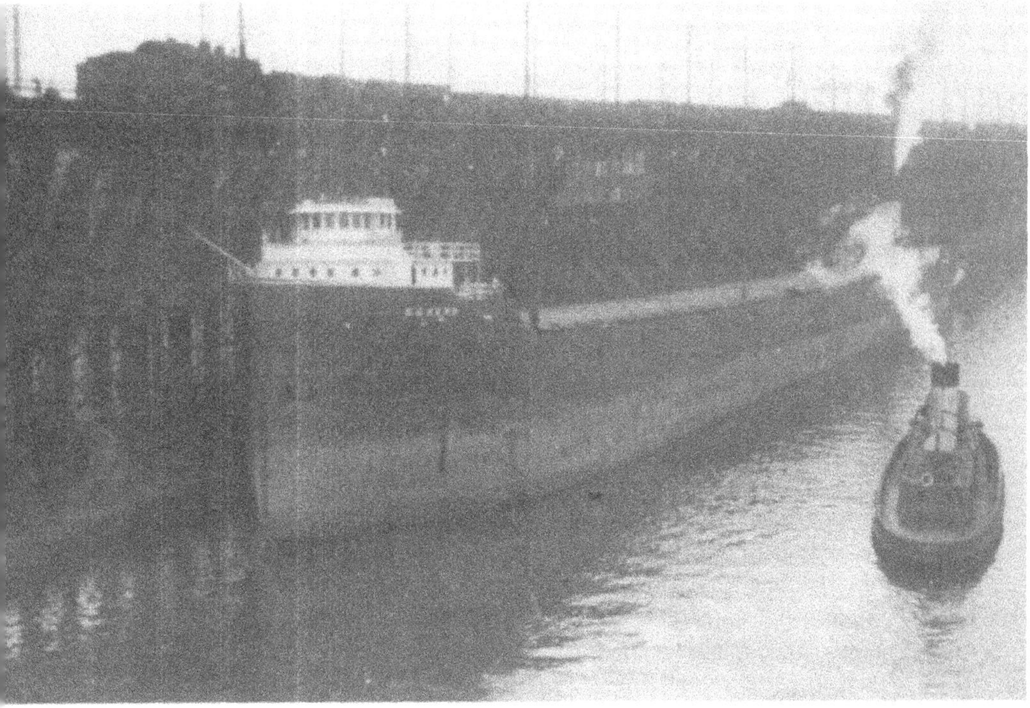

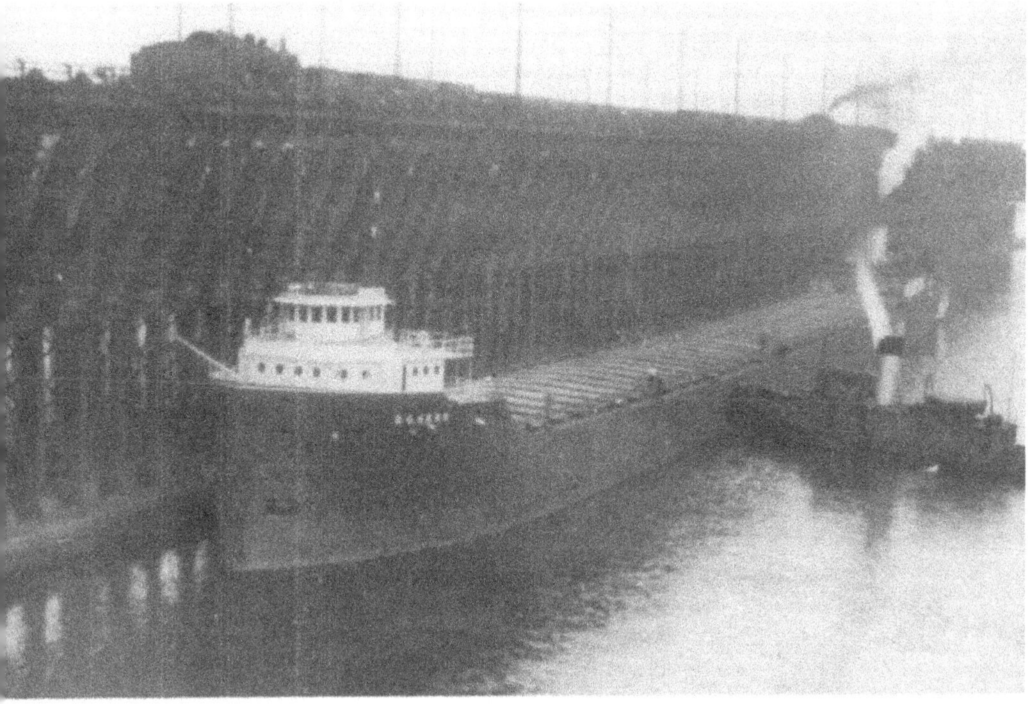

dump hatches on the bottom of each car, allowing the ore to drop into the pockets. After six carloads were dumped into each pocket, more ore cars were spotted on the dock. When the men were done, the *Kerr*'s entire cargo was either in the dock pockets or in cars atop the dock waiting to be dumped.

By 4:10 P.M., the *Kerr* was ready to load. Captain McElroy backed the ship away from the coal dock and, with help from the tug *Edna G.*, turned the steamer toward the south side of Dock 2. As the steamer approached the dock on its starboard side, deckhands dropped mooring lines down to men on the dock wall, who looped them over bollards. Winches on the *Kerr*'s deck hissed and clanked, drawing the ship tight to the dock.

Once the ship was secured, the *Kerr*'s wheelsmen took their places at the mooring winches while the deckhands gathered around the hatches. First Mate C. A. Wallace picked out a spot where he could watch the "tell-tale" that would indicate whether the ship was listing. Atop the dock, some men took their places at the controls that raised and lowered the chutes while others stood by the cars ready to dump them into the dock pockets. Finally, everybody was ready.

Precisely at 4:43:30, Watt and Wallace exchanged signals and the loading began. Ore chutes swung down into the *Kerr*'s open hatches and iron ore slid into the hold with a roar. Two minutes after loading began, ore was pouring into every one of the steamer's hatches. As each pocket emptied, dockworkers raced among the railcars spotted atop the dock, opening their bottom dump chutes to send more ore thundering through the pockets and directly into the *Kerr*.

The operation was running without a hitch when a mooring cable parted at the ship's bow. The *Kerr* began easing away from the dock, threatening to bring the adventure to a disastrous ending. Frantic shouts and curses brought loading to a halt. Within seconds, however, the *Edna G.* surged forward to push the steamer's bow back to the dock and hold it there. Only a minute passed before loading could resume.

More chutes swung down. More cars were emptied. More iron ore slid into the freighter's hold. As the *Kerr* settled lower and lower, it remained perfectly level. Watt's careful planning was paying off. Gradually the roar of moving rock ceased and the ore dock chutes began rising back into place. At exactly 5:00 P.M. all chutes were up and the *Kerr* was ready to cast off. The men of the Two Harbors ore dock and the *Kerr* had loaded 12,507 tons of ore in an incredible sixteen and one-half minutes, shattering the old record of 9,362 tons in twenty-five minutes. As dockworkers cheered, the *Edna G.* helped the *Kerr* back away from the dock and turn toward the open waters of Lake Superior. At 5:20 the steamer cleared the breakwall even as the deckhands were still closing the hatches.

The *Kerr*'s record load drew little attention from the public, but it won praise and admiration among lake and mining men. *Skillings Mining Review*, a mining trade journal published in Duluth, described the feat as "almost beyond belief": "The average person can hardly conceive of 12,507 gross tons being placed aboard a ship in sixteen and a half minutes, and as a matter of fact to have accomplished this great task required the most careful arrangement. Every move on the part of the men operating the dock and ship must be in perfect accord, and the slightest misconception of orders, or in the execution of such, would have disarranged the whole performance."[11]

The *Kerr*'s record-setting performance did not end in Two Harbors. Upon hearing of the feat on the upper lakes, officials of the Pittsburgh and Conneaut Dock Company in Conneaut took it upon themselves to set their own record. All was ready when the *Kerr* steamed into port on September 12. Dockworkers manning the unloading machines plunged into their work with a vengeance. They scooped out the last ore just three hours and five minutes later—a new unloading record.[12]

A promising note during these troubled years was U.S. Steel's acquisition in 1920 of the Michigan Limestone and Chemical Company. The Steel Corporation made Michigan Limestone a subsidiary and retained its president, Carl D. Bradley. The company operated a quarry near Rogers City, Michigan, that produced limestone which was of high quality and almost pure white. Steel mills added limestone to molten iron in the blast furnaces to carry away impurities as steel was made. The stone also was finding widespread use in making cement and in producing lime, which in turn could be used in everything from glass and paint to baking powder and ammonia.[13]

Michigan Limestone's operation included three ships: *Calcite, W. F. White,* and *Carl D. Bradley.* These vessels were revolutionary in their own right, representing the latest developments in self-unloading ships. The first ship designed and built to be a "self-discharging" vessel was the steamer *Wyandotte,* launched in 1908 for the Michigan Alkali Company. Four years later Michigan Limestone built *Calcite,* which was considerably larger than the pioneering *Wyandotte.* The *White* and *Bradley* followed over the next several years, adding considerably to the company's capacity to move and deliver its product to customers.

Each ship's hull was painted gray to hide the limestone dust that accumulated during loading and unloading.

The early self-unloaders functioned much the same as do their descendants today. The cargo hold is built with its sides sloping toward the center of the ship along the keel. Where the two sides come together, a series of steel gates can be opened which allow the cargo to drop onto a conveyor belt running the length of the ship beneath the hold. This belt carries the cargo to another series of belts or buckets, which carry it up on deck and onto another belt running through a long boom on deck. The boom is then swung over the ship's side to discharge the cargo onto a dock. The big advantage of self-unloaders is that they can deliver cargo directly to a customer's dock without requiring expensive shoreside unloading rigs.

Michigan Limestone would build several more self-unloaders in the years to come to serve the rapidly growing stone trade. After 1923 these ships would operate under the name Bradley Transportation Company. To U.S. Steel and people involved in the lake trade, they were simply known as "the Bradley boats." Although the Bradley fleet carried countless cargoes to Steel Corporation mills, it remained separate from the Pittsburgh Steamship Company and maintained its headquarters and a modest repair facility in Rogers City.[14]

Bradley Transportation Company was not the only nautical subsidiary of U.S. Steel that sailed independently of the Pittsburgh fleet. U.S. Steel formed a fleet of small towboats and barges about 1917 to carry raw materials and finished steel products to and from mills along the Ohio, Mississippi, and other rivers. By 1932 this so-called "Carnegie fleet" was the biggest on the rivers, consisting of 15 steamers and 440 barges carrying 10 to 15 million tons of cargo a year.[15] In 1908 the U.S. Steel Export Company bought nine ships to carry Steel Corporation products to foreign markets. Within a few years U.S. Steel formed Isthmian Steamship Lines to handle its overseas trade, eventually building twenty-eight ships for the fleet. Among them were *Steelmotor* and *Steelvendor*, built in 1923, and *Steel Electrician* and *Steel Chemist*, added two years later. These specialized ships were 258 feet long, enabling them to pass between the Great Lakes and the Atlantic Ocean through the series of small Canadian locks along the St. Lawrence River. They carried finished steel products from mills on the lakes, such as the U.S. Steel mill in Duluth, to ports like Waukegan, Illinois, and Cleveland. Other cargoes went to Montreal and American ports along the Atlantic coast. When winter brought Great Lakes shipping to a halt, the ships continued trading along the coast. They resembled conven-

The *James J. Hill,* originally part of American Steamship Company, was one of the Pittsburgh fleet's early giants. By the time this photo was taken, the *Hill* boasted the fleet's trademark "tin stack," as well as a newly enclosed upper pilothouse to protect crewmen from the elements. McKenzie Photo, Ken Thro Collection, courtesy Lake Superior Maritime Visitors Center.

tional lakers with their pilothouses forward and engines in the stern, but they were propelled by diesel engines—a rarity then—and used cranes mounted on deck to load and unload steel billets up to sixty-five feet long.[16]

After two years of recession, the U.S. economy rebounded sharply in 1922. Trouble continued on the lakes because of a miners' strike that cut coal shipments and from low rates that made grain shipping largely unprofitable. But the amount of iron ore carried was 42.6 million tons—nearly double the previous year's tonnage. The recovery gained speed in 1923 as the nation hungered for consumer goods, especially the increasingly sophisticated and popular automobiles. Modernization of the railroads and a spree of highway construction added to the demand for bulk cargoes shipped over the lakes. Ore tonnage climbed to 59 million tons. Stone hit 9.9 million tons—up from just 4 million in 1915. This commodity was relatively new to the lakes, yet it was already taking its place alongside iron ore, grain, and coal as a leading cargo. The increase was so noticeable that the Lake Carriers' Association compared stone's importance to that of grain as a cargo for lake ships.

As the pace of the economy quickened, so did that of the lake fleets. On June 29 the Pittsburgh fleet's *George W. Perkins* loaded 9,965 tons of ore in thirty minutes. No special preparations were made; the feat was simply a normal part of the trade. Ship traffic was so heavy by late summer that the Duluth, Missabe and Northern Railroad ore docks in Duluth loaded 103 ships during the week of August 20. On August 25 alone the docks poured 208,000 tons of ore into twenty-three vessels. Again, just as it had in 1917, the fleet received timely help in the form of new ships. The *Joshua A. Hatfield* and *Richard V. Lindabury* would be the last of the fleet's 600-footers. They featured a new arrangement of cargo hatches. Previous ships had hatches that were nine feet wide, but these two new ships' hatches were twelve feet wide. Also, the sides of the vessels' cargo holds were vertical rather than sloped in order to reduce the damage caused by unloading machines.

While the increase in ship traffic signaled better times for the fleets, it also meant a greater likelihood of collision, particularly in the crowded channels that connected the lakes to one another. On June 3 the Pittsburgh fleet's *Schiller* was lying at anchor off Point Iroquois at the upper end of the St. Marys River waiting for the fog to lift. Out of the gloom glided the steamer *Horace S. Wilkinson*, which slammed into the ore-laden *Schiller*, holing it below the waterline. Aboard the stricken steamer, crewmen raced to their stations as the master ordered the anchor raised. As soon as the hook cleared the bottom, he ordered the ship to get under way, pointing it toward the American shore in hopes of beaching his vessel. The *Schiller* finally settled to the bottom in forty feet of water, leaving its smokestack and pilothouse jutting above the surface.

The fleet began wrecking operations on June 8. For the next three weeks, salvage crews bolted together timbers to make a large wooden patch that divers fitted over the hole made by the *Wilkinson*. The ship's steel telescoping hatch covers, lying just below the water's surface, were removed and replaced by eight-inch wooden covers designed to be relatively watertight. With the vessel made tight, workers started up pumps and began clearing the engine room and cargo hold of water. Twice the operation had to be stopped; the first time when heavy seas battered the after cabin, and again when the after cargo hold bulkhead gave way. Nonetheless, the ship was afloat again by July 2, and the next day workers began raising steam on its starboard boiler. After the pumps and other wrecking equipment were removed at the Soo, the *Schiller* proceeded under its own steam to Cleveland, where its cargo was unloaded before the ship continued on to Toledo for dry-docking and repair.

Another busy season was predicted for 1924. The lake fleets expected to carry 55 million tons of ore, but a short, sharp recession put an end to those hopes. On June 13, Pittsburgh Steamship Company withdrew five of its smallest steamers and three barges from the ore trade. Over the next ten days it idled four more steamers and three more barges. At the end of July more steamers and the remaining eleven barges went to the wall. During July and August all the lake fleets combined laid up more than one hundred vessels.

While the season's tonnage figures and early layups caused rumblings through the fleet, the big news that year was that Harry Coulby—stalwart crusader for safety and implacable foe of unions—was retiring as president of Pittsburgh Steamship Company. Coulby was fifty-nine years old and his wife had recently died. He wanted to travel a bit, perhaps visit his hometown in England, and become more involved in Pickands-Mather and Company, where he remained a partner.

No one could dispute Coulby's legacy to the fleet, nor his unofficial title of "Czar of the Great Lakes." When he took over the Pittsburgh fleet in 1904, he began pushing for bigger ships, ordering the first 600-footer in 1906 with the *J. Pierpont Morgan*. When he retired, the fleet had thirty ships as big as or bigger than that first giant. His crusade for worker safety also was paying off. During his final year with the fleet, the steamers *Clemson, McGonagle, Farrell, Roberts Jr., Olcott, Buffington, Widener, Houghton, McLean,* and *Fairbairn* each recorded their third consecutive years with no injuries to crew members. His staunch opposition to unions had stifled them in the early years of the century and kept them at bay ever since. Whenever unrest threatened, Coulby was there to dole out generous pay raises to keep his men reasonably content and, as a result, make sailors on the Great Lakes the best-paid mariners in the world. As a powerful member of the Lake Carriers' Association, Coulby also had crusaded for better aids to navigation and wider and deeper channels. When he died in 1929, the association eulogized him as one of the most important men in the history of Great Lakes shipping: "In the development of waterways, bigger and still bigger ships, coordinated effort, all to the end that lake commerce would be fastest and cheapest in the world, his breadth of vision and unerring judgment commanded the most serious attention in any council."[17]

Replacing Coulby was his protégé A. F. Harvey, a man long familiar to the Pittsburgh fleet and its employees. Harvey was born in Cleveland and grew up in the city that was the heart of Great Lakes shipping. He graduated from Yale University in 1894 and immediately returned home to join Pickands-Mather and Company. When the Pittsburgh Steamship Company was formed, Harvey was hired as an assistant to President Augustus Wolvin. When Coulby took over the fleet, he brought Harvey back to Cleveland and promoted him to assistant general manager. The long apprenticeship began to pay off in 1916 when, at age forty-five, he was named the fleet's vice president. Now, with Coulby gone, he was elected president and a director of the Pittsburgh Steamship Company.

Despite his new post, Harvey was not as prominent as Coulby, and he kept a lower profile in the Lake Carriers' Association. He was in firm command of the Pittsburgh fleet, but largely carried on the policies and traditions established by his predecessor. Safety remained a top priority, both for men working on board ship and in navigation. With more than seventy steamers and barges in operation, accidents of all sorts were common. Harvey closely monitored these mishaps and continually circulated memos among the masters that summarized recent accidents. Always labeled "Confidential Letter to Masters," these circulars offer a glimpse of Harvey's professional demeanor with the captains. He used the letters to discuss how accidents could be avoided and, as often as not, to drive home a lesson about the need to follow company policies, maintain good shipboard discipline, or adhere to government navigation rules. He also used them to chide, praise, or absolve those involved in accidents. When an incident warranted relieving a master of his duties, Harvey told all the masters about it in the circulars—undoubtedly to control the inevitable rumors while at the same time providing an example of what happened to those who failed to exercise sound judgment.

Harvey could be strict or forgiving, but like Coulby, his word was law and his rulings in each memo were straightforward and firm. "I am very sorry to state that we were compelled to relieve Captain Blessing of the command of his steamer because of intoxication," he wrote at the end of a two-page memo in 1926. "During the last four or five years I have heard several rumors to the effect that Captain Blessing was drinking. I have had him in the office twice and told him frankly that I had heard these rumors but had been unable to prove them. He arrived at the Soo for Inspection in such a condition that the entire Inspection Committee reported he was more than 'half seas over.' Under the circumstances there was only one course open to me."[18]

Harvey did not fire masters often, but many surely felt the sting of being singled out after making a mistake that may have resulted from a mental lapse or just plain carelessness. On May 23, 1927, Captain J. F. Parke was in command of the steamer *Francis E. House* as it steamed up the east side of Lake Michigan near the treacherous Gray's Reef Passage. The first mate was on watch at about 5:00 that foggy morning when the ship ran aground.

Harvey's memo recounted in detail the courses, times, and landmarks involved in the *House*'s passage and subsequent grounding. He then rendered one of his typically stern judgments to make his point to the masters: "We all know that in running from Skillagalee to Gray's Reef Light Ship we are approaching some of the worst water on the Great Lakes. We also know that there may at any time be an unknown current of unknown strength and unknown direction in this vicinity. Under the weather conditions existing, it would seem to me that Captain Parke should have been in the pilot house himself a few minutes before arriving at Skillagalee instead of actually getting on deck as he did after they were about two miles past. At the time he came on deck the vessel was eight or nine minutes beyond Skillagalee and as this light could no longer be seen he gave himself no opportunity to really test visibility. I think he depended too much on his Mate at this time. Captain Parke left orders he was to be called ten minutes before ship would be abreast of Skillagalee. As a matter of fact he was not called until he was abreast and did not get on deck until the *House* was some two miles by."

"I feel compelled to hold Captain Parke solely to blame for this accident and to criticise his navigation severely," Harvey continued. "I was very much surprised to read his report because for many years Captain Parke has been considered and has been one of the most consistently careful and successful navigators who ever sailed for us." Despite Harvey's criticism, Captain Parke was in command of the *House* the following year, no doubt eager to keep his name out of future circulars.[19]

Just as Harvey used his memos to chastise careless veterans like Captain Parke, he also used them to absolve from blame and bolster the confidence of new men who blundered. Such was the case when the steamer *Maricopa* and barge *Smeaton* grounded on Lake Superior's north shore while upbound September 10, 1929, on a windy, rainy night. With no landmarks or lights visible, Captain G. T. Comins of the *Maricopa* was steaming by his stopwatch and miscalculated the distance he had traveled before making a crucial turn.

"In checking over the time, distance and courses run by Steamer *Maricopa,* we find that according to the Master's calculations, he should have been about four miles off the north shore," Harvey noted, writing as though he were the thoughtful headmaster checking over a student's solution to a mathematics problem. "Captain Comins, from lack of experience no doubt, did not realize that a light steamer would head reach while towing a barge with the wind northeast on steamer's quarter and running before it, which no doubt, brought the *Maricopa* four or five miles farther to the north when he hauled than he expected to be. For that reason, we do not believe Captain Comins should be too severely criticised for his navigation in this case. He is due a great deal of consideration for the manner in which he handled the steamer and tow after the accident."[20]

While he was sure to criticize poor seamanship, Harvey also readily recognized outstanding achievements by his masters and men. His first opportunity to do so came early in his tenure as president. In the fall of 1925, a Pittsburgh ship was involved in one of the most dramatic yet unheralded rescues in the history of shipping on Lake Superior. A substantial autumn gale was pummeling eastern Lake Superior on November 5 as Captain George Banker brought the steamer *Richard Trimble,* one of the fleet's 600-footers, upbound through the Soo Locks and into Whitefish Bay. Although his vessel's cargo hold was empty, Banker thought the steamer's 1,800-horsepower engine would enable him to handle the gale.

The *Trimble* left the shelter of Whitefish Bay late on November 5 and steamed into the open lake. Conditions were severe, but Captain Banker kept going. After several hours, the ship was about eighty miles from Whitefish Bay and plunging into seas whipped up by winds blowing more than fifty miles per hour. It was a rough night to be out on Lake Superior. From the *Trimble*'s pilothouse Banker could make out the lights of the *Hamonic,* a popular passenger steamer belonging to Canada Steamship Lines. The *Hamonic*'s masthead lights were swinging in wild arcs as the ship rolled in the trough of the waves, but it was not showing distress signals, so Banker kept the *Trimble* pointed into the wind. The situation continued to deteriorate, and at about 11 P.M. the wind blew the *Trimble* around and pushed it into the trough of the waves. Admitting defeat, Banker ordered his ship turned around and headed back to shelter.

The *Trimble* was running before the seas when the men in the pilothouse spotted the *Hamonic* again, this time blowing distress signals on its whistle and burning a red distress flare. Captain Banker promptly changed course to bring his vessel closer to the passenger ship. After battling wind and seas for about two hours, he brought his freighter to within three hundred feet of the wallowing *Hamonic*. Through the roar of the wind, a man on the passenger ship's stern shouted that its propeller was gone. The *Hamonic*, its passengers, and crew were helplessly adrift.

"At the time I went under the *Hamonic*'s stern . . . the sea was washing all over us," Banker reported. "We could do nothing in the night. It was snowing hard and for a while I could not see him. I could not hear anything in the pilot house on account of the wind. I checked down to stay as near to him as possible and went at various speeds down before the wind, thinking to get back to him sometime before daylight."

In the dim, gray light of the stormy morning, Banker turned the *Trimble* into the wind and began searching again for the *Hamonic*. He located the passenger ship at about 9:45 and began closing the distance between the two ships. By 10:30 A.M. the *Trimble* was close enough to the *Hamonic* for Banker to shout through his megaphone for the Canadian ship to prepare its towline. Aboard the Canadian steamer, its captain, standing atop the pilothouse clad in a long fur coat, waved and responded simply "All right!" The *Trimble* continued steaming past, preparing to turn in a broad circle so it could come alongside again to pick up the towing hawser. "The wind was fifty-five miles an hour and the seas were sweeping over us," Banker recalled. "The seas seemed bigger than they had been. When we would turn, we would roll our decks under [water]."

Taking a vessel's towline always called for caution, but with both ships battling high winds and waves, the maneuver became extremely dangerous. One miscalculation by Banker, one nasty roll, one unexpected wave could slap the two ships together, sending both to the bottom. The *Trimble* steamed off two miles before turning. Meanwhile, sailors aboard *Hamonic* tied about three hundred feet of three-quarter-inch line to a life preserver and cast it into the water for the *Trimble*'s men to snag. This light "messenger" line would be connected to a heavier line, which in turn would be tied to the towing cable. The men aboard the *Trimble* would need to snag the messenger line, then haul it aboard until finally getting the heavy hawser on deck. The life preserver remained near the *Hamonic*'s hull, however, so Captain Banker had to bring the *Trimble* perilously close to the helpless steamer. The first attempt failed; the *Trimble* steamed off again and turned to make another pass. The second attempt failed, then a third.

On his fourth attempt to snag the line, Banker daringly brought the *Trimble* within twenty-five feet of the rolling *Hamonic*. A sailor standing on the ore carrier's exposed deck amidships snagged the line with a grappling hook, and his companions quickly pulled it aboard and began hauling it aft to the *Trimble*'s towing bitts.

"The towline was a ten-inch line. I held the stern of the *Trimble* as best I could about seventy-five or 100 feet from the bow of the *Hamonic* while we made the towline fast," Banker said. "The towline led from his bow to our stern chock. Seas were coming over our stern and the men were in constant danger from these seas. Even after we got the line the seas came right over our stern. After we got the line made fast we had to maneuver very carefully because both vessels were jumping and rolling heavily. We had squalls of snow right along and the weather was very cold. Our thermometer was washed overboard from the front of our observation cabin and we were all covered with ice. The *Hamonic* put out about 500 feet of towline, and we got him down abreast of Whitefish Point at 6:21 P.M. November 6."

The *Trimble* brought the *Hamonic* into the bay and let go the towline off Iroquois Point. The *Hamonic* dropped anchor, badly battered but safe. His duty to his fellow mariners completed, Banker ordered the *Trimble* back on course and unceremoniously departed.

"Practically our entire time from 2 A.M. to 9 P.M., November 6 was devoted to the *Hamonic*. I was forty-eight hours without sleep and for thirty hours I didn't even have a meal," Captain Banker said. "All the men on the *Trimble* were concerned in the rescue of the *Hamonic*. All worked faithfully, some of them without dinner and all were pleased that we had rescued her."[21]

6

The Roaring Twenties

Great Lakes shipping, like America's economy, roared along in the late 1920s. The combined tonnage of iron ore, coal, grain, and stone topped 125 million for the first time in 1928 and surpassed 138 million the following season. These two navigation seasons were the best the lake fleets had enjoyed since the Great War. During these years the Tin Stackers mostly earned their keep hauling iron ore, but the fleet's smaller ships also kept busy carrying coal, stone, and even grain, along with such prosaic cargoes as pit iron, scrap turnings, cement clinkers, flue dust, and coke. Signs of prosperity were visible everywhere by 1929, such as in Detroit, where the newly opened Ambassador Bridge soared across the busy shipping lanes of the Detroit River to link the Motor City with Windsor, Ontario. It also could be seen in the Pittsburgh Steamship Company, which launched the new steamers *William G. Clyde, Horace Johnson,* and *Myron C. Taylor.* All three boasted an overall length of 604 feet. Things just could not have been better.

Everything started to change on October 24 when a sliding stock market suddenly tumbled into a frenzy of panicked selling on the New York Stock Exchange. "Black Thursday," as the day was quickly dubbed, was followed five days later by the infamous "Black Tuesday," when the bottom dropped out of the market. The crash reverberated across the nation and directly affected the financial holdings of an estimated 25 million Americans. But in the light of recent prosperity, it was perhaps only natural for people to dismiss the panic as an aberration. Shortly after the crash, Henry G. Dalton, a partner in Pickands-Mather and Company, told businessmen at Duluth's prestigious Kitchi Gammi Club not to panic over the stock market's tumble. "I take a confident view of the outlook for business and industry. I see nothing in the situation to cause apprehension." *Skillings Mining Review* took a warier stance two weeks after the crash: "It has been a great shock to the country, has

A brand-new *Horace Johnson* takes on ore in Two Harbors during its maiden voyage in 1929. The nation's healthy economy prompted the Pittsburgh fleet to build three ships that year. Hugh McKenzie photo, Ken Thro Collection, courtesy Lake Superior Maritime Visitors Center.

entailed much individual loss, but the most that can be said of its influence nationally is that it has created uncertainty."[1]

The uncertainty carried into 1930 as the economy shuddered and financial gloom continued to spread. On the Great Lakes, tonnage fell by about 25 million from the previous season. Most of the decline came in iron ore while grain and coal cargoes held their own. Still, the downturn was not as bad as the troubled years of 1920 and 1921. Factories ashore were cutting wages, but the bigger lake fleets, perhaps fearing a return of union organizers, kept sailors' pay at the same rate as in the past few years. Despite the downturn, the Pittsburgh Steamship Company completed two ships it had started building before the troubles began. *Thomas W. Lamont* made its maiden voyage on May 28, sailing from the American Ship Building Company's yard in Toledo to Duluth to load iron ore. *Eugene P. Thomas* soon followed suit. They were the only ships built on the Great Lakes that year—a distinction they would hold for a painfully long time.

America's economic troubles worsened in 1931. On the lakes, few ships sailed before June 1 and most were laid up by the middle of October. Ore tonnage was half that of the previous year, and nearly one hundred ore carriers never left their docks. Pittsburgh Steamship Company operated only forty vessels; the smallest and least economical ships and barges remained idle. Few fleets made a profit amid what the Lake Carriers' Association termed the nation's "industrial prostration." Most people did not think it could get any worse. James A. Farrell, president of U.S. Steel, asserted in a nationwide radio broadcast that "our worst experiences are behind us and . . . we are entering upon a period of gradually increasing trade activity."[2]

But 1932 was the worst year yet. Steel mills and factories around the Great Lakes and throughout the nation closed their doors. Demand for iron ore and stone was minuscule. There was demand for coal only because so many people used it for heating their homes. A. F. Harvey contemplated the unthinkable: running no ships at all. He finally decided to fit out nine steamers to sail from July 1 to November 1. That season the Pittsburgh fleet carried 173 cargoes; in 1929 it had carried 2,600. Iron ore shipments amounted to just under 4 million tons, a far cry from the 50 to 60 million tons moved during normal years. Harvey's decision to fit out a handful of ships was driven at least partly by the fleet's desire to protect its masters, mates, and engineers—the core of its skilled workforce. Still, that year many experienced masters

Visitors watch the ore chutes swing into place as the *Horace Johnson* loads its first cargo on its maiden voyage to Two Harbors. The telescoping-leaf hatch covers have been pulled open and are stacked neatly at the end of the hatches; the bright white paint inside the hatches has not even been scarred by red ore. Hugh McKenzie photo, Ken Thro Collection, courtesy Lake Superior Maritime Visitors Center.

accepted berths as mates, while mates sailed as wheelsmen and watchmen. Many unlicensed men in the Pittsburgh fleet did not sail at all. Typical among them was Guido Gulder of Duluth, who had gone to work as a young deckhand in 1929 and had sailed steadily the next two years. However, he spent 1932 working in a bakery and never set foot on a ship. For the first time in nine years, fleets in the Lake Carriers' Association cut wages. Deckhands, wheelsmen, watchmen, oilers, firemen, stewards, and porters saw 20 percent of their pay disappear—if they were fortunate enough to be sailing at all. "From a revenue earning point of view, 1932 was the most disastrous year encountered in the last quarter of a century," the Lake Carriers' Association lamented after the season ended.[3]

Just when it seemed things could not get any worse, they actually got better. Franklin Roosevelt was inaugurated as president early in 1933 and soon proclaimed a ten-day "bank holiday" to slow the draining of financial reserves. The United States abandoned the gold standard and Roosevelt banned gold exports. Congress passed the National Industrial Recovery Act, and people began talking about a "New Deal" worked out by the president's "brain trust." At the beginning of April the nation's steel industry operated at 14 percent of capacity; by the end of the month it was at 29 percent. As industry perked up, so did Great Lakes shipping. Ore tonnage shot up to 25 million tons—far below normal, but still a "tonic for depressed spirits."[4]

During these bleak years unions made great strides in organizing working men and women around the country. Much emphasis was placed on reducing the length of the workday from the typical ten or twelve hours to eight. At the same time, the International Seamen's Union, after a ten-year absence from the Great Lakes, made another attempt to organize lake sailors. The Lake Carriers' Association responded to both initiatives in 1933 by radically altering the labor system aboard members' ships. Previously, men served six hours on watch and had six hours off. Under the new plan, they served a four-hour watch followed by eight hours off. Voluntarily adopting the "three-watch system" beat government regulators to the draw, but it required fleets to add a third mate, third assistant engineer, and several additional crewmen to each ship to man the additional watch. On the older ships, in particular, this led to crowded living conditions for men such as deckhands and firemen.

Not long after adopting the three-watch system, the Lake Carriers' Association bowed to pressure and submitted a plan to the National Recovery Administration to allow sailors in the fleets to vote on whether they wanted union representation. The International Seamen's Union got its name on the ballot, but so did a "company union" proposed by the fleets.[5]

Ballots were duly distributed to shipmasters to pass out to their men. It was not exactly a secret vote. If the sailors wanted the company union to represent them, they asked for an "A" ballot and signed it. If they wanted no union, they asked by a "B" ballot, and if they favored the International Seamen's Union, they asked for a "C" ballot.

Offering the men two choices of unions doomed either union's chances of winning a majority. Leaders of the International Seamen's Union quickly protested the results to the National Recovery Administration. One of several affidavits included in the union's brief singled out the Pittsburgh Steamship Company for its role in the election. According to the affidavit, a sailor on one of the Pittsburgh ships said the master distributed to the men a set of "A" ballots he had received from the fleet office that were signed by A. F. Harvey. "If you want to remain in the employ of the Pittsburgh Steamship Company," the master allegedly told his men, "you should vote the ballot that has A. F. Harvey's name on it. If you want to support a lot of bloodsuckers, then vote the ballot of the Sailor's Union of the Great Lakes, but you will be sorry for it." After more electioneering, he supposedly added, "There is only one way for me to vote and that is the ballot with A. F. Harvey's name on it, and if you want to remain in the employ of the company, you had better do the same." The union's protest failed, and the Pittsburgh Steamship Company and other large fleets remained open shop organizations.[6]

The slow, painful climb back to normal conditions continued in 1934, although a severe drought caused a recession that was euphemistically termed a "pause in the recovery." A significant improvement in the economy, especially in the second half of the year, made 1935 the best season on the Great Lakes since the depression began. Still, fleets in the Lake Carriers' Association saw only 284 of their 397 vessels sail that season. The Pittsburgh fleet operated about 40 vessels in 1934 and 1935. Many sailors remained out of work even as ore tonnage crept back above 50 million tons.

Even though Great Lakes shipping was well below peak levels in the mid-1930s, running a fleet as large as the Pittsburgh Steamship Company remained a complex endeavor requiring more than twenty-five hundred men and the coordinated effort of numerous U.S. Steel subsidiaries. Making steel required a huge investment in mines, ships, and mills. To turn a profit, all facets of the operations had to run smoothly. A roaring blast furnace could not wait for a tardy ship or consume just any grade of ore. The right material had to be in the right place at the right time.

At the Pittsburgh Steamship Company, preparations for the next shipping season actually began before the current season ended. During the final weeks of navigation ships were assigned to various ports to lay up for the winter. Vessels requiring dry-docking or shipyard work were directed to the appropriate locations. Normally about 30 percent of the fleet laid up in South Chicago, Milwaukee, or Manitowoc, Wisconsin. The remaining vessels spent the winter in Lake Erie ports like Toledo, Cleveland, Lorain, and Conneaut.

By December 1 most ships were laid up for the winter. Repairs were planned in advance based on reports received from the masters and from a detailed inspection conducted every September as each ship passed the Soo. Much of the major work was done by shipyard crews, but many Pittsburgh sailors and engineers were hired over the winter to perform smaller repair jobs and maintenance work.

Not long after the last ships tied up, men in the fleet's Cleveland office began planning for the next shipping season. The Pittsburgh Steamship Company's entire season of operation was closely linked to those of U.S. Steel's ore mines, its railroads in Minnesota, its dock companies and railroads on the lower lakes, and the vast mill complexes of its various subsidiaries. U.S. Steel managers in New York and at steelmaking subsidiaries around the Midwest predicted how much steel they would make, what kind they would make, and how much iron ore and limestone they would need to make it. These figures were compiled for every mill and revised periodically throughout the winter. As the start of navigation neared in the spring, the Oliver Iron Mining Company, Michigan Limestone and Chemical Company, and the Pittsburgh Steamship Company were provided detailed statements showing the amounts and grades of iron ore and the quantity of limestone to be shipped that season. At the same time, each steel mill was given its assignment for the year so it would be prepared to receive ore and stone after the opening of navigation.

At fleet headquarters in Cleveland's Rockefeller Building, preparations for the next season began in earnest in January when the first estimates arrived for ore and limestone movement. The ships and barges needed for the next season had to be selected and, if needed, additional ships chartered from other fleets. By April 1 more definite estimates would be set for ore and limestone, and partial estimates would be available for coal. Other factors, such was water levels in connecting waterways and the length of the season as determined by spring ice, were figured into the calculations.

Final estimates received by the fleet office as the navigation season opened included not only the distribution of ore by grades among the various steel mills, but also the rate at which the railroads could clear ore off the docks, the rate at which deliveries could be made to lakefront mills, and the rate at which ore was to be stockpiled on the docks for use during the winter. After factoring in all these items, fleet officials drew up a weekly loading scheduled and passed it to the Oliver Iron Mining Company to govern its mining operations. Another schedule was given to the Michigan Limestone and Chemical Company and to the coal mining companies to govern their operations. Adjustments would be made to the schedules throughout the season to account for delays caused by storms and other unexpected interruptions.

The opening date for navigation depended on the weather. During much of March, fleet officials carefully scrutinized Weather Bureau reports regarding ice conditions on the lakes. No vessel movement would be initiated until ice in key areas such as the Straits of Mackinac, St. Marys River, and Whitefish Bay allowed free movement of ships. When an opening date was picked, about a dozen vessels were placed in commission that day, followed by an approximately equal number on each succeeding day until the entire fleet was under way. The staggered start ensured that ships would not jam up at loading docks or at the Soo.[7]

By 1937 the Pittsburgh Steamship Company was back to operating its entire fleet of seventy-five ships and barges. It also chartered thirty-five to forty-five vessels each week. That meant the fleet and chartered ships were carrying about 135 cargoes a week from ports on the upper lakes to Gary, South Chicago, and several ports on Lake Erie. Each trip generally took seven days as a ship spent three days running empty or "light" on its upbound journey to Lake Superior and three and a half days running downbound with cargo. A total of about half a day was spent loading and unloading. During a trip, a ship traveled an average of 1,750 miles, at least 300 of which were in the twisting channels of the St. Marys River, the fog-prone Straits of Mackinac, or the congested Detroit and St. Clair Rivers. As delays occurred in these waters and on the open lakes, fleet dispatchers rerouted ships and shuffled cargoes to ensure a steady supply of ore arriving at the mills.

Once the ships started sailing, the season quickly settled into a routine for the thirty-five to thirty-eight men and the occasional woman who made up the crews. On the forward end, the master and mates supervised navigation of the ship. The master did not stand a regular watch, but directed navigation whenever entering or leaving port, locking through the Soo, steaming in congested waterways, or sailing in foul weather. Each mate stood two watches a day in the pilothouse. In addition to his watches, the first mate also "ran the ship" by supervising the loading of cargoes and discipline of the crew. Each watch also had a wheelsman, who steered the ship while under way and ran the deck winches when in port; watchmen, who kept a lookout for other ships and occasionally relieved the wheelsman; and deck watchmen, who worked on deck and periodically "sounded" the ship's ballast tanks to report how much water they contained. At the bottom of the hierarchy were the deckhands, who opened and closed hatches, handled lines, cleaned the decks, and performed shipboard maintenance such as polishing brass and painting. With no provisions for overtime pay, everyone was prone to working long and unusual hours, but none more so than the deckhands. They typically would be roused early in the morning by the sound of a mate or steward shouting "Rise and shine for the Pittsburgh line!" They put in an eight-hour day, but then might be called out in the middle of the night to handle hatch covers and mooring lines while the ship loaded or unloaded cargo. When that was done they often had to hose down the deck and scrub coal or ore dust off the cabins before being allowed to snatch a few hours of sleep before going back to work.

All these men generally lived in the ship's forward end. The captain had his own cabin, which was located right below the pilothouse and which often doubled as an office. He also had a bathroom, which might be shared with guests occupying a stateroom. The first mate and second mates often shared a room while the third mate bunked with one of the wheelsmen. Watchmen and deckhands generally shared rooms belowdecks in the forecastle. On the older ships that might mean as many as six men in a room. In many ships these rooms were arranged around an area that had tables and chairs and a common washroom. Until the 1930s, many of the men on the forward end were hired by the master or first mate, and often followed him if he was transferred to a different ship.[8]

On the after end of the ship were the chief engineer, who supervised the engine room and all mechanical matters, and the first, second, and third assistant engineers, licensed men who each stood two daily watches at the throttle in the engine room. Working in the engine room on each watch was an oiler, who oiled the myriad moving parts of the triple expansion steam engine and assisted the engineer. Most ships had six firemen, with two working on each watch. Under normal conditions one man shoveled coal for the first half of the watch while the other passed coal from the bunkers to the fireroom floor. Then they would switch roles for the last half of the watch. On some ships they were assisted by a coal passer, who shoveled coal out of the coal bunker and dumped the ashes from the furnaces.[9] These men all lived in the deckhouse that sat on the stern of the ship. The chief engineer had his own quarters, but the oilers usually shared a cabin and all six firemen often crowded into another.

Also housed aft were the steward, who served as chief cook and supervised the galley; a second cook; and one or two porters, who helped in the galley, washed dishes, handled the bedding and linens, and cleaned up the master's office and made his bed. If a woman was aboard a ship, it always was as a cook, often working alongside her husband who was steward. These generally were the only jobs available to African Americans, not only in the Pittsburgh fleet but on virtually all lake freighters.

The galley crew had a big responsibility. The other men worked hard, shoveling coal into the furnaces, lugging wooden hatch covers, and scrubbing down the deckhouses. They developed huge appetites and expected them to be satisfied with food that was both plentiful and tasty. The steward had to meet the company's guidelines governing the cost of each meal while keeping the men satisfied with a robust breakfast, a big lunch, a hearty supper, and a supply of snacks for the overnight watches.

The sailor's life attracted many young men from cities around the Great Lakes. Typical among them was Guido Gulder, who grew up near the ore docks in Duluth and went on to spend his career working for the Pittsburgh Steamship Company. He started out as a deckhand—"decking" in sailor's parlance—and soon aspired to being a wheelsman. "Decking then paid $77 a month, so if you saved your money you could come off in pretty good shape, but wheelsmen earned $100 a month," he recalled.[10] Deckhands worked hard. The older ships and barges had heavy wooden hatch covers that had to be removed before loading, stacked at the end of the hatch, and then lugged back into place and covered with stiff canvas tarpaulins before getting under way. Newer ships had telescoping steel hatch covers that were pulled open by a steel cable running to a steam-powered deck winch. These hatch covers also had to be covered with tarpaulins. During cold weather the tarpaulins grew stiff with frozen spray, and once in port the deckhands would have to roll them up and lug them into the boiler room to thaw.

Some of the men aboard ship were family men, others were loners or single young men. Many shared a special camaraderie that resulted from spending so many hours together away from their families and homes. Some read or worked on hobbies. Others gathered in the messroom aft or in the dunnage room up forward to play cards—usually poker, cribbage, or hearts. Mail and newspapers were always a welcome diversion, picked up at the Soo or snatched up in a bucket from the little mail boat that came alongside each ship passing through the Detroit River. The men did their own laundry, dumping soap into a bucket and drawing off hot water before sitting down with a washboard to scrub socks, pants, and shirts. If a ship had a wringer washer down in the dunnage room, the men figured they had it pretty good.[11]

When their ship was in port, most men did not stray too far because loading often was finished in three or four hours—and departing vessels did not wait for absent crewmen. Some men stayed aboard because they had to help with loading, handle lines, run the winches, or stand a watch. Those fortunate enough to live nearby tried to get home for a few hours. Others might walk "uptown" to visit stores, a theater, or a tavern. South Chicago was known for its taverns and boardinghouses near the docks. Two Harbors had a drugstore open around the clock that was a popular destination for sailors. Some ports, however, were located in rough parts of town, which discouraged touring. "I never bothered to get off much except to go to the store and get some things," Gulder recalled years later. "It got to the point in Gary where it was OK to get off in the daylight but at night it was not so great; the same in South Chicago."[12]

During the 1930s, men who sailed the Tin Stackers were unlikely to spend much time attending Saturday night dances or Sunday morning church services. On Lake Erie and lower Lake Michigan the steel company docks did not unload ships on Sundays, so the fleet's dispatchers always tried to schedule vessels to be under way those days. That meant Saturday nights were hectic as dockworkers hustled to finish unloading ships before the weekend break. Monday mornings were equally busy as they cleared up the backlog that awaited when they returned to start a new week.

Overall, Pittsburgh Steamship Company enjoyed a good reputation among sailors, Gulder said, even though it was strict about "human behavior." Drinking was discouraged. Trouble between men was not tolerated. "They weren't very liberal on crewmen bringing friends on board, especially if the girls weren't related to any of the sailors," he added.[13]

The tug *Virginia* has *Sir William Fairbairn* under tow on Cleveland's Cuyahoga River. The destination of many Tin Stackers often was a steel mill or stone dock on a narrow, twisting river. Hugh McKenzie photo, Ken Thro Collection, courtesy Lake Superior Maritime Visitors Center.

Men sailing for the Pittsburgh fleet became most familiar with the half dozen American ports that loaded iron ore. Every time a Tin Stacker called at Two Harbors, Duluth, Superior, Ashland, Marquette, or Escanaba, it was playing its role in the long and elaborate process of supplying U.S. Steel's mills with the specific type of ore needed to produce a particular grade of steel.

Most of U.S. Steel's iron ore came from the iron ranges spread across northeastern Minnesota and Michigan's Upper Peninsula, where the Oliver Iron Mining Company gouged a wide variety of ores from open pits and underground mines. Ores varied in color, hardness, and purity. Most ore was a russet hue, but others were ocher-yellow, deep red, grayish blue, or dark brown. Some ore, such as that from Minnesota's Soudan Mine, was so hard it could scratch glass. Other ores were the consistency of red mud or topsoil. Many grades of ore had to be crushed and screened to remove rock and sand, or to make the ore a suitable size for loading.

Mine workers separated ores into groups according to their content of iron, phosphorous, silica, manganese, alumina, and moisture. The two major steelmaking processes were Bessemer and open hearth, so the two main classifications of ore were Bessemer and non-Bessemer. Oliver Iron Mining Company subdivided these classifications into grades. Ores from its Mesabi Range mines were labeled 1, 2, and 5 for Bessemer grades, and 3 and 7 for non-Bessemer grades. Ores from Minnesota's Vermilion Range were a Bessemer grade named Pioneer, a non-Bessemer grade called Frontier, and another grade called Vermilion Lump. Michigan's Marquette Range turned out the non-Bessemer Bedford grade, while the Gogebic produced Norden, another non-Bessemer. On Michigan's Menominee Range, Oliver mined a high-phosphorous non-Bessemer ore called Barton.

The process of loading a ship actually began when the Pittsburgh Steamship Company designated a vessel to proceed to a port to take on a particular grade and type of ore. That information was then telegraphed to the Oliver Iron Mining Company, which relayed it to one of its mines. At the Mesabi Range mines, an employee known as a grader then had up to three days to specify the type of ore to be mined and loaded aboard a train for delivery to the ore dock.

When a trainload of ore arrived at the mine shipping yard, samples were taken from the cars and sent to a laboratory for analysis. The train was then immediately sent on across northeastern Minnesota to sorting yards at Two Harbors or at Proctor, a few miles from the ore docks in Duluth. Before the train arrived, the laboratory analysis was completed and the results telephoned to another grader at the railyard. Upon the train's arrival, the grader, knowing the weight and chemical composition of each car, distributed them into lots known as blocks that contained the grade and tonnage designated for a particular ship.

The ore docks were owned by railroads. U.S. Steel's ore docks were located in Duluth and Two Harbors. By the 1930s they were owned and operated by the Duluth, Missabe and Iron Range Railway, which resulted from the merger of the earlier Duluth, Missabe and Northern and Duluth and Iron Range Railway. The railroad's dock No. 6 in Duluth was, at 2,304 feet, the longest ore dock in the United States. Atop the ore docks—eighty-four feet above the water—ran four parallel railroad tracks that received ore cars from the classification yard. Inside the dock were several hundred bins or "pockets" running down both sides of the dock. The railcars—each carrying 50 tons of ore—were pushed out onto the dock, where men opened their bottom dump chutes to empty the ore into the pockets. The contents of eight cars, or 400 tons, were dumped into each pocket. Each pocket had a long chute for loading vessels. Once a ship was alongside the dock, the chute would be lowered and the ore in the pocket dumped by gravity into the vessel's hold.

Loading a ship was not merely a matter of randomly dumping ore into its hold. A certain number of pockets in a dock were assigned to each ship to be loaded, depending on the capacity of the vessel. Since the arrival time of each ship was known several days in advance, ore could be shipped to the dock and dumped into the designated pockets to be ready for loading. Once a vessel was tied up alongside the dock, the first mate took charge of the loading, working with the dock crew to make sure cargo went into the ship evenly to avoid placing undue stress on the hull and to prevent the vessel from listing or sagging. Simultaneously dumping ore from a number of pockets ensured even loading along the length of the ship, and raising and lowering the chutes spread the ore across the width of the hold. As loading was completed, small amounts of ore were added where needed to trim the ship. Then deckhands shoveled any spilled ore into the holds and finished closing the hatches. If foul weather was a possibility, they covered the hatches with tarpaulins and secured them with steel bars to keep water from leaking into the hold. After that, they finished cleaning the deck.

Once a Tin Stacker was under way, it usually was destined for one of the ore receiving docks that U.S. Steel subsidiaries operated in six ports around the lower Great Lakes. The most tonnage was handled by the Pittsburgh and Conneaut Dock Company in Conneaut, where five huge Hulett unloaders and one bridge crane dug iron ore and limestone out of the ships and barges. Next-highest tonnage was handled by the Carnegie-Illinois Steel Corporation in Gary. It was a massive operation, with seven Huletts and seven bridge cranes. Carnegie-Illinois also had two docks in South Chicago with eleven Huletts and six bridge cranes. The National Tube Company dock in Lorain, the Pennsylvania and Lake Erie Dock Company in Fairport, Ohio, and the American Steel and Wire Company dock in Cleveland each had four to six Huletts. Conneaut, located seventy miles northeast of Cleveland near the Ohio-Pennsylvania border, loaded ore aboard trains of U.S. Steel's Bessemer and Lake Erie Railroad for shipment to inland mills in the Pittsburgh and Youngstown districts. The docks at Gary, South Chicago, Lorain, and Cleveland sent ore directly to adjacent steel mills.

From 1899, when George Hulett installed the first one on the Carnegie dock in Conneaut, until the early 1980s, the dominant means of removing iron ore from a ship or barge was the Hulett unloading machine. Huge, ungainly, immensely practical, the Huletts were erected in batteries along the edge of the dock. The first ones ran on steam; later models were powered by electricity. Mounted on rails, the 1,500-ton machines could roll along the dock to position themselves even with a vessel's hatches.

The Hulett's dominant feature was its eighty-foot arm that held a vertical leg which could be extended out over a ship. At the bottom of the leg was a bucket capable of digging into the iron ore and filling itself with seventeen tons of rock. Just above the bucket, a cab was built into the leg for the machine's operator. The operator would lower the leg into a vessel's hold and chew up a bucketful of ore. Then he would raise the leg out of the hold, move it back over the dock, and empty its load into a receiving hopper on the machine's superstructure. The vertical leg would then return to the cargo hold for another bite. As the amount of cargo inside the ship dwindled, the Hulett operator could pivot the leg on its vertical axis to reach ore that was not directly beneath the open hatches. In the final stages of the operation, men with shovels and brooms descended into the hold to help scoop the final tons of ore into the Hulett's bucket.

Once the Hulett operator emptied his bucket into his machine's receiving hopper, the ore was discharged into a small "larry" car that weighed the load. Then the larry car traveled overhead to a position above a railroad car and dumped the ore into the car. No further weighing was necessary as the loaded railroad cars were made up into a train for shipment to the steel mill. If the ore was destined for a storage yard, the Hulett's larry car ran out on a cantilever extension of the machine, farther back from the dock than the railroad tracks, and dumped the ore into a receiving pit or trough. There a 575-foot-long bridge crane, which resembled a cantilevered bridge on tall legs, picked up the ore one bucket at a time and transferred it to a storage site. Because of the many grades of ore that were handled, great care was used in stockpiling ore so that particular grades could be located weeks or months later when needed.

Conneaut's five Huletts, their arms bobbing up and down like pecking chickens, could unload an average of 3,750 tons an hour in 1937. That meant they could empty the biggest Tin Stackers in about three and a half hours. In July 1937 they unloaded more than 1.7 million tons. In their top day that month, they ran twenty-two hours and twenty minutes and unloaded 83,000 gross tons of ore—roughly the equivalent of cargo carried by six of the fleet's modern ships.[14]

Encouraged by the prospect for continuing improvement in the economy in 1937, the Pittsburgh Steamship Company resumed its program of modernizing the fleet. Following the pattern set by Gus Wolvin and Harry Coulby years before, A. F. Harvey ushered in a new era in marine engineering. On February 16, 1937, the American Ship Building Company submitted a proposal to the fleet to build a new ship for

$1.37 million and a duplicate ship for another $1.3 million. Great Lakes Engineering Works also bid for the work, and each yard won contracts to build two ships—the first new vessels to be built on the Great Lakes since the *Lamont* and *Thomas* in 1930.[15] The Pittsburgh fleet announced that the new ships would be named *William A. Irvin, Governor Miller, John Hulst,* and *Ralph H. Watson.*

First of the ships to be launched was the *Irvin,* named for the president of U.S. Steel Corporation. On November 10, 1937, William A. Irvin watched proudly as his young wife, Gertrude, smashed a bottle of champagne across the ship's bow and the vessel slid into the water at the American Ship Building Company's yard in Lorain. Among those attending the ceremony were A. F. Harvey; Benjamin Fairless, the Steel Corporation's newly elected president who would shortly replace Irvin; Adolf H. Ferbert, vice president of the Pittsburgh fleet; J. N. Rolfson, the ship's first master; and William Bourlier, its chief engineer.

The new ships were loaded with all the innovations that shipbuilders had developed in recent years but which had gone unused because of the depression. Each was more than 610 feet long overall and could carry more than 14,000 tons of cargo. They had eighteen hatches instead of the usual thirty-two. Each hatch had a one-piece steel cover that was lifted on and off by a low, horizontal hatch crane traveling on rails along the deck. These watertight covers meant tarpaulins were no longer needed on the hatches, saving time and eliminating injuries caused by the big sheets of canvas catching the wind and knocking down deckhands. A passageway, soon dubbed "the tunnel," ran the length of the ship belowdecks, enabling men to safely walk from one end of the vessel to the other in foul weather. Electricity rather than steam powered the anchor windlass, deck winches, auxiliary pumps, and steering gear. In the boiler room, automatic stokers carried coal from the bunkers into the furnaces, ending much of the firemen's backbreaking work. The crew's cabins were made entirely of steel, heavily insulated against heat and cold, and "vermin proof." Entrances to individual cabins were inside the deckhouse to shield men from the elements. Electric welding—another innovation—was used extensively in the ships to join steel plates in the hoppers, tank tops, coal bunkers, and other parts of the hull.[16]

The biggest innovation, however, was found in the engine room. Instead of a triple expansion steam engine, each of the four new ships was propelled by cross-compound steam turbine engines. In simple terms, steam flowing through a turbine turns a multibladed fan that generates power. Turbine fans spin at thousands of revolutions per minute, so some means is needed to translate that energy into the seventy-five to ninety revolutions per minute that is practical for a ship's propeller.

Turbines were not new to the Great Lakes. The Bradley fleet had used turbine engines in the *T. W. Robinson* in 1925 and the *Carl D. Bradley* in 1927. In the Bradley vessels, the turbines generated electricity to power huge electric motors. The motors, in turn, were geared to turn the propellers both forward and astern. In the four new Pittsburgh ships, the turbines were linked to double reduction gears that tamed the energy and turned the propeller shaft at a manageable speed. As compound engines, they passed steam from a turbine in a high-pressure chamber to a turbine in a low-pressure chamber. Because turbines cannot be reversed, these geared turbines were given a separate turbine set to turn in the opposite direction. This turbine simply idled until needed, when it could be put into gear to drive the ship astern. The 2,000-horsepower turbines gave the four new Tin Stackers a speed advantage of three-tenths of a mile per hour over other vessels in the fleet. It was a small increase, but factored in over the course of the season it could mean a measurable increase in tonnage carried by the new ships.[17]

The *Irvin* and *Miller* offered distinctly different profiles from their two sister ships because each had an extra deck in the forward cabin that contained four luxurious staterooms and an observation lounge for guests. A deckhouse just aft of the first cargo hatch contained a dining room for guests and a special galley to serve them. During the summer months, extra cooks and porters were carried on the forward end to serve the well-to-do people invited aboard by the Steel Corporation.

The economic optimism that prompted A. F. Harvey to order the four new ships proved to be premature. The 1938 shipping season was a dismal one for iron ore tonnage—almost as bad as 1932. The Pittsburgh fleet put only thirty-six ships into operation. Nonetheless, the *Hulst* made its maiden voyage on May 21. The *Miller* and *Irvin* followed in June, and the *Watson* finally got under way on September 1.

The first half of the 1939 shipping season showed no signs of being any better than the previous year. Traffic picked up considerably in the second half, however, as events in Europe once again took a grim turn toward war. By 1940 demand was rapidly growing for iron ore, stone, and coal, and the backlog of cargoes meant ships of the Pittsburgh fleet and others would sail late into the fall.

The weather in the first week of November was unusually mild for the Great Lakes. Sunday, November 10, was a day that evoked everything pleasant about Indian summer as Tin Stackers and other ships plodded along under clear skies. The following day dawned brightly, too, but even as the sun rose in spectacular hues, the U.S. Weather Bureau was posting storm warnings as a powerful weather system raced out of the High Plains and across the Midwest.

On southern Lake Michigan, the *Miller* was departing Gary bound for Calcite to load limestone. On its way out of port the ship passed the *Peter A. B. Widener,* which was starting to unload iron ore. Not far away was the *Thomas F. Cole,* heading to Lake Superior for ore. On the lake's northern end, the *George W. Perkins* and *A. F. Harvey* were downbound. Up on Lake Superior that morning, the old *Malietoa* headed toward the Soo as Captain Forrest Pratt pondered his ship's barometer. The pressure was dropping at an alarming rate and seemingly showed no inclination to stop. He tapped the glass and exclaimed, "Jesus, this is going to be bad one!"

The terrible storms of 1905 and 1913 had developed relatively slowly, giving some ships time to seek shelter. But the tremendous storm that struck the Midwest and Great Lakes on November 11, 1940, seemed to explode from the atmosphere as it raced in from the west. Along the Mississippi Valley, winds rose to sixty-five miles per hour, bringing heavy rain and then sleet as the temperature dropped forty-five degrees within hours. Unseasonably heavy snow followed close behind. Whipped into a blizzard, the storm stranded motorists and killed dozens of unprepared duck hunters in Iowa and Minnesota. The horizon turned into a low mass of black clouds as the storm approached. It raced across western Lake Superior, dropping barometric pressure to all-time lows in Duluth and Houghton, Michigan.[18]

The wall of wind, rain, and flying debris raced across Illinois and Wisconsin. Ceremonies set for 11 A.M. to honor the armistice that halted the Great War came to an abrupt end as the storm roared through town after town. In downtown Chicago, massive advertising signs crumpled while utility poles and trees fell in ranks. Aboard the *Widener* in Gary, crewman Jesse Cooper watched in astonishment as the wind toppled railcars onto their sides. The high winds and massive change in air pressure briefly lowered the harbor's water level by seven feet, leaving the *Widener* sitting on the bottom.[19]

Out on the lake, shipmasters saw the ominous wall of clouds on the horizon and ordered their vessels turned into the wind. The onslaught whipped the lake into a hellish froth, sinking Interlake Steamship Company's *William B. Davock* and Sarnia Steamships Ltd.'s *Anna C. Minch*. Both vessels went down with all hands off Pentwater, Michigan. Not far away the Canadian steamer *Novadoc* was blown ashore and mercilessly pummeled by the surf.

Aboard the *Perkins,* Captain Ed Baganz, in his first season as a master for the Pittsburgh fleet, desperately sought shelter as winds rose to eighty miles per hour and waves grew to enormous heights. After enduring the seas for several hours he finally managed to anchor the ship behind Garden Island. The *Perkins* narrowly avoided Lansing Shoal only after lighthouse keepers there flashed their house lights on and off to alert Baganz that he was steaming toward shoal water.[20] Captain Robert Parsons pointed the *Cole* toward the western shore of Lake Michigan and spent the next several hours plunging into the worst seas he had ever encountered. Waves ten feet high rolled down the spar deck and battered the after cabins. The steward's heavy stove came adrift in the galley, smashing equipment and destroying walls before finally battering itself to pieces.[21] The *Harvey* was off Point Betsie when the storm struck. Battling heavy seas and snow, Captain E. K. Male made the difficult decision to seek shelter in the shoal-lined waters off South Manitou Island. "I turned around off Point Betsie . . . and ran before it but the sea came aboard and broke in four doors: The after hall, dining room, oilers' room and Galley. Did considerable damage to the wood work and loosened a number of rivets in the deck amidships," he reported later.[22]

Corpses were just beginning to wash ashore north of Pentwater when people around the Great Lakes began comparing the Armistice Day Storm to the great maelstrom of 1913. The Pittsburgh Steamship Company was fortunate this time. Its worst casualty was the *Cole.* Along with damage to its stern cabins, the ship's hull had flexed so hard while bucking the waves that hundreds of rivets had sheared off, dangerously weakening the vessel and forcing it to proceed immediately to port for repairs. On many other Tin Stackers, sailors soberly reflected upon the fact that good seamanship, stout ships, and, in some cases, incredible luck had kept them from sharing the fate of the *Davock* and *Minch.* Aboard the *Malietoa* on eastern Lake Superior, only Captain Pratt's skillful seamanship had prevented disaster as the ship steamed blindly through a blizzard while desperately seeking shelter in Whitefish Bay. Guido Gulder, now a third mate, had spent the storm working the hand-operated sounding machine—essentially a partially automated lead line that told masters the depth of water beneath their ship. Many other ships had sought shelter in the bay, but the *Malietoa* managed to avoid all of them. The memory of sailing blindly through the storm remained etched in Gulder's mind even when he recalled the day fifty-six years later. "Every once in a while we'd hear a boat go by," he said, "but we never saw any of them."[23]

7

Tin Stackers Go to War

After struggling for a decade to recover from the Great Depression and its aftershocks, America's economy finally regained solid footing because of the specter of war looming over Europe. Tension on the continent had been growing for years as Adolf Hitler's Germany struck an increasingly belligerent pose. Sacrificing Czechoslovakia to Germany was supposed to appease him, but it did little to ease anyone's fears. The storm finally broke on September 1, 1939, when Germany unleashed its blitzkrieg on Poland. Great Britain and France quickly declared war on Germany, and another European conflagration seemed imminent.

The outbreak of war sent a jolt through America's economy and prompted a late surge in vessel traffic on the Great Lakes. Only 60 percent of lake freighters were operating in the first half of the season, but dozens more were brought out in the fall. The lake fleets carried nearly half the season's ore tonnage after September 1.

In the spring of 1940, Germany sent its juggernaut rolling across Europe. Belgium, the Netherlands, and France fell before the onslaught. Britain's army was sent reeling across the English Channel after abandoning most of its armaments in the fields around the French town of Dunkirk. Germany marched on, gaining control of Greece, Yugoslavia, Hungary, Romania, and Bulgaria.

Germany's onslaught galvanized the United States. Shedding years of isolationist complacency, it belatedly launched a program to modernize the Army and Navy as a safeguard against Germany and an increasingly bellicose Japan. Orders went out for new ships, airplanes, trucks, and guns. Great Britain, now the last bastion of freedom in Europe, desperately bought all the weapons it could afford. The arms of war were made of steel, and around the Great Lakes the mills gained strength and called for more iron ore.

photo was taken, the ship's original hatch covers had been replaced by one-piece covers, and a hatch crane had been added to the deck to handle them. Atop the pilothouse is a diamond-shaped antenna for the radio direction finder, a navigation aid adopted by the fleet beginning in 1926. William Roleff photo, courtesy Northeast Minnesota Historical Center.

Rearmament was a tonic for America's economy. Recovery meant more jobs and a return to normal life for millions of people. For industry, it meant more production and profits. For the Pittsburgh Steamship Company, it meant ships coming out of layup, a longer navigation season, and growing orders for cargoes of iron ore, coal, and limestone. For the men who sailed Steel Trust ships, the end of the depression came in the form of a job and a paycheck; of more days spent sailing; of going back to work as a mate instead of as a wheelsman or deckhand or any other berth that was available.

The return to prosperity was welcome in the United States. Still, the underlying cause of the recovery dawned slowly on many people. For Guido Gulder, it came as an epiphany as his ship was tied up in Lorain and he curiously eyed row after row of new steel tubing stacked on the dock. "I remember asking a guy on the dock at National Tube where all the pipe was going. He said, 'That's not pipe. Those are shell casings.' It was all being loaded aboard Canadian canallers and going to Canada for shipment to England."[1]

Amid the tumult of a busy navigation season came word that the man who had guided the Pittsburgh Steamship Company through its toughest times was stepping down at the end of the year. A. F. Harvey was sixty-nine years old and had worked for the Pittsburgh fleet every day of its thirty-nine-year existence. In his sixteen years as president he had carried out the fleet's goals of effectively moving iron ore, building more efficient ships, and keeping out the unions. More importantly, he

The addition of the *William A. Irvin, Governor Miller, Ralph H. Watson,* and *John Hulst* in 1938 introduced steam turbine technology to the Pittsburgh Steamship Company. Instead of an engine room dominated by the towering triple expansion steam engine, the *Governor Miller* had two turbines, visible here at the bottom of the photo. U.S. Steel photo, Ken Thro Collection, courtesy Lake Superior Maritime Visitors Center.

had successfully steered the fleet through the greatest economic disaster in the nation's history. The work had taken its toll, however, and Harvey would be dead within a year.

As Harvey's replacement, the Steel Corporation selected fifty-seven-year-old Adolph Henry Ferbert, another longtime employee who had worked his way up through the ranks of the fleet's front office. A. H. Ferbert, as he was known, was born in Cleveland in 1883. As a young man he took a clerk's job at the Upson-Walton Company, a prominent supplier of equipment to Great Lakes fleets. He joined the Pittsburgh Steamship Company in 1904 to get a better job as a stenographer. Two years later he earned promotion to the furnace order department, and by 1912 he was a vessel dispatcher. Ferbert spent the next decade in the exacting job of keeping the ships moving. He planned vessel traffic, juggled schedules, and issued sailing and loading orders. His work attracted the attention of Harry Coulby and A. F. Harvey, who began grooming him for a leadership role. In 1922 they made Ferbert the fleet's traffic manager. Just two years later, Coulby retired and Harvey moved up to take his place. That left the vice president's post open, and Ferbert was chosen to fill it. Contemporaries described him as progressive and interested in the well-being of his officers and crews. It was often pointed out that A. H. Ferbert took great pride in personally knowing each of the fleet's shipmasters.[2]

The insatiable demand for iron ore grew ever larger in 1941 as the federal government inaugurated an organized program of defense production. The Pittsburgh Steamship Company announced in March that it was placing orders to build five ships, taking advantage of a new federal program offering tax breaks to encourage construction of vessels. Despite heavy ice on Lake Superior, the first ships laden with ore departed on April 7—the earliest start to the season since 1916. Within the first week, seventy-one ships from various fleets were scheduled to load at the ore docks in Duluth and Two Harbors. The Pittsburgh fleet had sold only eight ships and twelve barges during the depression years, so it still had plenty of 400-footers like the *Zenith City* and *Queen City* that could be pressed into the ore trade even if they lacked the efficiency of the newer ships. With every available ship under steam, ore tonnage on the lakes rose to a record 80 million that season.[3]

By the time the United States entered the war in December, Great Lakes shipping was already in high gear. Iron ore shipping was immediately recognized as a crucial defense industry. The Roosevelt administration organized the Office of Defense Transportation and gave it authority over all forms of transportation. The office set a goal of 88 million tons of ore to be shipped on the Great Lakes in 1942. The men running the lake fleets, concerned that the government might take control of their ships, formed a committee and various subcommittees to carry out the directive. Donald C. Potts, the Pittsburgh fleet's traffic manager, took a hand in drafting proposals that largely prohibited U.S. lake freighters from carrying anything other than iron ore. Other measures called for an early opening to navigation and easing restrictions on how much cargo a ship could carry.[4]

The 1942 season was far different from anything Pittsburgh sailors had ever experienced. Ore cargoes began moving in March. Ships arriving in lower lake ports over the weekends started unloading on Sundays for the first time since 1918.[5] Big ships like the *D. G. Kerr* began loading more ore than they were designed to carry, and the effect was immediately noticeable. "It handled just like a log," one wheelsman recalled years later.

The rush to get steamers under way early that spring resulted in a rash of accidents related to poor sailing conditions and ice. *MacGilvray Shiras* was carrying coal to the fuel dock at Detour, Michigan, when it ran aground in western Lake Erie at about 4:30 on March 26. The steamer had been poking along as the master and mate tried to spot a navigation light that had not yet been put into service for the season. *Thomas F. Cole* lost a blade off its propeller on March 28 as it bucked heavy ice approaching the Detour fuel dock in the St. Marys River. After receiving a new blade at the Soo, the *Cole* continued into Lake Superior. There Captain M. H. MacBeth found himself in danger of being blown ashore on Whitefish Point. To escape, he had to batter the *Cole* into hard, blue ice up to two feet thick. Somewhere in the river or on Lake Superior—probably both—the *Cole* sustained extensive damage to the plates and frames in its bow, requiring the vessel to be drydocked in Lorain.[6]

The season's second week was even worse than the first. Late on April 4, the *William A. McGonagle* and *William B. Schiller*, both carrying badly needed iron ore, joined a pack of downbound ships trying to push their way through an extensive ice field in Whitefish Bay. About 9 P.M. the wind shifted to the southwest and gradually picked up to a brisk twenty miles per hour. A cold, miserable rain began to fall, reducing visibility. Early on April 5 the wind shifted to the northwest and increased to thirty miles per hour. Captain Arthur Dana of the *McGonagle* realized the ice field surrounding his ship was starting to drift toward the low shores of Ile Parisienne. He tried to turn the *McGonagle* left, then right. Failing that, he ordered full speed ahead in an attempt to push his way past the island. It was hopeless. A prisoner of the ice, the *McGonagle* was pushed ashore on the island shortly after 4 A.M.

Not far away, Captain C. J. Brinker found himself in the same situation aboard the *Schiller*. As soon as he noticed the ice pack was carrying his ship toward Ile Parisienne, Brinker tried backing out of the morass. Unable to get free, he tried turning left and right several times before ice jammed his rudder, leaving his ship helpless. The vessel grounded broadside on the island less than an hour after the *McGonagle* had fetched up.

The *Schiller* remained stranded for two days before the powerful wrecking tug *Favorite* and the smaller tugs *Roen* and *Louisiana* were able to fight their way through the ice. After dumping 600 tons of ore over the side, the steamer was light enough for the tugs to pull it free. The salvage flotilla finally freed the *McGonagle* on April 9, but only after it jettisoned 2,300 tons of ore. Both steamers limped southward with punctured ballast tanks. In one of his confidential letters to the fleet's shipmasters, A. H. Ferbert exonerated captains Dana and Brinker. Their ships, like many others, had been helplessly caught in the moving ice. They simply were unfortunate to be near the island when the wind picked up.

Misfortune also plagued other Tin Stackers that spring, but none as severely as the *Eugene J. Buffington*. The ship was plowing through the ice field in Whitefish Bay on April 7 when it sustained heavy damage to its bow. Numerous bow plates had to be replaced while nearly as many were removed, run through a rolling mill, and riveted back into place. The ship was hardly out of dry dock when, at the end of April, it damaged its bow again by hitting a pier at the Soo.

Repairing the *Buffington* was becoming an expensive habit for the fleet. But the worst came June 21, and it nearly ended the ship's career. The *Buffington* was downbound on northern Lake Michigan, passing among the treacherous islands and shoals that have claimed scores of vessels over the years. Throughout the day, Captain G. C. Chambers ordered his mates to deviate from the standard courses set by the Lake Carriers' Association because "he wanted to cut corners to make good time."[7] Despite these unusual circumstances, he spent little time in the pilothouse, instead passing along his orders to the first and second mates during their watches.

By evening First Mate C. A. Penzenhagen determined that the *Buffington* would pass within a mile of Boulder Reef. He alerted the captain, then turned over the pilothouse to Third Mate J. J. McDonald, who began his watch. McDonald thought the ship would pass uncom-

The desperate need for ships to move ore during World War II prompted the Pittsburgh fleet to launch an extraordinary effort to salvage the *Eugene J. Buffington* after it ran aground on Boulder Reef in Lake Michigan in 1942. To help mend the ship's broken back, workers bolted steel plates to the hull to hold the bow and stern together. Courtesy Lake Superior Maritime Visitors Center.

fortably close to the reef, but would still be safe. He was wrong. At 6:34 P.M. the *Buffington* rumbled over Boulder Reef about 1,000 feet west of the buoy marking the hazard. The shock of striking the reef in just twenty feet of water broke the *Buffington*'s hull amidships. The steamer ground to a halt. Its bow and stern sagged downward, severing all steam and water lines between bow and stern. After alerting the fleet office by radiotelephone, the crew could do nothing but wait until help arrived.[8]

The Pittsburgh fleet quickly rallied to save the stranded ship. Great Lakes Towing Company's wrecking tug *Favorite* and its lighter barge *T. F. Newman* were hired and arrived at the scene late on the twenty-second. As soon as the wrecking tug tied up to the *Buffington,* its crew went to work methodically stripping the pilothouse and forward cabin of equipment in preparation of the salvage effort. Heavy seas periodically battered the ship, interrupting salvage work and causing extensive damage to the forward and after cabins. Five days after the stranding occurred, the steamer *J. Pierpont Morgan* tied up alongside the stricken ship so the *Newman* could begin using its crane to transfer iron ore from the *Buffington*.

Once salvage crews had the *Buffington* ready to move, the *Clarence A. Black* was tied alongside the stricken ship to accompany it on the trip to the shipyard in South Chicago. Remarkably, the *Buffington* proceeded to the yard under steam from its own boilers. Courtesy Lake Superior Maritime Visitors Center.

By July 1 all the ore had been lightered. A diver sealed underwater openings in the hull while workers built watertight cofferdams inside the ship to seal off the break. Heavy-duty pumps were brought aboard to begin clearing the ship of water. When the hull was relatively dry, the salvage crew began riveting and welding heavy steel plates over the break in the hull to tie both ends of the ship together. Finally, and perhaps most remarkably, the *Buffington*'s own boilers were brought back to life in late July. With the steamer *Clarence A. Black* tied alongside, the *Buffington* proceeded under its own power to the American Ship Building Company's shipyard in South Chicago.

Salvaging the *Buffington*—just getting it off Boulder Reef and down to South Chicago—cost $133,741. The thirty-three-year-old ship still needed extensive and costly repairs. A. H. Ferbert took a personal interest in the matter, calling the fleet's agent in Chicago for details about how much American Ship Building Company would charge to put the *Buffington* back together. The agent relayed the concerns of "the boss" to the fleet's marine superintendent. "From his [Ferbert's] conversation," the agent wrote, "I gathered that if these costs were too excessive, it might be decided to scrap her."[9] A survey of the *Buffington* resulted in a nine-page itemized list of needed repairs, which American Ship Building estimated would cost $587,848. It was a steep price to pay, but 600-footers were too valuable to scrap just then. Ferbert authorized the work.

Ferbert dealt sternly with the *Buffington*'s officers. He scored Captain Chambers as "very negligent" for deviating from the standard course without reason and for failing to properly supervise the mates' navigation through a treacherous area. "In the face of such negligence," Ferbert wrote, "there was nothing left for me to do but ask for Captain Chambers' resignation." First Mate Penzenhagen, who earlier had been recommended for promotion to master, was criticized for turning over command of the ship to the inexperienced third mate even after recognizing that the vessel was steaming perilously close to the reef. "After careful thought," Ferbert continued, "it is our opinion that the First Mate used very poor judgment, and had he been looking out for his

With the *Buffington* in dry dock, shipyard workers go to work replacing a sizable portion of its keel and bottom shell amidships. This photo, shot inside the ship's cargo hold, shows men standing on the bottom of the dry dock after removing much of the damaged steel. The vessel's side and bottom frames are clearly visible. U.S. Steel photo, courtesy Lake Superior Maritime Visitors Center.

own position and the interest of the Company, he would have stayed in the pilot house until he saw the ship safely by Boulder Reef, or saw to it before he turned her over to the Third Mate that the master was in the pilot house to look after the navigation of the ship. We therefore feel obligated to discharge the First Mate." Third Mate McDonald fared somewhat better because of his good record and lack of experience. Although criticized for negligence, he was demoted rather than fired.[10]

Losing several vessels early in the season hurt the fleet, but the tonnage was made up in later months as five new ships entered service. Begun before the war, the ships arrived in time to help the fleet during the critical year when iron ore tonnage on the lakes rose to an all-time high. The new ships were named *Benjamin F. Fairless, A. H. Ferbert, Leon Fraser, Irving S. Olds,* and *Enders M. Voorhees,* but the trade press quickly nicknamed them "super freighters" and "super dupers" because they were bigger than anything sailing the lakes.

During the months before American involvement in the war, fleet officials had worked out the details for construction of the five new ships. They essentially decided to build duplicates of the *Ralph H. Watson* and *John Hulst* but on a bigger scale. On a set of specifications for the *Watson* and *Hulst* dated March 4, 1938, someone in the fleet's construction office simply penciled in a new length and beam for the five ships. Throughout the specifications, they also beefed up the size of equipment and piping to support a longer, wider vessel. A separate memo listed several design improvements to be made to coal bunkers and ventilation in the after cabins and in the engine room. The ships would be the first to use mechanical fans to force air through the vessel. Smaller details also got attention. The memo concluded with a notation about the placement of toilets in the crews' cabins: "Please put artillery in bathrooms around the corner from doors so as not to be the most prominent thing seen if doors be left open."[11]

As modified, the plans called for each of the new ships to be a whopping 639 feet long overall and 67 feet wide. Their width drew particular interest from the fleet and from shipyard officials. On December 27, 1940, Captain William Meister, the Pittsburgh fleet's superintendent of construction, met representatives of American Ship Building Company and Great Lakes Engineering Works to discuss whether to make the vessels 70 feet wide. Their final recommendation was that 67 feet would be more practical because "loading and unloading is very detrimental if beam is more than 67'" and "70' beam boat would be harder to handle in a seaway." Other objections noted that a wider ship would be heavier and more expensive to build.[12]

Along with being big, the new ships were powerful. Each vessel's turbine engine produced 4,000 horsepower—double that of the first turbine-powered ships built just four years earlier. The bigger engines gave the new class of ships a speed of sixteen miles per hour, about four miles an hour faster than anything else in the fleet. The new freighters quickly produced results. On one trip in 1942 the *Voorhees* took on 17,180 gross tons of iron ore in Two Harbors to break the loading record for that port. *Fraser* broke the record in Duluth by loading 17,033 tons.

Iron ore tonnage continued to climb to amazing levels. In 1940 the lake fleets carried 63.7 million gross tons. That shot up to 80.1 million the following season and to 92 million in 1942. In 1943, a crucial year when the outcome of the war remained in doubt, Great Lakes freighters carried a record 94.5 million gross tons of iron ore. More than a quarter of that tonnage was hauled by the Pittsburgh Steamship Company. The fleet's total of 25.2 million gross tons was its greatest single-season tonnage up to that time.[13]

More help for the fleet arrived in 1943 in the form of three new ships named *Sewell Avery, George A. Sloan,* and *Robert C. Stanley.* These were among sixteen ships the U.S. Maritime Commission had ordered built in Great Lakes shipyards. The vessels were then offered to various fleets at a modest price plus the exchange of equivalent tonnage in obsolete ships. Fleets were allowed to lease back their obsolete ships and continue operating them until the end of the war, when they would have to be turned over for scrapping. In exchange for the trio of modern vessels, the Pittsburgh fleet agreed to trade in half a dozen ships that were nearly fifty years old: *Zenith City, Robert Fulton, Queen City, Rensselaer, Pentecost Mitchell,* and *Herman C. Strom.*

The new ships were named the Maritime class. Taking advantage of new technology accelerated by wartime urgency, much of their steel was welded together rather than riveted. In other ways, however, these ships represented a step back in technology, which was necessitated by wartime shortages of materials. Each of the new ships was 621 feet long and 60 feet wide. Turbines were at a premium and most production was reserved for the Navy's warships, so *Stanley* and *Sloan* were built with triple expansion steam engines, *Avery* with a double-compound Lentz steam engine.

Wartime production pressures also may have been responsible for a structural problem that quickly became apparent in the *Stanley.* The ship was crossing Lake Superior during a storm on November 10 when a crack developed across its spar deck and spread down both sides of the hull. As the ship flexed in the heavy seas, the crack widened until it was a gaping four inches across in spots. In a desperate attempt to keep the ship from breaking up, crewmen ran steel mooring cables between the deck winches at the bow and the stern and then tightened the cables to reduce the flexing of the hull. The *Stanley* survived, but it did not sail again until the crack was repaired and the hull strengthened.

While the Great Lakes fleets were preoccupied with the challenge of meeting the unprecedented demand for iron ore, they were still keeping a wary eye on organized labor. The bigger fleets, including Pittsburgh Steamship Company, did not like what they saw. Beginning in 1939 and continuing through the early years of the war, the National Maritime Union renewed its effort to organize the sailors. It succeeded in several cases, winning the right to represent men of the Interstate Steamship Company, International Harvester, Inland Steel, and Bethlehem Steel fleets. Encouraged by their success with these smaller fleets, union officials decided the time was right to go for the biggest prize of all—the Pittsburgh Steamship Company. They launched an organizing drive, sending representatives to ports all over the lakes to seek out sailors of the Pittsburgh fleet. By November the union had enough new members to ask the National Labor Relations Board to hold a representation election. The board agreed and announced the election would take place in June 1944.[14]

Pittsburgh Steamship Company had successfully resisted the efforts of organized labor for forty years, so A. H. Ferbert and other fleet officials were not intimidated by the prospect of another union election. They had six months before the men would cast their ballots, and they put that time to use by waging a widespread, persistent, and controversial campaign against the National Maritime Union.

Soon after the ships began sailing in 1944, the fleet started sending sailors mail aimed at alienating them from the National Maritime Union. One piece was a reprint of a speech made in Congress denouncing the union. The men also received two letters from A. H. Ferbert urging them to vote against union representation. One letter played on the tradition of engineers, mates, and masters hiring the same men each year and taking them along as they were promoted through the fleet. "You should carefully consider some of the issues which have been referred to you by the Union," the letter stated. "One is 'rotary hiring.' To make sure you understand what this means . . . it means that you are entitled to return to the same ship in the spring that you laid up the previous fall, but if you are following an engineer or mate, and you want to work with him, and he is promoted (and given a berth on another ship), he cannot hire you."[15]

The anti-union campaign also played on another cherished tradition of the lakes: the substantial cash bonus each man received if he remained with the same ship for much of the season. In a tongue-lashing of a union sympathizer, one Pittsburgh officer said, "You CIO people, you're going to cause these men to lose their bonus. These men get a bonus at the end of the season. You're going to come up here and try to start overtime, and the Pittsburgh fleet isn't going to give them bonuses."[16]

The most controversial move occurred when every Pittsburgh sailor received in the mail a pamphlet the union had published titled "The NMU Fights Jim Crow." Several years earlier the National Maritime Union had vigorously fought discrimination against African-American sailors working aboard American ships on the high seas. However, the union studiously avoided mentioning this topic during its organizing campaign on the Great Lakes, where African Americans and other racial minorities traditionally were limited to the jobs of cook, porter, or, occasionally, fireman.

The Pittsburgh Steamship Company denied mailing the pamphlets, and a later investigation by the National Labor Relations Board found no evidence to the contrary. Nonetheless, some of the fleet's officers used the pamphlets as cudgels in their attacks on the union. One union sailor claimed a Pittsburgh shipmaster told him, "If these lake boats are organized they will have a bunch of niggers down here and you will have to work with them." The captain went on to say that the union was making a mistake "by showing all the pictures of the Negroes working with white crews," and added, "the men on the lakes aren't ready for that sort of thing."[17]

The election was held in the second week of June. The National Maritime Union did fairly well, but still lost. Of 1,619 sailors who cast ballots, 720 voted in favor of union representation while 899 voted against it. The union filed charges of unfair labor practices with the National Labor Relations Board, claiming Pittsburgh Steamship Company interfered with the sailors' right to organize. The board largely upheld the charges, and it ordered the Pittsburgh fleet to stop discouraging membership in the union. The damage already done to the union's reputation made the board's ruling a hollow victory, however, and the Pittsburgh fleet remained free of the National Maritime Union and organized labor.[18]

The demands on transportation continued through 1944 and 1945 as the U.S. war effort reached full stride. America's shipyards turned out thousands of warships and freighters of every size. Factories built tanks by the shipload and turned out enough new bombers and fighter planes to equip growing air forces around the globe. As the nation produced more ships, tanks, and planes, it needed more young men to fill the ranks of the ever-expanding Army and Navy. Earlier in the war, men under age twenty-six who were engaged in jobs considered vital to the war effort were not drafted. Now, the armed forces were clamoring for more men, and the Selective Service was casting its net wider than ever before, ending deferments that previously were ironclad.

Nowhere was the problem more acute than on the Great Lakes. The key to winning the war was moving supplies to the front, and virtually all those supplies moved by ship. Younger sailors had been receiving deferments because of the importance of iron ore shipping. But men who knew how to navigate a ship or run an engine were desperately needed everywhere, and a growing number of sailors in the Pittsburgh fleet and others found themselves being drafted. Captain R. M. Crawford, the Pittsburgh fleet's shore captain, sent a letter to shipmasters early in the year warning of the problems they would face in hiring a crew that spring: "We wish to impress upon you the fact that it is becoming increasingly difficult for us to obtain selective service (SIC) deferments even for experienced single men without dependents between the ages of 18 and 25. It is practically impossible to obtain deferments for inexperienced men in this age group. Draft boards are much more lenient with beginners who are over 25 years of age and married."[19]

Young men were the backbone of the Great Lakes fleets. They were the men who worked as deckhands, watchmen, wheelsmen, firemen, and assistant engineers. As the labor pool shrank, the Pittsburgh fleet resorted to hiring a growing number of older men and young boys to man its ships.

Edmund Siegrist turned sixteen in November 1944 and tried to join the Navy using a forged birth certificate. When that failed, the youngster from Bay City, Michigan, decided to apply for seamen's papers. When ships began sailing on the Great Lakes that spring, he headed to Detroit. After spending two days hanging around the Lake Carriers' Association hall, he was called for a deckhand's job on the Pittsburgh fleet's *Maunaloa*. The mail boat at Detroit took him alongside the steamer as it passed through the Detroit River, and he climbed a Jacob's ladder to the deck. "Your age was OK," he recalled years later, "as long as you passed the test to get your papers."

Siegrist was quickly introduced to the life of a deckhand. Aboard the *Maunaloa,* six deckhands bunked in one room. The ship had eleven hatches, and each hatch had eleven heavy wooden hatch covers. Deckhands used a wooden club with a hook in the end to catch the handles on the hatch covers and lift them on and off. It took two deckhands to carry one hatch cover to the side of the deck where they were stacked during loading and unloading. After the hatch covers were put back in place, the deckhands had to cover them with tarpaulins and secure them with wedges and clamps to keep the hatches watertight. "Our first trip back we ran into a hell of a storm," Siegrist said. "Water coming over the deck knocked those wedges right out. We had to get out on deck at 3 A.M. to put those wedges back in."

After two trips that season the *Maunaloa* was sold to a Canadian company. Much of the ship's crew, including Siegrist, ended up on the *B. F. Affleck,* where his apprenticeship continued. The deckhands painted, mopped floors, and washed the decks. After the anchors were used, they went down to the dunnage room with wire hooks to lay out the anchor chain in its locker so it would play out properly the next time it was used. Some of the lessons learned were hard to forget. At one port Siegrist was watching another deckhand loop the mooring line over a bollard. The man was handling the line by the loop instead of using the handle woven into the wire rope. As the mooring line tightened, the man's fingers were caught inside the loop and the pressure instantly amputated his fingers. It was the sort of mistake experienced men did not make.[20]

Although they were outsiders in a way, the wartime workers generally fit in well aboard the ships. As a flat-footed eighteen-year-old, Wendell Barrow was not considered fit for military service, but aboard the *Henry H. Rogers* he was welcome. "We had a few guys who'd go out and get drunk, but most were pretty nice. In my presence they made me feel right at home," he said. "The engine room and forward end crews were separate. Sometimes you got into a little hassle with each other but usually not. It wasn't a big deal; we all ate at the same table." Barrow, who had sailed aboard freighters on the Atlantic Ocean, was impressed with the *Rogers* and the Pittsburgh fleet. "It was a well-equipped ship—well taken care of; well fitted out."

The influx of new men aboard the ships meant a dilution of the culture of safety that the fleet had nurtured for so many years. On one trip through the Soo Locks, Barrow was riding the "bosun's chair"—a small seat attached to a line running through a boom, which was used to lower crewmen over the side to handle mooring lines. As a joke, the men handling the line on the bosun's chair lowered Barrow so far that his feet dragged in the water as the ship approached the lock wall. If Barrow had hung there long enough he would have been crushed, but his crewmates quickly hauled him back up to a safe height. "The old man [captain] saw that and I thought he was going to have a heart attack," Barrow said. "He didn't like that goofing around." On another trip the ship was leaving the dock at Lorain when a tardy crewman arrived. While the ship laboriously turned around, the crewman ran to a bridge spanning the harbor channel. As the *Rogers* passed, the man leaped to the ship's deck. "The third mate was the safety officer and he just about had a stroke," Barrow said.[21]

On May 8, 1945, Germany surrendered to the Allied forces. Three months later Japan followed suit. The world's greatest armed conflict was over. News of Japan's capitulation set off a wave of celebrations across the country, many of which were recorded in photographs and movies that became historical icons of the age. No one recorded the hundreds of smaller but equally heartfelt celebrations that took place that day on the ore carriers which ultimately made victory possible. "We were in Lake Superior when Japan surrendered," Edmund Siegrist recalled. "I'd never seen Superior that calm. We just ran out on deck and began dancing, screaming and hollering. There was no place else to go."[22]

When news came of Japan's surrender, the nation paused briefly, as if uncertain what to do next. Work stopped at many mills and docks around the lakes. "We just sat there for a couple of days until they started to unload us," Guido Gulder said. "They had to figure out how to switch from war production to peacetime."[23]

Government supervision of Great Lakes shipping ended on December 31. By the start of the 1946 navigation season, the country was making the transition to a peacetime economy, struggling a bit like a man trying on a suit that he had not worn for some time. The end of hostilities left workers free to demand better wages and benefits that had been deferred for several years. In the United States, coal miners in Appalachia and iron ore miners on the "Old Range" in Michigan and Wisconsin went on strike. Steelworkers and sailors in Canada also walked off their jobs. The season began slowly. By July 1 only 12 million tons of iron ore had been shipped because the coal miners' strike left many ships without fuel. Nonetheless, by the end of the season the fleets had carried nearly 60 million tons of ore—the third-largest peacetime ore tonnage. Pent-up demand for consumer goods and plenty of money earned by workers during the war meant the country would not suffer through an economic recession as it had after World War I.[24]

At this point, the Pittsburgh Steamship Company remained the dominant fleet on the Great Lakes. It had built four modern ships at the end of the depression and added eight more during World War II—more new tonnage than any other lake fleet had added during the same period. Following the war it shed the obsolete, inefficient ships that were part of agreement to buy the *Avery, Sloan,* and *Stanley.* It also sold off the venerable *Mataafa* and the last of its barges, the *John Fritz* and *John A. Roebling.* Now only 10 of the fleet's original 112 vessels remained. The fleet's 62 ships could carry a combined total of 776,064 of the gross tons commonly used to measure iron ore cargoes. In contrast, in 1905 the fleet could carry 513,651 gross tons per trip with 102 ships and barges. The Pittsburgh fleet's next largest competitor was the Interlake Steamship Company, which had 36 ships. Cleveland Cliffs Steamship Company had 22 vessels, and the Hutchinson and Company fleet had 21.[25]

The vast sums of money poured into weapons development during the war resulted in numerous advances in aircraft, engines, medicine, and communications. Among the first of these wartime technology breakthroughs to yield a peacetime benefit was radar. The Pittsburgh Steamship Company was always quick to adopt new communications devices and "safety appliances" as they were introduced on the lakes. In 1922 the fleet purchased its first gyrocompasses, which used a gyroscope instead of a magnetized needle to determine true north. Gyrocompasses were more accurate than magnetic compasses and could not be affected by the iron deposits so common around Lake Superior. Four years after the first gyrocompasses were adopted for use, the fleet began adding radio direction finders to its ships. Every lighthouse on the lakes was equipped with a transmitter that endlessly broadcast its identity through a code of dots and dashes. By using a ship's radio direction finder to take a bearing on three transmitting stations, a shipmaster could determine his location. Within a year every ship in the Pittsburgh fleet had a radio direction finder in its pilothouse and a distinctive loop- or diamond-shaped antenna mounted on the pilothouse roof.[26]

Another advance was the introduction of radiotelephones into the fleet in 1937. Guido Gulder was third mate on the *Buffington* when it got its first radiotelephone, and he recalled the master's concern that the ship follow the fleet's strict policy of using the devices only when necessary. Consequently, no one aboard the *Buffington* got much practice using the radiotelephone. One day the master spent several minutes impatiently dialing and redialing as he tried in vain to contact a nearby ship. Frustrated, he finally gave up and stomped out of the pilothouse. As soon as he left, the wheelsman quietly confided to Gulder, who was on watch, that the master had been dialing the *Buffington*'s number instead of that of the other vessel.[27]

As helpful as the earlier devices had been for improving scheduling and safety, it was radar that held the most promise. Following the war, the Lake Carriers' Association became interested in radar primarily as a means of reducing the risk of collision in rivers and channels. The Pittsburgh Steamship Company joined five other fleets, the Lake Carriers' Association, the U.S. Coast Guard, the Federal Communications Commission, and six radar manufacturers in agreeing to test prototype radar sets aboard Great Lakes ships during the 1946 season. Donald C. Potts, an official in the fleet's operations department, and R. C. Stanbrook, from its engineering office, were put in charge of the project for the Pittsburgh fleet.

Radiomarine Corporation of America, a division of RCA, installed a radar set aboard the *Ferbert*. Other manufacturers installed sets aboard vessels from the other fleets, including *John T. Hutchinson, Ernest T. Weir, William G. Mather, George F. Rand,* and *Frank Armstrong*. Throughout the season, the vessels tested different types of radar sets and varying wavelengths to determine their effectiveness in detecting other ships, showing a ship's position, and navigating in rivers and harbors. The tests led to several improvements in the radar sets, and recognition that Great Lakes ships needed specially designed equipment to enable vessels to avoid collisions and maneuver in narrow waterways. The final report also called emphatically for adoption of radar by all lake freighters. Its usefulness became apparent to the Pittsburgh fleet in 1947 as Captain Frank Davenport repeatedly was able to keep the *Ferbert* moving in rivers and on the open lakes while other vessels were forced to drop anchor because of fog.[28]

By 1950 radar was installed in the pilothouse of every vessel in the Pittsburgh fleet. The new sets were big and bulky by today's standards, and special sets had to be developed to fit into the cramped pilothouses of the smaller ships. Generally, the newest ships received radar first, although on one ship the fleet delayed installation because the master was so old that shoreside officials believed he could not adapt to it. When he retired at the end of the season, the ship got its radar.

Radar quickly became a useful tool, but it was still new enough to have serious limitations. No one saw it as a miracle cure-all for the hazards of navigation. "The weather bothered them [early radar sets] a lot," said Bill Wilson, who was aboard the *Voorhees* when radar was installed in 1948. "If it got foggy, you could maybe pick up [objects] six miles away. We hardly navigated with it then."[29]

Like the two previous years, 1948 was a good one for Great Lakes shipping and the Pittsburgh Steamship Company. Ships hauled nearly 83 million gross tons of ore down the lakes that season—tonnage comparable to the busiest of the war years. The Pittsburgh fleet had carried its share, but it had done so without the guidance of A. H. Ferbert. The fourth man to guide the fleet and the last one who was present during its early days died May 23 at the Cleveland Clinic at age sixty-five. "Long a proponent of the progressive viewpoint," the Lake Carriers' Association noted, "he was keenly interested in the well-being of officers and members of crews on his company's vessels and took an active part in maintaining and improving the good working conditions for which the Great Lakes bulk cargo fleet has long been proud." Taking Ferbert's place was Walter C. Hemingway, a New Englander and Yale graduate who had worked with U.S. Steel in various capacities for forty-one years. Before coming to the Pittsburgh fleet he was general manager of Federal Shipbuilding and Dry Dock Company in Kearny, New Jersey. Donald Potts became vice president of operations, a post that made him the apparent heir to the presidency. Both men had served their apprenticeship during the darkest days of depression and war. They both would face their own challenges, however, in the form of a new war, a new labor battle, and a question over the very future of Great Lakes shipping.[30]

8

A Billion Tons of Ore

In 1950 the Pittsburgh Steamship Company was nearly a half century old. With sixty-two ships, it remained the largest fleet on the Great Lakes. Its twenty-four hundred employees maintained a sharp edge of efficiency, enabling the fleet to routinely carry up to 25 million tons of iron ore each season. Since 1901 the fleet had hauled an estimated 1 billion tons of iron ore while surviving devastating storms, economic chaos, and the ravenous appetite of global war.[1] Now, at the dawning of the first new decade following World War II, it seemed the Pittsburgh fleet was finally returning to normal operating conditions. No one could foresee that the years ahead would be every bit as challenging as those just passed.

On the surface all was well, for these were prosperous times in the United States. The steel industry was at its zenith. Americans were clamoring for automobiles, household appliances, and other goods made of steel. In cities across the country, new highways, bridges, and skyscrapers consumed vast stocks of the metal. Meanwhile, the steel industries of Europe and the Far East were still recovering from the devastation of World War II and provided little competition in the marketplace. Then, in the summer of 1950, the United States joined other members of the United Nations in sending troops to defend South Korea from invasion by North Korea. The nation's steel industry moved to a wartime footing once again, and demand shot up for iron ore.

Behind prosperity, however, lurked the shadow of doubt concerning the Great Lakes iron ore industry. The Mesabi Range—backbone of Lake Superior iron ore production—had been mined for more than fifty years, producing a staggering amount of ore. Concerns were growing over how much high-grade ore remained and how economically it could be mined. U.S. Steel Corporation, through its Oliver Iron Mining Company, still owned or controlled an estimated 85 percent of the best ore reserves in Minnesota. Yet Oliver was spending $20 million to build a plant near Virginia, Minnesota, that could concentrate the iron found in a plentiful low-grade ore called taconite. Other mining companies were experimenting with taconite or already moving ahead with plans to process the ore into a usable product. Several companies, including U.S. Steel, also were exploring for ore or developing mines in Labrador, Venezuela, and the Dominican Republic. Many experts questioned whether taconite could feasibly take the place of high-grade ore, or whether steelmakers would be better off simply mining ore in other countries and bringing it back to the United States in huge seagoing ore carriers.[2]

At the same time, steel companies were questioning the long-term viability of their Great Lakes fleets. Vessels built during the boom years early in the century were now several decades old. Many needed substantial improvements to continue operating economically. Others were too small to ever again sail efficiently in the ore trade.[3]

Despite uncertainties over the industry's future, the continuing strong demand for iron ore set off a boom in ship construction. The first large ship built on the Great Lakes after World War II was Inland Steel's *Wilfred Sykes*. When the ship was delivered to its owners on January 12, 1950, it represented a substantial stride in construction on the lakes and immediately took the title of the first "modern" ore carrier. The *Sykes* was built specifically to carry 20,000 tons of iron ore on each trip from Duluth to Indiana Harbor, Indiana. It was 678 feet in overall length and 70 feet wide, making it the second-longest U.S. merchant vessel and the first U.S.-flag Great Lakes freighter with a 70-foot beam.[4] The length and width of the vessel were the maximum dimensions considered feasible for existing navigation channels and docks. The new ship also was one of just a handful of lakers to burn oil in its boilers. For its crew, the *Sykes* made such radical changes as giving each licensed officer a private cabin and bunking just two unlicensed men to a cabin.

In the wake of the *Sykes*'s construction, other fleets announced plans to build new ships. Several Victory ships and T-2 tankers built during World War II also were purchased for conversion to Great Lakes ore carriers. In the summer of 1950, Walter Hemingway announced that Pittsburgh Steamship Company would construct three new ships: *Philip R. Clarke, Arthur M. Anderson,* and *Cason J. Callaway*. Two of the building contracts would be awarded to American Ship Building Company, and the third would go to Great Lakes Engineering Works at River Rouge, Michigan. The vessels were to be completed in time to

Construction of the *Arthur M. Anderson* began in January 1951 in Lorain, Ohio. After seven weeks, shipyard workers had built a sizable portion of the amidships section. U.S. Steel photo, courtesy Lake Superior Maritime Visitors Center.

operate during the 1952 season. At the same time, Irvin L. Clymer, president of Bradley Transportation Company, announced that his fleet would award a contract to Manitowoc Ship Building Company to build a 666-foot self-unloader named *John G. Munson*.

The trio of ships being built for the Pittsburgh fleet represented yet another new vessel design. The fleet designated them the AAA class, which was simply a continuation of its system for differentiating ships according to size. When several other fleets built ships from the same design, the shipping industry named them the Pittsburgh class. The first of the new class was the *Clarke*. Designating it Hull 867, the American Ship Building Company laid down its keel in December 1950 at the company's shipyard in Lorain. The three ships were launched between late 1951 and early 1952, and all made their maiden voyages in 1952.

Somewhat more modest in size than the *Sykes*, each of the new Pittsburgh ships was 647 feet in overall length and 70 feet wide. Their design was an expanded version of the Maritime-class ships built during the war. Hull streamlining introduced by the Maritime ships was refined in the new plan, and the bow and stern were modified after extensive testing of models. The hull was designed to be asymmetrical near the stern to improve the flow of water into the propeller. The rud-

By early 1952, the *Anderson*'s hull was complete and the vessel was ready for launching. U.S. Steel photo, courtesy Lake Superior Maritime Visitors Center.

der also was offset slightly to perform more efficiently in the water streaming back from the propeller. Increasing the size of the hull enabled the ships to carry about 19,000 tons of iron ore—2,000 tons more than their next largest fleetmates.

The new vessels also were the first Pittsburgh ships to use alternating current electrical power and the first built with a system to treat sewage before pumping it overboard. Their boilers were fired by oil instead of coal. These produced enough steam to power a Westinghouse geared turbine engine that produced 7,000 horsepower and could drive the ship at sixteen miles per hour. The new power plants enabled these ships to complete a round-trip in just over five days, compared to the six or seven days required by their older fleetmates.

The Bradley fleet's *Munson*, slated to operate in the stone trade, was built and launched during the same period as the new AAA-class vessels. With its 666-foot overall length and 72-foot beam, the ship was the lakes' largest self-unloader. Its coal-fired boilers powered a 7,000 horsepower engine that drove the ship at sixteen miles per hour.

The combination of speed and size meant the Pittsburgh fleet's three new ships not only could carry more cargo than any other ships in the fleet, but they also could make several additional trips a year. This was realized in 1953 when the *Anderson* made forty-six round-trips and carried a single-season record of 866,855 tons of cargo. In

August alone the fleet carried 4.1 million tons of ore—the first time it had broken the 4-million-ton mark for a single month. Of that amount, 358,000 tons was hauled just by the *Anderson, Callaway,* and *Clarke.* Again, new ships were leading the fleet toward its goal of efficiency— carrying the most tons of ore with the fewest ships and men.

While the new ships were under construction in 1951, the Pittsburgh fleet began a program that received much less attention in the press but which certainly carried important long-term implications. Over the next ten years, the fleet planned to repower a major portion of its ships, removing their triple expansion steam engines and replacing them with either diesel or turbine engines. By now, fuel costs had become a major expense. A typical 600-foot freighter burned 320 to 350 tons of coal during each seven-day trip. With the price of coal often reaching more than $9 a ton, this meant a fuel bill well over $3,000 per trip. More efficient engines—especially those that burned cheap fuel oil—were seen as an important means of cutting costs.

Fleet officials decided to begin the effort by experimenting with the *Eugene W. Pargny* and *Homer D. Williams,* sister ships built in 1917,

Thirteen months after construction began, the *Anderson* splashes to life with flags flying and a shipyard crew aboard ready to secure the vessel once in the water. U.S. Steel photo, courtesy Lake Superior Maritime Visitors Center.

The newly completed *Anderson* steams along during its trial run on August 7, 1952. The vessel's "billboard" lettering on the hull proclaims the Pittsburgh fleet's new identity as a division of U.S. Steel. U.S. Steel photo, courtesy Lake Superior Maritime Visitors Center.

to determine which type of engine was most efficient. Both ships were sent to the shipyard in Lorain, where their triple expansion steam engines were removed. The *Pargny* received a diesel engine built by the Baldwin Lima Hamilton Company, while the *Williams* was repowered with oil-fired boilers and a steam turbine made by General Electric Company. "We intend to follow closely the performance of these two ships next year," Walter Hemingway said, "and the type of future repowering of our many ships will depend upon their performances."[5]

Amid the hoopla that accompanied construction of the new ships and the repowering of the old ones came two significant milestones in the fleet's history. First, alterations in the corporate structure of U.S. Steel changed the Pittsburgh Steamship Company from a subsidiary to a division of the Steel Corporation. Beginning January 1, 1953, the fleet was renamed Pittsburgh Steamship Division. This completed a reorganization that had begun a year earlier when the Bradley Transportation Company was merged with U.S. Steel. Titles to the Bradley ships had been transferred to the Steel Corporation with no change in the vessels' operation.[6] Second, and most significant, in 1953 organized labor penetrated the anti-union armor that had shielded the Pittsburgh fleet for decades. Surprisingly, it was the United Steelworkers of America that finally succeeded where so many maritime unions had failed.

The Steelworkers could trace the roots of their victory back to World War II. The National Maritime Union had never recovered from the drubbing it took in 1944 when it attempted to organize the Pittsburgh fleet. So in 1951 the union announced it was giving up its efforts to organize the Great Lakes fleets. Union officials ceded their jurisdiction over the sailors to the Steelworkers union. It was an unusual move—a maritime union bowing out in favor of a "land union"—but it made sense from a strategic standpoint. The Steelworkers already represented thousands of men in ore mines and steel mills, including many of those owned by U.S. Steel. The union could muster widespread support for the sailors, and U.S. Steel could not conduct an anti-union war on the lakes without jeopardizing labor relations ashore.[7]

The Steelworkers chartered Local 5000, Great Lakes Seamen, and launched a vigorous campaign to sign up members and hold elections on representation. Before the 1951 season was half over the union had organized the fleet owned by Jones and Laughlin Steel. The union then set its sights on the Pittsburgh fleet. Management resisted strongly for two years, but by 1953 enough men had signed up with the union to prompt the National Labor Relations Board to call an election. This time there was no Harry Coulby or A. F. Harvey to browbeat the men into submission. Nonetheless, the vote was close. Of 1,337 men eligible to vote, 1,309 cast ballots. The final tally showed that 700 voted in favor of representation by the Steelworkers, 501 voted for no union, and 92 voted for the Seafarers International Union. In April 1954 the National Labor Relations Board certified the Steelworkers union to bargain on behalf of unlicensed men employed by the Pittsburgh Steamship Division.

Contract negotiations got under way and proceeded quickly. On July 16 the two sides agreed to a contract setting forth terms for pay, holiday benefits, and grievance procedures as well as addressing broader issues such as living conditions, safety, and employee seniority. Harkening back to its early battles against unions, the fleet secured a clause in the contract that clearly established the ship's master as a higher authority than union work rules. "Nothing contained in this Agreement is intended to or shall be construed to restrict in any way the authority of the master as established by law," the contract stated. "Unlicensed personnel shall comply with all lawful orders of their superior officers, and the refusal of an employee to work as directed on any day shall be grounds for discharge." For its part, the union secured a pay scale ranging from $1.27 an hour for porters, coal passers, maintenance men, and wipers up to $1.58 an hour for wheelsmen and $1.66 for boatswains. The contract formalized the traditional season bonus, setting it at 10 percent of a man's total wages for the season if he had been employed before September 15. Overtime pay at time and a half was established for anyone working more than eight hours in a workday. Deck crewmen were exempted from chipping and painting on Sundays. The contract also delved into the finer points of shipboard life, establishing meal times, requiring electric fans in men's rooms, and mandating a change of bed linens once a week and bath towels twice a week. More battles lay ahead, but organized labor had returned to the Pittsburgh fleet, and this time it was back to stay.[8]

In 1954 the Pittsburgh Steamship Division operated 59 ships, which carried 26.4 million tons of ore. That was a remarkable change from the fleet's first year, when 112 ships and barges carried 10 million tons of iron ore. Although the number of ships had steadily declined over a half century, the Pittsburgh fleet remained the biggest bulk cargo operation on the lakes. Keeping it running was a complex task, even for those who never set foot on a ship.[9]

The Pittsburgh fleet employed more than one hundred people during the mid-1950s in its main office on the fifteenth and sixteenth floors of the Rockefeller Building in Cleveland. These employees were responsible for a mind-boggling array of tasks—planning, testing, recording, reporting, ordering, checking, screening, and tabulating—that kept the ships sailing. Every day of the week during the navigation season the office staff was in contact with the ships and docks by telephone, radiotelephone, teletype, and telegraph.

Key to the shoreside operation was the traffic department. During the navigation season, the fleet moved iron ore at an average rate of more than 7,000 tons an hour. Coordinating this southward stream of rock was the traffic manager and his staff of nine. Their duties broke down like this:

Traffic manager: Throughout the season, the traffic manager worked with colleagues in U.S. Steel's Raw Materials Section to plot the steel mills' need for ore, stone, and coal.

Chief dispatcher: The chief dispatcher was in direct charge of vessel movement. He transmitted orders to start vessels moving in the spring and continued throughout the season with orders on when and where to load and take on fuel.

Assistant dispatchers: Working with the chief dispatcher were two assistant dispatchers who plotted the ships' movements and their destinations six to seven days in advance. Much of their work was based on telegraph reports they received twice a day from ship reporters at the Soo and in Detroit. They also worked with the fleet's vessel agents in Duluth, Chicago, and Conneaut.

Working as an assistant dispatcher was demanding work. During the peak years of World War II, Pittsburgh fleet dispatchers controlled the movement of one hundred ships a week. Donald Potts worked his way up to chief dispatcher in 1938 based largely on his uncanny ability to keep track of the fleet's vessels—a skill often envied by his co-workers.

Furnace order supervisor: The supervisor of the furnace order department maintained a weekly schedule of the needs for twenty-one different grades of ore and advised the dispatchers on the need for each grade.

Coal clerk: The coal clerk and his assistant plotted the need for coal and made arrangements for ships to carry coal cargoes.

Location clerks: Finally, the location clerks each day plotted the movement of every Pittsburgh ship as well as every downbound ship the fleet had under charter. It also was their job to field the telephone calls from sailors' wives, girlfriends, and family members seeking information about the location of individual vessels.[10]

Another key part of shoreside operations involved supplying the fleet's vessels with food, mechanical parts, clean linens, and mail. When the fleet was formed, the notion of getting supplies to each of its 112 vessels was a daunting prospect, yet having every ship purchase its own food in every port would be inefficient and costly. Fleet officials instead chose to supply their own vessels at two strategic locations which nearly all of them would frequently pass—Conneaut and the Soo.

In 1903 the Pittsburgh fleet purchased Lake Erie Supply Company in Conneaut to provide food to its vessels calling there. After the Steel Corporation was formed, the old "Carnegie port" became one of the busiest for the Tin Stackers, making it an ideal supply depot. The fleet's primary supply point, however, was in the St. Marys River just below the locks. During its peak years, the fleet's Soo Warehouse and its supply boats served dozens of passing ships a day in an operation unique to the Great Lakes and possibly the world.

Shortly after its formation in 1901, the fleet built a warehouse on the Kemp Brothers Coal Company dock three-quarters of a mile below the Soo Locks. From there a ninety-eight-foot former ferryboat named *Superior* began hauling supplies to passing vessels of the Steel Trust fleet. That vessel was replaced in 1917 by the *Frontier,* a 140-foot passenger ferry that had operated near Buffalo as *Niagara Frontier.* Affec-

The *Ojibway*, the Pittsburgh fleet's supply boat at the Soo since 1946, approaches a vessel in the St. Marys River below the Soo Locks. U.S. Steel photo, courtesy Lake Superior Maritime Visitors Center.

tionately called "the Gutwagon" by sailors, the *Frontier* operated around the clock with two crews of eleven men, each working a twelve-hour shift every day during the sailing season. *Frontier*'s commodious hull enabled it to carry a large supply of food that could be dispensed without returning to the warehouse. For many years, a butcher even worked on board so that stewards from passing ships could step aboard for a few minutes to select the cuts of meat they wanted to serve in their galleys.

The original Soo Warehouse was replaced in 1937 with a bigger building. A decade later the fleet built the *Ojibway*, a sixty-foot diesel-

Crewman aboard the *Richard V. Lindabury* prepare to unload a pallet of supplies swung aboard from the *Ojibway*. U.S. Steel photo, courtesy Lake Superior Maritime Visitors Center.

powered boat with a small crane for hoisting pallets aboard passing ships. The awkward-looking craft was designed specially for its trade, and it was meant to sail in the sheltered waters of the St. Marys River. After the *Ojibway* was completed in Ashtabula, one of the supply boat crews sailed the new vessel to the Soo in a rough passage that became known in warehouse lore as "the cruise of the Pitching Pickle."

The Soo Warehouse in the 1950s was operating at its prime, typically serving about twenty-five vessels a day. It provided ships with everything from fresh strawberries to 9,000-pound anchors—nearly all of it delivered at any time of day while ships were under way. Primary customers of "the store" were vessels of the Pittsburgh Steamship Division, but it also served ships from the Bradley Division, Interlake Steamship Company, Great Lakes Steamship Company, and the Hutchinson fleet.

The heart of the supply operation was the two-story warehouse. In a typical year it handled a mind-boggling volume and array of goods. A ship's master, engineer, or steward could order from among six thousand items. They could get television sets, chairs, shower curtains, towels, anchors, life jackets, navigation lights, lumber, and cement. They could order twenty-five different types of oil, twenty types and sizes of chain, fifty-five types of manila and wire rope, fifty kinds of soap, two hundred types of packing and gasket material, thirty types of paint and scrub brushes, and countless pieces of hardware and small parts used aboard ship.

It was in food, however, that the warehouse distinguished itself. A ship's steward, planning his menu a week in advance, could choose from among 500 grocery items—and there was always plenty to go around. Each day a local creamery delivered 900 gallons of milk to the warehouse. Each week the warehouse's butchers cut and distributed 20,000 pounds of beef, pork, chicken, and turkey. Each year the warehouse shipped out 100,000 pounds of flour and 50,000 pounds of butter and coffee.

At the same time it was picking up supplies, a ship also could drop off damaged items that could not be repaired on board. Throughout the season, the warehouse's blacksmiths kept busy making or repairing fire tools used in the boiler rooms and renewing the wheel chains that turned the rudders of the older vessels. In a sail loft located on the warehouse's second floor, workers using heavy-duty sewing machines repaired carpets and made and repaired canvas hatch tarpaulins, lifeboat covers, parcel bags, sea anchors, and laundry bags.

In filling a ship's food and supply order, the warehouse crew followed a well-practiced routine. As an upbound ship approached the

Soo Locks, the *Ojibway* came alongside to pick up order forms filled out by the steward and engineer. The forms were dropped off at the warehouse, where workers began filling the order. The warehouse's first floor was dominated by 125 large bins—one for each ship it served. Also located on the first floor were meat coolers, a deep freeze, the grocery department, and a "green room" for storing vegetables. Once an order was received, men stacked the requested dry goods and supplies on pallets or "skids" inside the appropriate bin. Perishable food items were set aside in the coolers to await delivery.

A couple days later, when the vessel reached Whitefish Bay on its downbound trip, it alerted the warehouse by radiotelephone. Inside the warehouse, the men on duty recorded the ship's location on a blackboard, knowing they had almost exactly four hours before the vessel arrived. The warehouse men completed the ship's order, packed everything on pallets, and moved the pallets to the warehouse dock to be picked up by the *Ojibway*.

Aboard the *Ojibway*, the workday consisted of a hectic and endless series of runs between the warehouse dock and passing ships. On deck, a crewman's cry of "Around the bend!" heralded the approach of each steamer requiring service. As the ship passed the warehouse, the *Ojibway* pulled alongside and tied up. Using their vessel's deck crane, the *Ojibway*'s crewmen could hoist the pallets aboard the ship in about five minutes. The ship's crew gathered to unload the pallets so the empties could be quickly swung back aboard the supply boat. At the same time, mail and newspapers were put aboard the ship and soiled table and bed linens exchanged for fresh ones. Then the *Ojibway* cast off its lines to return to the warehouse for another ship's load.[11]

Although the number of ships served by the Soo Warehouse had declined to less than twenty-five by 1997, it remained an integral part of the fleet's operations. Eight buildings provide 54,000 square feet of warehouse space for the $1.2 million inventory maintained during the navigation season. A state-of-the-art computer system tracks inventory and orders. Along with handling supplies, the warehouse stores wrecking equipment and acts as distributor for companies that manufacture lubricants, machinery, and chemicals used aboard ship. The warehouse's fifteen employees run a twenty-four-hour operation throughout the navigation season. The *Ojibway* continues to supply passing ships with food, freight, and laundry service. It also takes crew members, passengers, and maintenance workers to and from ships, and it handles garbage removal. In addition, ships at regional ports are served by trucks dispatched through the warehouse.[12]

Throughout the 1950s, shipboard life for Pittsburgh sailors continued much as it had in earlier years. Sailors who were not on duty still amused themselves by playing cards or reading, but some now also enjoyed painting, assembling plastic models, or playing shuffleboard on courts they painted on deck. They also had a new amusement called television to help pass the time. Ships throughout the fleet began sprouting TV antennas atop their deckhouses to improve reception whenever the vessel neared Detroit, Chicago, Cleveland, or Milwaukee. Food was still a significant fringe benefit of a shipboard job as well as providing a time for the men to socialize. Occasionally, geese or ducks became tired or disoriented while flying over the lakes and landed aboard a ship. The unfortunate ones ended up in the galley as an added course of wild game, as did fish that were caught for amusement by men trolling off the fantail. Sometimes in hot weather a steward would hold an outdoor wiener roast and serve food to the men on the spar deck as they played musical instruments, joked, or talked as the sun went down.[13]

Occasionally, the monotony of shipboard life was broken by moments of excitement or danger. When that happened, it often was good training and a sense of duty that guided men's actions. Frank Stuller, second mate on the *Ralph H. Watson,* was checking the ship's draft marks as it unloaded on July 18, 1958, at the National Tube Company dock in Lorain. As he was working he saw deckwatchman James Eckert fall backward off the boarding ladder leading from the dock to the ship's deck. Eckert landed hard on the dock and tumbled into the water. Without hesitating, Stuller leaped into the water, grabbed the stunned Eckert before he sank, and pulled him to safety. In another instance the same year, the *John Hulst* was steaming through heavy fog off Gray's Reef in northern Lake Michigan when Captain F. L. Palagyi spotted five motorboats clustered together, their operators obviously lost. Palagyi contacted the boats by radiotelephone, organized them into a single-file line behind the *Hulst,* and shepherded them to the Mackinaw Bridge, where the boaters could regain their bearings and return to safety.

Some parts of shipboard life, however, were different these days, reflecting the tense relations between the United States and the Soviet Union. A sailor who set up a ham radio operation in his cabin extolled the virtues of his hobby, pointing out that it also would prove useful if an atomic attack destroyed radio stations ashore. All ships in the Pittsburgh fleet took part in Operation Sky Watch, an Air Force program that encouraged sailors on the Great Lakes to watch the skies and report any large aircraft which might be Soviet bombers. In three years of participation, half the calls made to the Air Force came from Pittsburgh ships, although apparently none of the aircraft reported ever proved to be Soviet.[14]

Aboard a few ships in the fleet, the sailors were joined throughout the summer by special guests of the Pittsburgh Steamship Division and U.S. Steel. Each week, a party of company officials, customers, or the socially well-connected brought their families and friends aboard the *William A. Irvin* and *Governor Miller* to enjoy a round-trip between the lower lakes and Duluth or Two Harbors. By now, the fleet's public-relations men preferred to use the nickname "Silver Stackers" rather than Tin Stackers, and the two ships with guest quarters certainly reflected this high-brow moniker. Guests slept in luxurious rooms in the forward deckhouse and ate in the private dining room located on deck. An extra cook, steward, and porter rode the ship just to cook for the guests, serve them drinks, and keep their rooms tidy. Top-rate food was brought aboard for the guests, and they could request virtually anything for dinner, including lobster. One of the benefits for the crew was that leftovers and unused portions of this lavish fare often found their way to the mess tables of the officers and men. "You really ate when the passenger season was on," recalled Ed Finnigan, who served as a mate aboard *Governor Miller*. "The sky was the limit."[15]

The presence of guests put pressure on the master and crew to perform well, keep a neat ship, be reasonably genial, and generally limit the amount of cursing used to get jobs done on deck. Crewmen never knew when one of the guests might turn out to be among the most powerful people in American industry. For instance, William A. Irvin, former president of U.S. Steel and a vice chairman of the board, was fond of riding his namesake and took a special interest in its operation. He liked to talk with deckhands and oilers and often posed questions about the ship to the mate on duty in the pilothouse.

The carrying of guests was a long-standing practice of the fleet, and it sometimes spawned stories that became part of company lore. Among them was the story of when automaker Henry Ford and William Livingstone, longtime president of the Lake Carriers' Association, rode the *James A. Farrell* in July 1913. During the trip Ford wandered out on deck each day at dawn in his robe and bare feet, sometimes chatting with the crew, other times just admiring the sunrise. When the ship's steering engine failed, Chief Engineer Mike Toner invited the automobile magnate to inspect the mechanism. Ford, who began his career as a mechanic, was delighted to be the first to spot a sheared pin that had disabled the engine. Throughout the trip Ford and Captain Fred Bailey kept up a good-natured debate on whether automobiles would ever replace horses as the primary means of land transportation. Bailey continually expressed his doubts about the automobile's practicality. As Ford was leaving the ship, Bailey caught his arm and whispered, "That's a marvelous machine you have there, Mr. Ford, but I'm still not sure it'll ever replace the horse and buggy." The following winter, a few days before Christmas, Bailey stepped out of his house in Vermilion, Ohio, to find a new Ford automobile parked at the curb—a gift from his shipboard guest of the previous summer.[16]

During their trips, guests often spent their time playing shuffleboard, flying kites, skeet shooting, hitting golf balls into the water, or enjoying refreshments in their private lounge. One activity aboard the *Irvin* that became a tradition during the 1950s was for guests to use verse when recording their impressions in the guest log. Over the years they filled the log with pages of tortured poetry as they tried to outdo the humor and rhymes set down by previous guests.[17]

Consider this rhyme from a trip in 1950:

> We came aboard as the Huletts dug
> their last big bite of ore;
> And lined the rail as the diesel tug
> Was about to demonstrate what it
> was there for.
>
> The ship avoiding all the rocks,
> The captain met us with a smile,
> And showed us his geranium box,
> Informing us his name was Colin
> Campbell Carlisle.
>
> The galley, the haunt of Willis, Dave
> and James
> With most fantastical cuisine,
> The fanciest appetite it tames
> But on this voyage there's no chance
> of your keeping lean.
>
> We watch the locks at the well
> known Soo
> Lift the carriers heavy and light,

> For the iron trade a rendezvous
> One spot where government
>   ownership is all right.
>
> Happy those few who may enjoy
> A trip that so delights the senses.
> One can no week the better employ
> Especially with the company paying
>   all the expenses.

Other erstwhile poets resorted to parodying the classics, such as Clement Moore's "Twas the Night Before Christmas." Their version of the closing verses included:

> Captain Whit waved goodbye, to
>   his crew gave a whistle,
> And the *Irvin* receded, like the
>   down of a thistle.
>
> But we heard him exclaim 'ere he
>   hove out of sight,
> "Gad, now they have gone; I guess
>   they're alright!"

The *Irvin*'s food was the subject of many verses—virtually all of them alluding to culinary excess not often enjoyed ashore:

> The food is marvelous never ending;
> Any dish you may desire;
> Willis will cook and feed you full;
> Without a crane you can't retire.

If their poetry was any reflection of their emotions, most guests were sorry to leave the ship at the end of their voyage. They said it in many ways, including this:

> This wonderful trip we shall never
>   forget
> The beautiful ship and the food we et
> While flying is great and sometimes
>   unnvervin'
> It can't top sailing on the *William
>   A. Irvin.*

But even amid the comforts of guest rooms, good food, and new ships came occasional reminders that sailing on the Great Lakes could still be a hazardous profession. One such occurrence in 1953 involved several ships from the Pittsburgh fleet, and provided ample evidence that having a fleet of well-maintained ships sailed by able men was the best way to avoid accidents on the lakes.

Before dawn on May 10, the steamer *Henry Steinbrenner*, owned by the Kinsman Transit Company, departed Superior with 7,000 tons of ore consigned for Bethlehem Steel. Across the narrow tip of western Lake Superior, the Pittsburgh steamer *William E. Corey* was loading at Two Harbors and preparing to leave in a few hours. The wind was moderate from the northeast when Captain James Robinson took the *Corey* out past the breakwall about 11 A.M. An hour later, however, Robinson received radio reports warning of a southeast storm with winds expected to range up to thirty-five miles an hour.

Also leaving Two Harbors that day was the steamer *D. M. Clemson* under the command of Captain A. M. Everett. The wind was already blowing at twenty-five miles an hour when the *Clemson* sailed at about 12:30 P.M. Four hours later the steamer was bucking waves near Devil's Island in the Apostles. Uneasy over the rising winds and waves, Everett ordered out his deckhands to put more fasteners on the hatch covers. Then he warned the engine room and galley to rig for heavy weather.

The storm grew in intensity during the afternoon, far surpassing the original forecast. By 6:30 P.M. the *Clemson* was rolling through heavy seas whipped up by a fifty-five-mile-per-hour wind. The shrieking winds climbed to eighty miles an hour by midnight as the temperature plummeted. A little after 3 A.M. a wicked gust of wind and an extraordinarily large wave boarded the *Clemson* right over the bow, shredding the heavy canvas "weather cloth" stretched across the front of the texas deck to keep spray off the pilothouse. The impact was so severe that it bent the steel stanchions holding the weather cloth.

Aboard the *Steinbrenner*, what had begun as a typical voyage through a spring storm had turned into a deadly menace. The ship's hatches were not covered by tarpaulins, and they were taking water through the night under the continuous pounding by waves washing over the deck. As the fifty-two-year-old ship labored off Isle Royale, it grew increasingly sluggish and began to ride lower and lower in the water.

About 5 A.M. on the eleventh the wind began to ease. Aboard the *Clemson,* Captain Everett ordered a slight increase in speed and then went below to rest. Forty miles to the west, aboard the Tin Stacker *D.*

G. Kerr, Captain G. A. Lehne exchanged small talk with his third mate in the pilothouse. It was odd, he remarked, how at times they were in calm seas while the wind howled overhead, yet at other times the wind seemed to drop out of the sky to pummel the ship.

A short time later, men on watch in the pilothouses of the *Clemson*, *Kerr*, and *Corey* overheard the *Steinbrenner* radio for help from the Canadian steamer *Hochelaga*, which was only twelve miles away from the troubled ship. Aboard the *Clemson*, Captain Everett was summoned from his bedroom. He immediately signaled the engine room for full speed and ordered the ship swung to the north toward the *Steinbrenner*'s position. Aboard the *Kerr* and *Corey* the same orders were given at almost the same moment. The *Steinbrenner* made a few more frantic radio calls, then—silence. By 6:30 A.M. the ship was gone.

Around western Lake Superior a host of American and Canadian ships altered course and raced toward the *Steinbrenner*'s last known position. The *Clemson* and *Corey* slammed into wave after wave as they pounded northward within sight of one another. They were several hours away from the site of the sinking, but their masters kept pushing, knowing they might be the only hope for men in the water. After some time, the two ships were joined by the *Joseph H. Thompson*, a converted World War II troopship now carrying ore for the Hanna fleet. Captain Everett and Captain Robinson held a radio conference with the *Thompson*'s master. They agreed to stay about four miles apart to increase the chances that one of the ships would spot survivors.

Shortly before 11 A.M. lookouts on the *Thompson* spotted wreckage. The Hanna ship had overtaken the *Corey* and arrived first on the scene. The *Wilfred Sykes* joined them a short time later. The ships had raced along for hours. Now they had to check their speed and delicately maneuver into position to rescue men who were nearly helpless aboard lifeboats and life rafts.

About 11:20 A.M. the *Sykes* and *Thompson* hove to windward of a raft containing survivors. As the *Thompson*'s crew plucked several men off the raft, the *Corey* maneuvered in to spread storm oil to help calm the waves. Meanwhile, the *Clemson* was several miles to the west bearing down on a lifeboat with seven men aboard. Captain Everett plowed past the boat, stopped, then carefully backed his ship. In a fifty-mile-per-hour wind and rough seas he gently put the lifeboat in the shelter of his ship's lee. Crewmen on the *Clemson* hoisted the survivors aboard by rope. The men were taken to the captain's room, where they were given hot baths, warm food, and dry clothing. When all apparent survivors were aboard, the *Clemson, Kerr,* and *Corey* formed a line abreast and began searching for bodies.

The men aboard the *Clemson, Kerr,* and *Corey* that day were later presented with ten-carat gold medals from the Pittsburgh Steamship Division for their role in the rescue. The following year, the Lake Carriers' Association presented a large bronze plaque to each ship in honor of the teamwork and seamanship used in the rescue. Donald Potts, who had become the fleet's president following Walter Hemingway's death on March 31, 1952, offered thanks on behalf of the men. "We take special pride in receiving this award for, as a division of U.S. Steel, we have worked hard to make safety, cooperation and teamwork conscious attitudes of our people," he said, "and we like to feel that previous training and conditioning in safety and teamwork were there for the men to draw on in a time of need."[18]

The end of the Korean War in 1953 brought a precipitous 36 percent drop in iron ore tonnage during the 1954 navigation season. While most sailors were preoccupied with this short-term setback, two events took place that heralded major long-term changes for all Great Lakes fleets.[19]

On May 13, President Dwight Eisenhower signed Public Law 358, giving final approval for the United States to join Canada in building the St. Lawrence Seaway. For decades, small oceangoing ships had been able to enter the Great Lakes through the series of locks running along the Canadian side of the St. Lawrence River. The new seaway would have far fewer locks and would be capable of handling far bigger ships. The public and the press focused much attention on the potential for general cargo ships to sail from Europe straight to Midwestern cities like Cleveland, Chicago, and Milwaukee. But the real potential impact lay in the possibility that ships up to 730 feet long would be able to haul iron ore from new Labrador mines to Canadian and U.S. mills.

The second major event occurred in Two Harbors on the first day of the shipping season. Columbia Steamship Company's steamer *Reserve* loaded 17,784 tons of taconite pellets made by the new Reserve Mining Company at its preliminary taconite concentrating plant in Babbitt, Minnesota. "Not only was this cargo the largest single consignment of taconite which has been shipped but it marked the first time that a season had been opened by the prepared mineral," the Lake Carriers' Association reported.[20]

By this time steel companies involved in Minnesota and Michigan mining were making serious commitments to developing taconite as a replacement for the high-grade ores nearing depletion. Taconite is an extremely hard rock containing 20 to 35 percent iron ore—far less than the amount contained in high-grade natural ores, and much too little to use in its natural state. The material is abundant on the Mesabi Range, but because of its hardness it is difficult to mine. Taconite formations must be blasted to dislodge the rock and break it up into pieces small enough to transport.[21]

Scientists working for the University of Minnesota and several mining concerns had been experimenting with taconite since 1914. They faced myriad problems in reducing the rock to a powder, separating out the iron, and concentrating the iron into a product suitable for steel mill furnaces. Experiments over the years resulted in a wide range of potential products containing about 60 percent iron and shaped like briquets, nodules, and pellets of varying sizes.[22]

The first commercial production of taconite began on June 21, 1922, at a plant in Babbitt, owned by a small group of investors incorporated as the Mesabi Iron Company. The plant produced a shapeless concentrate called sinter containing 61 percent iron. That fall the first load of sinter was shipped by rail to Two Harbors and loaded aboard a ship bound for the Ford Motor Company's steel mill at River Rouge near Detroit. Two years later the venture failed, the victim of high production costs and lingering technical problems.[23]

Experimentation and research continued, however, led primarily by the University of Minnesota's Mines Experiment Station. Refinements in production during the early 1940s attracted the attention of many mining and steel companies that were seeking greater ore reserves. Oglebay Norton Company had acquired the old Mesabi Iron Company lands in 1939 and, along with Republic Steel and Armco Steel, formed Reserve Mining Company to examine the feasibility of making a taconite product. Three years later Pickands-Mather joined four steel companies in forming Erie Mining Company for the same purpose. Both projects moved slowly, but Reserve took an early lead. In 1945 Oglebay Norton began buying land near a small inlet on the Minnesota shore of Lake Superior called Silver Bay. Six years later the company decided to move ahead and build a plant there to produce taconite pellets using rock mined in Babbitt and shipped in by rail.[24]

By 1955 taconite pellets were moving down the lakes in ships. "For the first time shipments of beneficiated taconite were an appreciable factor in Lake Superior operations," the Lake Carriers' Association reported. "At year end it was announced by Reserve Mining Company that 343,136 gross tons of the product moved by lake during 1955." Reserve completed its Silver Bay taconite plant in 1956 and that year shipped 3.5 million gross tons directly from there. Erie Mining Company followed, beginning to ship its taconite pellets in 1957 from a new port on Minnesota's shore called Taconite Harbor.[25]

Oliver Iron Mining Company cautiously embraced taconite in 1944 when it opened a laboratory in Duluth to study the product. Six years later the company broke ground near the Mesabi Range town of Virginia for an experimental taconite plant called Pilotac. The following year it began developing a taconite mine called Extaca near Mountain Iron, Minnesota. The first cargo of taconite sinter produced by Extaca was loaded aboard the *William B. Schiller* in Duluth on May 3, 1954. The 12,152 gross tons of material, produced by the plant over the winter, represented the first full shipload of sinter carried on the Great Lakes. As part of its experiments, Pilotac even handled some overseas ore. In 1955 the steamer *William P. Palmer* loaded a cargo of Venezuelan iron ore in Conneaut for delivery to Duluth. It was then taken to Pilotac to determine whether the fine-textured ore could be made more suitable for U.S. Steel's mills. Oliver finally announced plans to build a full-scale commercial taconite plant near Mountain Iron, but did nothing more. For the foreseeable future, most ships from the Pittsburgh Steamship Division would continue hauling natural ore.[26]

Continuing economic prosperity helped maintain a high demand for iron ore. In the spring of 1956, Great Lakes fleets were anticipating a good season. The season's outlook was equally bright—albeit for a much different reason—in the headquarters of the United Steelworkers of America and the Marine Engineers Beneficial Association (MEBA). The Steelworkers were preparing for a massive strike against the nation's steelmakers, and they fully expected it to succeed. MEBA was gearing up for a campaign aimed at organizing engineers of the Pittsburgh Steamship Division. While they expected a tough battle, they also were confident of victory.

At the union's convention in March, MEBA officials had decided to resume their organizing efforts on the Great Lakes after a ten-year absence. They were encouraged by the success of Steelworkers Local 5000 and enticed by the prospect of expanded lake shipping when the St. Lawrence Seaway opened. Union leaders voted to assess members $10 each to amass a $120,000 war chest for what they assumed would be a three-year campaign. They chose to take on their biggest opponent first, reasoning that once the Pittsburgh fleet fell into the union camp, the smaller fleets would readily follow.[27]

Just as MEBA began its organizing campaign, however, it ran into the nationwide Steelworkers strike. At midnight on June 30, Steelworkers around the Midwest walked off their jobs in the steel mills and iron mines. Stockpiles of ore were quickly loaded and shipped down the lakes, and by July 7 most ships were being laid up. A total of twenty-eight ships from the Pittsburgh fleet steamed into Duluth and Superior to wait out the strike. Unlicensed men were laid off, but mates and assistant engineers remained aboard and continued to receive their pay.

As the ships began tying up, MEBA organizers realized that they had been presented with the ideal opportunity to contact potential members in the Pittsburgh fleet. Officials of the Masters, Mates and Pilots Union also seized this opportunity and began a membership drive among mates on the Tin Stackers. By the end of July many men had signed pledge cards and begun attending union meetings. Under the Taft-Hartley Act, the assistant engineers and mates were not protected from employer retaliation because they were considered to be part of company management. Nonetheless, they were unhappy after spending three years watching unionized firemen and deckhands work forty-hour weeks and get overtime pay while licensed officers worked fifty-six-hour weeks at straight time. Many licensed men felt the Pittsburgh Steamship Division was arrogant and high-handed in the way it treated them.[28]

Once again the Pittsburgh fleet launched a counterattack against a union organizing campaign. This effort, however, appeared to be considerably less pernicious than those of past years. Donald Potts refused to meet with union representatives. Instead, a two-page letter dated July 30 was sent from Potts to the fleet's engineers. Although it was a form letter, each man's name was typed in at the top and each letter concluded with Potts' signature—or a good imitation of it—in ink. In the letter, Potts likened the union to a "new neighbor" that eventually would force the engineers to choose between being part of the fleet's management team or being "just another group of organized workers." Potts went on to write:

"Without unduly extolling the virtues of belonging to a management team, I want to remind you of the respect and recognition it brings. As officers of the Pittsburgh fleet, you are working supervisors, carrying more responsibility and working closer to the heart of your company than your unlicensed shipmates. We have steadily increased our communication to you, then visited with you aboard your ships, soliciting your ideas and suggestions to make this exchange of ideas flow both ways.

"This is the way toward proper individual recognition of each man's abilities and his specific performance. I seriously doubt that individual recognition is something that you can sit back and demand or legislate through group effort.

"Certainly all of you have been aware of the integrated program of several years to recognized the supervisory responsibility of fleet officers, and our continuing efforts to bring all of you together as a management team. Even so, I think most of you would be astonished at the time and effort top-level management people spend in trying to adapt your ideas and desires to the greatest benefit of our total operation. We are well aware that because we are all human, these efforts are met with a variety or reactions. Still—you can't argue with the facts.

"When iron ore requirements fell in 1954 and all Great Lakes shipping experienced a temporary dislocation, as highly valued management people, you were kept on the job long into the fall. Some of you were transferred to other ships to prolong your employment. Now, in 1956, your ships are tied up by a national steel strike, but again your employment continues.

"To make sure you are placed properly for the most harmonious operation, your departmental people keep in mind the location of your homes, personality traits, and personal desires when assigning you to a ship. Shore representatives regularly visit your ships to listen to your comments and counsel you on your problems.

"Among other privileges are trips for your wives and youngsters during the season, Vacation Pay, the Licensed Officers' Sick Leave Benefit Plan, a separate insurance plan and an excellent non-contributory pension supplemented by a contributory plan over and above the regular plan and many others of almost equal importance.

"Engineering officers on twenty-eight fleet ships now have private rooms and baths. Seventeen of these ships have been re-arranged through our quarters improvement program since 1950 at a cost of better than one million dollars. After-end quarters on five more ships will be re-arranged in this manner during the next two years. Target date for completion of the program is 1966.

"The present licensed officer salary plan, announced in 1954, permits the officers who perform outstandingly to receive more than the old method of payment would allow. When the plan went into effect, every officer received at least the same per diem rate that he received at the end of the 1954 season. Several officers received merit increases. No officer's rate was cut, but every officer's earning potential was increased. This plan was designed to recognize superior performance. As we forecasted in 1954, the program has increased our payroll costs throughout 1955 and the first half of 1956. The largest percentage of our officers are being paid at rates that are the same as, or higher than, Lake Carriers' rates; and nearly half are paid rates higher than Lake Carriers'.

"With the arrival of M.E.B.A, I think you can expect considerable fanfare and fuss over relatively minor issues. Certainly you can look forward to glowing promises that will probably never materialize.

"I am certain, however, that in a matter which concerns you as closely as this one might, you will think it carefully through and avoid sacrificing a future in management as the general course of our business progresses.

"I think we all know that advancement, security and recognition are not quantities we can demand, but qualities of life that each of us must earn for himself. That's the nature of the world we live in—and may it ever be so."[29]

The Steelworkers' union ended its nationwide strike on August 2. After a month of idleness, all the steel companies were eager to get their mills running as quickly as possible. Seeing this, MEBA officials took a gamble. They called a strike for the next day against the Pittsburgh fleet if it refused to recognize the union as the engineers' bargaining agent. They were not sure how their new members would react or whether other unions would support the walkout. Predictably, Potts refused to recognize the union. Its leaders went ahead with the strike and were greatly relieved when engineers aboard Pittsburgh ships in Duluth walked off their jobs at 5 P.M. on August 3. Organizers immediately sent telegrams to union officials in other ports. In Milwaukee, Conneaut, and Lorain, engineers on Tin Stackers tied up in those ports went ashore and set up picket lines. Within a day all of the fleet's fifty-eight ships were idled while vessels from other fleets were rushing to meet the renewed demand for ore.[30]

Potts countered the strikers' demands by proposing a return to work and an election to be conducted by an impartial third party. The men would be offered a choice of voting for no union, for MEBA, or for the Great Lakes Brotherhood of Engineers, which was a largely social organization suspected by many union men of being a company creation. The engineers rejected the proposal. The mates, who received a similar offer, responded by walking off the ships on August 7 and joining the pickets.[31]

Unlike the strikes called during the fleet's first fifty years, this one got strong support from other unions. Steelworkers Local 5000 refused to let its members sail Pittsburgh ships. Members of the Brotherhood of Railway Clerks and the Brotherhood of Railway Trainmen, who worked on the Duluth, Missabe and Iron Range Railway, refused to cross picket lines set up at the ore docks in Duluth and Two Harbors. Even if the Pittsburgh fleet had been able to hire enough strikebreakers to sail its ships, it could not have gotten anyone to load them.

After two weeks the Pittsburgh Steamship Division began to weaken in its resolve. Potts sent a letter to the strikers offering them a 12 percent pay raise and changes in shipboard conditions "aimed at improving our human relations and the way we work together." The engineers and mates rejected the offer. Finally, on August 24, Potts agreed to an election. It was not a voluntary capitulation. U.S. Steel was impatient to resume operations and put considerable pressure on Potts to resolve the dispute.

In the election, engineers and assistant engineers could vote for MEBA or no union. The mates and stewards could vote as separate units for the Masters, Mates and Pilots Union or no union. At the fleet's insistence, shipmasters were excluded from the vote because they were considered management employees. The unions agreed to the stipulation because most masters opposed unions, so including them in the vote could reduce the chance of union success. In an election held August 28, MEBA won the right to represent the engineers, and the Masters, Mates and Pilots Union won the right to represent the mates. Only the stewards rejected union representation.[32]

Shortly before noon on September 6 the *Arthur M. Anderson* backed away from the cluster of twenty-four Pittsburgh ships that had been tied up at the Duluth, Missabe and Iron Range Railway ore dock in Duluth since July 1. It departed light for Two Harbors, the first Pittsburgh ship to get under way since the strike ended. That evening, steaming out of Duluth's harbor at fifteen-minute intervals, were the loaded *Thomas W. Lamont, Henry H. Rogers, William E. Corey, Thomas F. Cole, Eugene J. Buffington, George W. Perkins, John W. Gates, Horace Johnson, B. F. Affleck, William P. Palmer,* and *Richard V. Lindabury*.[33]

Contract negotiations between the unions and the Pittsburgh Steamship Division began in early September. The first days of talks were futile. Having resisted unions for fifty years, men from the fleet office had no experience in labor negotiations and were ineffective against the experienced union bargainers. After several days U.S. Steel sent its own negotiators to Cleveland to speak on behalf of the fleet. Within two days the Pittsburgh fleet signed three-year agreements with each union. The new contracts included pay increases as well as the addition of paid holidays and vacations, a seniority schedule, and a 10 percent season bonus. The biggest benefit, however, was the addition of time-and-a-half pay for any mate or assistant engineer working more than eight hours in a day or forty hours in a week. Overtime, combined with the 5.5 percent pay raises, enabled the men to increase their income about 40 percent if they worked fifty-six hours a week.[34]

In its newspaper, MEBA trumpeted its triumph over the mightiest union foe on the Great Lakes: "Licensed Engineers of the Pittsburgh fleet won, under their contract, a high degree of consideration and recognition as Licensed Ships' Officers. The kind of recognition to which they were rightfully entitled by the standing of their profession." Speaking less formally, several mates told a labor researcher that they risked joining the union and striking because of grievances that had gone unrecognized and unresolved over the years. They spoke of winning job security and being assigned to ships according to seniority. But after much talk one man concluded: "I don't think those were the most important things we got. What we won on the lakes is freedom of speech. Now, if we don't like the way the master or the company treats us, we can say so."[35]

9

A New Leader, A New Look

Following the troubled 1956 navigation season, Donald Potts was transferred to the staff of U.S. Steel's executive vice president of engineering and raw materials, where the company said he was to work on "special assignments." He retired in 1958, ending a career that began in 1913 when he joined the fleet as a clerk in the traffic department. On March 1, 1957, Charles R. Khoury became president of the Pittsburgh Steamship Division. The soft-spoken, red-haired Khoury was a real sailor. He was born in New Haven, Connecticut, and attended private schools in Boston. As a teenager he went to sea on a three-masted bark and served as seaman, deck officer, and master before coming ashore in 1940 to manage the Terminal Shipping Company in Baltimore. A few months later he entered the Navy and remained on active duty through World War II and the Korean War before finally retiring in 1953. With a lifetime of nautical experience, Khoury seemed the perfect choice to step in as acting vice president of operations for U.S. Steel's Isthmian Steamship Company. The "acting" portion of his title was dropped in 1955—the same year he was promoted to rear admiral in the Naval Reserve. When U.S. Steel sold the Isthmian fleet in 1956, Khoury worked briefly as assistant to the vice president of raw materials, but soon was appointed head of Pittsburgh Steamship Division.

In taking over his new job, Khoury freely acknowledged he was not an expert on Great Lakes shipping. His job was to prepare the fleet for the future and to rebuild the relationship with employees that had been shattered by the previous season's strike. "We have a team made up of every man in the Division, a team that has done and will continue to do some remarkable things," he said in an article printed in the company's magazine. "If I could meet each one of them face to face, I'd say that I'm very glad to be here at Pittsburgh Steamship. I don't mean to drum at this word team, but when you sit in the top chair of a company,

The *Arthur M. Anderson* gets an early start to the 1956 season, steaming into Duluth on April 15. Wes Harkins photo.

you can see the whole picture. You can see why each man is needed to stand his watch or check his gauges or write his report. I am proud to be a member of that team, of your team, and I would like to tell everyone personally that I want all of us to grow and prosper together."[1]

Khoury quickly led his management team into action. By the end of 1957 he was proposing to build twelve new ships at a cost of more than $100 million. Preliminary design plans called for ships that would be 730 feet in overall length and 75 feet wide. Carrying 25,000 tons each trip at a speed of seventeen miles per hour, the new class would be bigger than anything else on the lakes. Khoury hoped to begin building the ships by 1964. When they were finished, he could retire twenty-nine older ships without diminishing the fleet's tonnage capacity. More significantly, he would be able to carry the same amount of cargo while employing seventeen fewer crews of increasingly expensive unionized labor.[2]

U.S. Steel's board of directors rejected Khoury's building plan. Questions still remained about U.S. Steel's future on the Mesabi Range. In addition, the federal government in 1959 began reviewing plans to build a new lock at the Soo capable of handling ships up to 1,000 feet long and 100 feet wide. Building a dozen ships that were 730 feet long did not seem like a good fit under the circumstances.

The 1950s, which had begun on such a promising note, ended in tumult. The threat of another Steelworkers strike at the start of the 1959 navigation season prompted the Pittsburgh Steamship Division to fit out all its fifty-three ships rather than just the forty-two most efficient ones. The fleet had to meet U.S. Steel's demanding tonnage and stockpile requirements by the strike deadline of June 30. This was accomplished despite the increased costs of operating older vessels and a series of vexing delays at loading and unloading facilities. The steel strike began on June 30, but the main walkout did not occur until July 14. The bitter dispute idled the steel industry until early November, when President Eisenhower invoked the Taft-Hartley Act to force the strikers back to work. By the middle of the month the Pittsburgh fleet was back in operation, but now ships encountered severe cold, heavy ice in Duluth's harbor, and frozen ore in railcars and docks. Khoury kept fifty ships sailing until mid-December. In a radical departure from its usual procedures, the fleet carried 1.3 million tons of ore in December. When *Eugene W. Pargny* sailed with the final cargo on December 20, it marked the latest date that iron ore had ever been loaded aboard a Great Lakes freighter.

As the 1960s began, the Pittsburgh Steamship Division and other Great Lakes fleets were adjusting to the recent opening of the St.

Lawrence Seaway, a two-year slump in iron ore tonnage, rising labor costs, and a substantial growth in iron ore imports. These pressures set the tone for the Pittsburgh fleet's annual meeting on March 22 and 23 at the Hotel Carter in Cleveland. Khoury told the assembled shipmasters and chief engineers that he expected stiff competition from ore imported through the seaway. He then announced that the fleet would carry up to 1 million tons of ore in 1961 from the Quebec-Cartier Mining Company in Shelter Bay on the St. Lawrence River. Another 2.5 million tons of cargo could be expected in 1962. This would be the first significant new port of call for the Pittsburgh fleet since its founding. Khoury also revealed that U.S. Steel was slowly moving ahead with plans to build a taconite plant in Minnesota that could be operational by 1965.

Then, standing at the front of the room before the Pittsburgh fleet's house flag, Khoury sternly chastised the men for the fleet's poor performance the previous year. Accidents, time lost, and costs were all up. Ships were even producing more smoke than in the past, raising the ire of people in cities around the lakes. He called for the men to demonstrate leadership in training and inspiring their men to cut costs, reduce the number of accidents, and decrease the amount of time wasted.

Robert H. Lucas, Khoury's assistant, stepped in to continue the tirade. In the previous season Pittsburgh ships had lost 11,000 hours of productive time—10,000 of those hours represented underloading, delays en route, excessive time spent in port, and delays due to accidents. The fleet overspent its budget by $1.1 million. Only four unlicensed men successfully tested for licenses needed to move up in the ranks. "I am aware of the fact that there are some in this audience who performed at standard, had no damage, trained and inspired their crewmen and had a lost-time, accident-free season," Lucas said. "To you gentlemen, I can only say by way of solace, we know who you are, and we won't soon forget you. To the rest of you I say, get off the dime: Help us, help yourselves."[3]

Then Khoury stepped in to drop his bombshell. "Lake Superior iron ore is in a life-and-death struggle to retain the ever-shrinking markets in which it can successfully compete," he warned. "To my mind, the only way to meet this challenge during 1960 is to establish the most practical yardsticks of optimum or maximum performance for our facilities and peel off every possible element of waste and excessive costs in our operations." He then announced that the fleet would institute performance standards for each ship. This "standard cost sys-

The *Alva C. Dinkey* steams up the St. Marys River at Mission Point on June 16, 1957. Although the weather that day was perfect, clouds were gathering in the Pittsburgh fleet office as management grappled with rising labor costs, growing competition from overseas ore, and renewed calls for bigger, more efficient vessels. Wes Harkins photo.

tem" set guidelines for ships to meet in numerous categories of costs. Meeting the cost standards, Khoury said, was every master's objective for the season. He made it clear this was essential to the future of the fleet as well as to each man's job.

The new standards were based on extensive studies setting the average costs, times, and speeds of each ship in the fleet. They covered such items as the cost of lubricants needed in the engine room, fuel consumed on each trip, food costs per meal, average sailing speeds, and the amount of time needed to sail from port to port, from harbor entries to the docks, and to load and unload various types of cargoes. Each master was given a set of standard costs for his ship. He would be required to track the costs of his vessel, compare them to the standards, and then investigate and solve any problems that caused his ship to exceed the standards. "Train your subordinates in cost control just as you would train them in navigation and maintenance," Lucas said. "The time has come to do away with the concept of ships' officers as lord mayors dispensing favors and decrees from on high and conceive them to be much more what they are: as presidents of small companies who must know every job to be done, taking a personal and an economic interest in every phase of their 'company's' operation."[4]

Captain Thomas F. Harbottle, the fleet's marine superintendent, drove home the point that from now on masters and chiefs would be judged as managers as well as mariners. "The program is new to Pittsburgh Steamship but it is not new to the Corporation. You may pass it off as just another one of management's gadgets for cutting costs. Gentlemen, I can assure you the program is not just a passing fancy. It is a proven method of evaluating management's effectiveness, and it will be used in the division to evaluate your contribution and effectiveness to the division's overall objectives." He added ominously, "I can assure you that in addition to myself, there will be others looking over your shoulder and following your course very closely."[5]

In addition to renewing the fleet's attention to controlling costs, the meeting also signaled a return to the days of closely monitoring safety. The fleet's safety record had slipped to dead last among U.S. Steel's eighteen divisions. Over the winter it also had spent about $1 million making repairs that could have been avoided if ships had been handled properly. Again, it was Lucas who delivered the pointed message. "Gentlemen, our safety record in the Pittsburgh Steamship Division is lousy! We've been low man on the safety totem pole in the United States Steel Corporation so long that I'm beginning to believe that you like it there or else you would have done something about it before now." He urged them to hold training sessions on the proper and safe way to perform shipboard tasks, to supervise men more closely as they worked and while they were off duty, and to firmly control drinking by crew members.[6]

The new requirements laid out at the 1960 annual meeting reinforced the unique nature of a modern shipmaster's job. According to tradition and law, he was the ultimate authority aboard ship. Any error or accident was his responsibility, even if he was sound asleep when it occurred. On the other hand, shipmasters increasingly were required to be, as Robert Lucas put it, presidents of small companies. They essentially were plant managers who had to control production costs and maintain safety standards. Unlike their shorebound brethren, however, shipmasters had to guide their "plants" through the Soo Locks and make sure they did not run aground in fog.

Joining this elite brotherhood required men to dedicate their lives to learning the mariner's trade. It was a long process. A man began as a deckhand, then earned promotions by learning new skills. He became a deck watchman, then a wheelsman. To earn a third mate's license or "ticket" he had to take an arduous Coast Guard test. From there came more tests to earn his second mate's license, then his first mate's license, then his master's license. Throughout this process, he also had to meet the standards of his employer. In the case of the Pittsburgh fleet, those standards were high and the tests to meet them were frequent.

Many masters in the Pittsburgh fleet began their careers as teenagers. Captain Bill Wilson, who sailed for the fleet for forty-four years, followed a path that was fairly typical. As a sixteen-year-old in tiny Highbridge, Wisconsin, in 1945 he had few prospects for a job after high school. His father, however, knew a man who was a second mate for the Pittsburgh fleet. At his father's request, the mate put in a good word on Wilson's behalf. Not long after that Wilson received a telegram telling him to report to the *Enders M. Voorhees* when it docked at Two Harbors. His father drove him there, arriving at the ore dock gate about 3 A.M. The teenager climbed the boarding ladder to the *Voorhees*'s deck and within a few minutes was at work handling lines as the ship was winched along the dock. The ship soon departed. Because Wilson had not yet put in his eight hours, he was put to work painting after the deckhands hosed down the deck.

A couple years later, after earning a promotion to wheelsman, Wilson began learning about steamboating under the tutelage of his mates and the *Voorhees*'s gruff old master. The master was quick to hand down the first lesson. "When I'd see a boat coming in the river, I'd get nervous," Wilson recalled. "He saw me looking out the window at one [ship] once and he said 'Don't you worry about that boat. You just steer that course I gave you. If we're too close to that boat it's my fault. You'll never be a wheelsman if you're looking out a window.'"

His first assignment as third mate came in 1950 aboard the *John W. Gates*. Wilson often stood his four-hour pilothouse watch, then went to work on deck learning skills such as how to splice a cable or how to cut glass for a porthole. After eight hours on deck, he returned to the pilothouse for his next watch. He also continued to receive lessons from the *Voorhees*'s master. The man had a sharp tongue, Wilson recalled, but was a highly skilled steamboatman. Without radar, the master could run the St. Marys River in the fog, proceeding slowly, watching until he saw the navigation buoys, then making his turns by degrees. One time it took thirty hours to make what was normally a four-hour trip, but the *Voorhees* had kept moving while other ships had anchored. One season the *Voorhees* made forty-two trips into Two Harbors without ever using the tug *Edna G.*, which was stationed there to help vessels maneuver in the small harbor. "That guy could do it," Wilson said. "You learned by watching that."

Along with learning shiphandling skills, Wilson also picked up less tangible but equally valuable talents. He learned the secrets of interpreting a barometer. By watching its movement during foul weather he could tell whether he was on the front side of a storm or behind it. He learned to decipher the mysteries of optical illusions. Sometimes ships or landmasses seemed to grow and stretch until they appeared unnaturally tall. When that happened on the eastern horizon it meant weather was coming from that direction—always a bad sign. Wilson also learned to use the weather as his ally. Wind and atmospheric pressure from passing weather systems could temporarily raise the water level at the Soo. A clever shipmaster could load a few hundred extra tons of cargo and then time his arrival at the Soo to coincide with the high water, enabling his vessel to safely pass through the locks with its additional burden.

Another lesson to be learned involved shipboard etiquette. In the 1940s men had to wear sport coats to eat in the officers' dining room. The rule was only enforced, however, when the master was present. On hot days, as soon as he left, the other men quickly peeled off their coats and slung them over the backs of their chairs. By the 1950s officers had uniforms and hats to wear on watch. The officers' dining room of that era boasted china, tablecloths, and linen napkins.

The most important protocol for any man to learn was how to behave in the presence of the shipmaster. "When I first started sailing, you stayed away from the master," Wilson said. "You went through your mates. You didn't talk to the master unless he talked to you." The starboard side of the deck was the side the master used when walking fore and aft. All the other men used the port side unless weather made it too risky. This unofficial but firm rule came about partly because sailors did not want to walk up the starboard ladder to the pilothouse. That route forced them to walk across the roof of the master's cabin, risking his ire if their footsteps awoke him. While most shipmasters were genial men, some were tough old birds. If a master took a dislike to someone, there was little the victim could do. "You just took it. The captain was the king," Wilson said. "If you got fired by a captain, you were done. You had to go to a different fleet."

Each year the masters conducted an appraisal of their mates for the fleet's personnel office. Men were evaluated in areas such as ability to learn and to study, appearance, cheerfulness, courtesy, judgment, initiative, loyalty to the fleet, and truthfulness. Earning the final promotion from first mate to master was the hardest step to accomplish. For many years a mate seeking his own command had to pass muster by a group of about ten Pittsburgh shipmasters known as "the Apostles." Would-be captains needed to sail with three of the Apostles and earn their approval before being deemed eligible to command a Tin Stacker.

After his promotion to master, Wilson enjoyed a fair degree of autonomy from the fleet office. That freedom was tempered by the pressures inherent in a job bearing so much responsibility. He had to keep a close eye on costs, time, and mileage. His ship's performance was compared to that of others in its class. Each master received detailed monthly reports showing how his vessel was doing in meeting its average costs. The obsession with costs, however, could turn into a disincentive for the men, Wilson said. If a master consistently performed better than his ship's average, sometimes the standard was raised. Eventually the average might become, for instance, a perfect docking, which was unrealistic to expect every time.

Sometimes the pressures of the job came from the shipmaster himself. Wilson said he never was pressured by the fleet office to sail if it might endanger his ship, but he personally felt a need to keep his vessel moving even in marginal weather. "The first boat I had, the marine superintendent said to me, 'I'll tell you something: The hardest thing for you to do will be going to anchor.' He was right. I couldn't just sit at anchor and not be concerned. I figured I was getting paid to move ore. So I was always finding a way to better myself so I could get in there."[7]

For the Pittsburgh fleet the 1962 navigation season brought with it the end of one era and the beginning of two new ones. The previous year the steamers *Isaac L. Ellwood* and *John W. Gates* had been scrapped. This season the *William Edenborn* and *James J. Hill* were sold. The four ships were the fleet's last vessels dating back to the formation of the Pittsburgh Steamship Company in 1901.

The first of the new eras began when excavation commenced for a new Poe Lock at the Soo. Dimensions of the lock had been increased to 1,200 feet long and 110 feet wide in recognition that 900- to 1,000-foot ships eventually could be built to increase competitiveness and cut transportation costs for the lake fleets.

The second new era began August 14, 1962, when *Arthur M. Anderson,* under command of Captain G. J. Olsen, steamed out of Conneaut. Its sailing orders were, as always, short and straightforward: "When unloaded, proceed to Port Cartier, load ore for Gary." Instead of turning northwest as Tin Stackers had done for decades, the *Anderson* turned northeast to a course heading of forty-nine degrees. In the pilothouse, wheelsman Burt Johnson feigned difficulty with the ship's wheel while Captain Olsen jokingly said, "By golly, she doesn't want to head this way." The *Anderson* proceeded through the Welland Canal and out the St. Lawrence River, the first Pittsburgh ship bound for the ore dock at Port Cartier, Quebec. It arrived there on August 17 to load processed ore for the Gary mill.[8]

Although it began a year later than Khoury had predicted, the Pittsburgh Steamship Division became the first U.S. fleet to carry ore from the new Labrador mines back into the Great Lakes. The company modified thirteen ships to operate in saltwater. Fresh water for the engine was carried in a ballast tank. Extra tanks for drinking and wash water were installed for crew use. At the same time, the fleet assigned several first mates to ride Canadian ships on Lake Ontario, the Welland Canal, and the St. Lawrence River until they qualified to test for their Coast Guard licenses covering that region. After that, the mates became "pilots" for their own fleet, boarding Pittsburgh ships bound for Port Cartier much as the regular but more expensive St. Lawrence Seaway pilots would do for other fleets. Under their supervision, more Pittsburgh masters and mates qualified to navigate those waters and to test for their licenses. For about the next decade, the most efficient Pittsburgh ships—*Anderson, Callaway, Clarke, Fraser, Fairless, Ferbert, Voorhees, Olds, Avery, Lamont, Thomas, Pargny,* and *Williams*—would be calling for ore in Port Cartier.[9]

Pressure continued to mount on the iron ore industry during these years. The Duluth, Missabe and Iron Range Railway decided in 1963 to shut down its Two Harbors ore docks and move all its ore through Duluth. The Two Harbors docks had shipped more than 21 million tons of ore as recently as 1953, but with natural ore tonnage erratic and taconite production several years away, the railroad and the Steel Corporation decided to save money by consolidating dock operations. It was the first year since 1885 that no ore moved through Two Harbors.[10]

More major changes were in store for the fleet. On the first day of January 1964, Edwin H. Gott, executive vice president of U.S. Steel Corporation, announced a reorganization plan aimed at streamlining the company. As part of the plan, the Pittsburgh fleet's division status was dropped and it was completely absorbed into the U.S. Steel Corporation. It was renamed simply the Pittsburgh Fleet. Charles Khoury was promoted to vice president-lake shipping with his office at U.S. Steel headquarters in Pittsburgh. With his new title came a corresponding increase in responsibilities. He now was in charge of all movement of materials on the Great Lakes as well as intercoastal shipment of semi-finished steel products. His duties included overseeing the Bradley Transportation Line, which was renamed the Bradley Fleet. The Pittsburgh Fleet's headquarters remained in Cleveland and Bradley's office

The *Benjamin F. Fairless* approaches a saltwater ship in a narrow stretch of the St. Lawrence River. The *Fairless* was among thirteen Tin Stackers that journeyed to Port Cartier, Quebec, to load ore for U.S. Steel mills. The Howard Weis Collection, Copyright Doris Sampson.

stayed in Rogers City, but there was a clear implication that this was a temporary arrangement.[11]

As part of the streamlining, the Pittsburgh Fleet also acquired the much-traveled *Clifford F. Hood*. This 434-foot steamer was a one-ship fleet for the American Steel and Wire Company, another U.S. Steel subsidiary. It was launched in 1902 as the *Bransford*. Pittsburgh Steamship Company acquired it in 1915 and renamed it *John H. McLean*. In 1943 the fleet transferred the vessel to American Steel and Wire, which renamed it *Clifford F. Hood*. The following year, shipyard workers added two deck cranes with seventy-five-foot booms and cut away enough of the deck to provide three large hatches. The ship went to work hauling semifinished steel products, ingot molds, scrap steel, and processed manganese ore to various U.S. Steel docks around the lakes. From time to time it also hauled iron ore, coal, and limestone. Now back on the Pittsburgh roster, it occupied an odd spot as the fleet's only crane ship.[12]

Throughout the 1960s, the Pittsburgh Fleet devoted a considerable amount of time to exploring more ways to save money. One means was to renew its experiments with diesel engines. Its first test had been made in 1951 when the *Eugene W. Pargny* was repowered with a diesel engine while its sister ship *Homer D. Williams* was given a new steam turbine. The *Pargny* experienced a lot of trouble at first as the engine's manufacturer and the fleet struggled to find the right match between engine and ship. They also had to learn how to deal with the heavy residual oil the engine used as fuel. Once the engineers worked out these problems, however, they realized that the *Pargny* consumed half the fuel needed to power the *Williams*. The experiment clearly showed that diesels engines were cheaper to buy, install, and fuel than triple expansion steam engines or steam turbines. In the Pittsburgh fleet, the days of the steam engine were numbered.[13]

The fleet took its next step toward diesel propulsion during the winter of 1962-63 when it sent the steamer *Eugene P. Thomas* to the American Ship Building Company's yard in Lorain to be repowered with a 2,250-horsepower Nordberg diesel engine. Nordberg, a Milwaukee company, had provided engines for several Great Lakes ships, including the car ferry *Arthur K. Atkinson*, the self-unloader *J. R. Sensibar*, and the Bradley fleet's *Calcite II*. The *Thomas* underwent sea trials April 26 and was ready for duty at the beginning of the navigation season.

More conversion work was slated for the end of the 1964 season, when the thirty-five-year-old *Thomas W. Lamont* was sent into Fraser Shipyards in Superior. Workers there removed the ship's boilers and its

2,200-horsepower triple expansion steam engine and replaced them with a V-12 3,200-horsepower Nordberg diesel engine. Diesel generators were added to supply electricity, and two automatic oil-fired boilers were installed to provide steam for cabin heat, deck winches, the steering engine, anchor windlass, and the whistle. The only outward change in appearance was the ship's smokestack. The new stack, essentially just a large exhaust pipe and muffler, was shorter and somewhat sleeker than the old one.

While the *Lamont* was in the shipyard it also became the first Pittsburgh ship to receive a bow thruster. The bow thruster was a propeller mounted in a tunnel running the width of the ship which could push the bow from one side or the other. The device gave the *Lamont* enough additional maneuverability for it to rarely need tugboats, enabling it to save time in port and to avoid tug charges.

The 604-foot *Lamont* underwent sea trials on April 30, 1965. Captain Herold Wissbeck took his newly revitalized command through the Duluth ship canal and out into Lake Superior under the watchful eyes of Khoury, Lucas, and Captain Harbottle. Joining them were dozens of shipping representatives, shipyard workers, and manufacturer's technicians. The vessel motored along the Minnesota shore at fourteen miles an hour—the fastest it had ever sailed. Although the main engine was operating, the engine room telegraph stood in the "stop" position. New engine controls allowed the master or mate on watch to operate the engine from the pilothouse. Instead of signaling for engine speeds of "slow," "half," or "full," the master could determine the exact number of propeller revolutions he wanted for any situation and monitor them with a tachometer. As the ship sailed along, engineers from Fraser Shipyards, Nordberg Engine Company, and the Pittsburgh Fleet office carefully brought the new engine to full power. When a malfunction occurred because of improper adjustment, a siren wailed and a revolving red light flashed a warning in the noisy engine room. After the ship cruised back into Duluth, Captain Wissbeck skillfully docked it with a few minutes of work with the bow thruster. "It's easy when you have the equipment," he modestly told his guests.[14]

Projects like the *Lamont*'s repowering were just a small part of the engineering work that went into the fleet's operation. The Pittsburgh fleet in the 1950s and 1960s maintained an engineering office staffed by men capable of handling virtually any building, repair, or maintenance job that might arise among its ships.

The engineering office was divided into the machinery, hull, and design departments. Men in the hull and design departments were hired from shipyards and had backgrounds in naval architecture. Staffing the machinery department were seven men who came from the ranks of the fleet's engineers. For more than thirty years one of them was Bruce Liberty, who joined the Pittsburgh fleet in 1951 as an oiler aboard the *Henry H. Rogers*. He rose quickly through the ranks, earning his chief's license at age twenty-six even though he was still years away from attaining that rank in the fleet. The Pittsburgh ships then were divided into three classes according to size. Assistant engineers worked their way up through the classes. By the time a man was promoted to chief engineer he had worked in ships of all ages and sizes and had dealt with triple expansion, turbine, and diesel engines. The superintendent of machinery offered Liberty a shore job in 1956, and the young man entered a world which was very busy and much different from that of a shipboard engineer. The fleet operated fifty-nine ships built between 1898 and 1952. While the vessels were divided into classes, individual ships within the classes were all different from one another. "For me, it was a chance to get experience involving every facet of the industry," Liberty said. "We had so many ships that there was always someone doing something: drydocking, casualties or something."[15]

The men of the machinery department worked on maintaining, repairing, and improving the mechanical operation of the ships. They budgeted and planned for the big repowering projects and for the smaller jobs involving boilers, propellers, and auxiliary equipment. They conducted performance studies and examined various ways to reduce smokestack emissions. A large part of their job involved travel. The men followed ships from port to port to work on problems or to prevent them. They rode every ship to "check the plant" to see whether the equipment was in balance for engine performance, fuel economy, and smoke abatement. The trips gave them a chance to share their knowledge with the engineers and to point out problems that had arisen aboard other vessels. Sometimes they were even called upon to substitute for a chief engineer if one became ill. "If they were short a chief, they'd say, 'Bruce, grab your license and bring her down,'" Liberty recalled.

The system worked well. Other companies had fleet engineers who handled mechanical and personnel matters. The Pittsburgh fleet, however, had a fleet engineer who handled personnel matters while the mechanical department focused strictly on engineering problems. Because the men from the mechanical department were not involved in granting promotions or making ship assignments, they were well received by the engineers, who perceived that the men were there to help. In addition, the fleet was self-insured so there was no need to hide problems. "There was no third-party insurance person involved, so when I got to a ship there wasn't any reason for the crew to disguise fault."

A big part of the mechanical department's job involved assembling the fleet's "winter list." Engineers on each ship constantly maintained a list of work that needed to be done during the next winter layup. At the same time, the machinery department kept track of work that needed to be done to upgrade equipment, add new equipment, or meet new safety regulations and pollution laws. They broke these tasks into three priorities: "must jobs," "desirable," and "improvements." The first category consisted of jobs that had to be done no matter what happened. The second category included work that could be done if time and money allowed. The final category involved lower-priority projects that might improve a ship's performance.

For many years, each ship underwent inspection in September during one of its trips through the Soo. The ship stopped while fleet officials descended on it to inspect such wide-ranging items as management techniques and cleanliness. At the same time, someone from the mechanical department compiled a list of technical problems. When the inspection was finished the ship's officers presented their list of problems to the office staff. These two lists were combined, priorities assigned, and costs determined. The list moved up through the fleet office. Some jobs won approval, while others were challenged or rejected outright. After the work was reviewed, a final list of projects was made.

Ships needing hull work were scheduled for shipyards over the winter. Other work was done as vessels sat in their layup berths. Crewmen were employed during the winter to perform the easier jobs, while the more complex work was contracted out. On the triple expansion engines the engineers often made their own repairs, while a manufacturers' representative or a specialized crew generally was needed for jobs involving the more complex turbines and diesels.

Throughout the years, Liberty said, the Pittsburgh fleet stressed safety and good upkeep. The fleet's safety director looked for hazards on ships and put needed safety improvements onto the winter work list. Ships were sold or scrapped not because they were run down but because they were too small to be competitive. The ships were built to operate forty to sixty years, but many lasted much longer. Most were in better mechanical shape when scrapped than when new because of refits and updating done over the years. "I don't think there was a better maintained fleet anywhere in the world than what U.S. Steel was doing," Liberty said. "You didn't have unlimited funds, but the money was there to carry out the tasks that needed to get done." His words were echoed by Ralph Bertz, who spent forty-one years with U.S. Steel and ended his career as head of engineering for the fleet. "Anything we said had to be repaired got repaired," Bertz said. "You had to defend your decision, but if I said it had to be done, it got done."[16]

The mid-1960s brought many more changes to Great Lakes shipping. Demand rose for all four major commodities: ore, stone, coal, and grain. Escalation of the war in Vietnam reduced the availability of young men needed to sail the ships. Fleets continued to scrap obsolete vessels, while lack of government incentives or tax breaks meant no fleets could afford to build new ones. Instead, they worked to make existing vessels more efficient by adding automatic boiler controls, bow thrusters, or diesel engines. During this time the Pittsburgh Fleet began selling off its vessels that were built as part of the famous *Gary* and *Morgan* classes. The former flagship, *William E. Corey,* was sold to a Canadian firm in 1963 along with *Elbert H. Gary.* The next year the fleet sold *Henry C. Frick* and *George W. Perkins,* followed in 1965 by *George F. Baker, Thomas Lynch, Norman B. Ream,* and *J. Pierpont Morgan,* namesake of the *Morgan* class. These ships were fifty years old and, with their fifty-six and fifty-eight-foot beams and twenty-seven-foot drafts, simply not as efficient as the fleet's longer, wider, and deeper vessels.

Changes also were taking place ashore. After shipping nearly 300,000 tons of ore in 1965, the remaining ore dock in Ashland, Wisconsin, ceased operation after the last mine closed on the Gogebic Range. Late in the year, however, the new Eveleth Taconite Company went on-line, adding another company to the list of those shipping taconite pellets from Lake Superior ports. More good news followed in 1966 when the Duluth, Missabe and Iron Range Railway reactivated its Two Harbors ore docks. On May 20 the *William J. Filbert* steamed out of port with the first cargo to move through the docks in three years. Two Harbors's reopening was another sign of taconite's growing importance. Three new taconite plants were scheduled to begin production in 1967, and the Two Harbors docks would be needed to handle some of that tonnage.[17]

Changes continued during the 1967 navigation season. The U.S. Steel Corporation consolidated its Great Lakes shipping operations on July 1 by merging the Bradley Fleet into the Pittsburgh Fleet. The move resulted from an ongoing review of company operations. U.S. Steel's needs had changed along with those of the marketplace, and maintaining two fleets with separate office staffs no longer made economic sense.[18]

Under the new arrangement, Charles Khoury remained in Pittsburgh as vice president–lake shipping. Management of the Bradley vessels was transferred to the Pittsburgh Fleet's office in Cleveland. The combined operations were officially named United States Steel Corporation Great Lakes Fleet. Robert Lucas, general manager–lake shipping for the Pittsburgh Fleet, moved to Cleveland. Joseph Parilla, operations manager for Bradley, also moved to Cleveland to assume the role of operations manager. Loran Hammett, the Pittsburgh Fleet's operations manager, was named general traffic manager.[19]

The former Bradley vessels retained their distinctive gray hulls, but the large letter "L" that adorned each ship's smokestack was removed and replaced with the silver-and-black colors of the Pittsburgh fleet. The ships also gave up their house flags, which carried the same markings as their smokestacks. "All carriers under the new service will fly U.S. Steel's Great Lakes Fleet house flag, the official USS emblem superimposed on a blue field," the *Skillings Mining Review* reported.[20]

Although they both served U.S. Steel Corporation, the Pittsburgh and Bradley fleets over the years had remained largely separate from one another. The Bradley vessels had been based in Rogers City since the fleet was formed in 1912 to serve the Michigan Limestone and Chemical Company. Although U.S. Steel purchased Michigan Limestone in 1928, the Bradley self-unloaders remained in Rogers City, where Michigan Limestone operated a quarry, crushing plant, and loading docks. It was a tightly knit company, with most sailors coming from Rogers City or small towns nearby. Although Pittsburgh ships frequently loaded stone in Rogers City, few men who sailed the Tin Stackers knew any of the Bradley sailors.[21]

During the 1950s the relationship between the two fleets slowly began to change and they began working together more often. In 1956 the Pittsburgh fleet transferred the steamers *Myron C. Taylor* and *A. F. Harvey* to Bradley. Christian F. Beukema, president of Michigan Limestone, said Bradley needed the ships to meet the growing demand for limestone used in making steel, cement, and chemicals and for construction.[22] The *Taylor* was sent to the Christy Corporation's shipyard in Sturgeon Bay, Wisconsin, on June 1. Shipyard crews immediately

The *Myron C. Taylor* steams out of Duluth harbor on May 21, 1956, carrying its last cargo as a straight-decker. It was transferred to the Bradley Transportation Company and on June 1 went into the Christy Corporation shipyard in Sturgeon Bay, Wisconsin, to be converted to a self-unloader. Wes Harkins photo.

began cutting down the ship's ballast tanks and installing hoppers in the cargo hold. Four months and eleven days later the vessel emerged as a self-unloader. It was the fastest such conversion on record. The ship carried the latest technology for discharging its own cargo. Pneumatically controlled gates in the bottom of the hoppers opened to drop stone onto two forty-eight-inch-wide rubber conveyor belts. The belts carried the stone forward to a bucket elevator, which lifted the cargo up to a hopper that fed it onto the ship's 250-foot conveyor boom. As part of the project, the after deckhouse and forward cabins were lengthened to allow expansion and modernization of the crew's quarters. The *Taylor* came out of the shipyard in the fall and proceeded to Michigan Limestone's new plant at Port Dolomite, Michigan, to take on its first cargo as a self-unloader. The *Harvey* went into the yard owned by Defoe Shipbuilding Company in Bay City, Michigan, to undergo the same process. When it emerged it carried the name *Cedarville*.

The new ships were hardly incorporated into the fleet when the Bradley steamer *Carl D. Bradley* broke up and sank near Gull Island in northern Lake Michigan during a severe storm on November 10, 1958. Although built the same year as the *Cedarville*, the *Bradley* apparently was in far worse condition. Much attention was given to several personal letters written by Captain Roland Bryan, who died along with thirty-two other men in the thirty-five-member crew. In one letter he wrote, "This boat is getting pretty ripe for too much weather . . . I'll be glad when they get her fixed up." Subsequent investigation by the Coast Guard showed the *Bradley* was scheduled for extensive maintenance work on its cargo hold and that it had run aground twice during the 1958 season. Neither grounding was reported, leading to speculation that the mishaps had caused structural weaknesses in the ship that had gone unnoticed.[23]

The Pittsburgh fleet transferred *William G. Clyde,* another 600-footer built in the 1920s, to Bradley in 1960. The *Clyde* replaced the venerable *Calcite*, which was Bradley's first vessel. At 436 feet in length, the old self-unloader was no longer economically viable. The *Clyde* was converted to a self-unloader during the winter of 1961 by the Manitowoc Ship Building Company and renamed *Calcite II.* Three years later it was repowered with a 3,200-horsepower Nordberg diesel engine and given bow thrusters.

Tragedy touched the Bradley Fleet again on the morning of May 7, 1965, when *Cedarville* was rammed by the Norwegian freighter *Topdalsfjord* as the vessels attempted to pass each other in the fog-bound Straits of Mackinac. The *Topdalsfjord*'s sleek bow, reinforced

The *John G. Munson* unloads stone into a massive pile on a Duluth dock. By now the ship bears the silver stack and "USS" emblem of U.S. Steel's Great Lakes Fleet. Ken Thro photo, courtesy Lake Superior Maritime Visitors Center.

for breaking ice, plunged into the *Cedarville*'s port side at the seventh hatch at 9:45 A.M. The collision ripped a twenty-foot gash in the self-unloader's hull. Captain Martin Joppich first stopped the *Cedarville* and dropped anchor. After surveying the damage, he ordered his engine started again and the anchor raised, then turned the vessel toward shore in an attempt to beach it near Mackinaw City. As the ship plowed toward shore it slowly rolled over and sank. Ten of the *Cedarville*'s thirty-five-man crew died. A Coast Guard investigation determined the *Cedarville* was traveling at nearly full speed when the collision occurred, even though visibility was only about three hundred feet. Captain Joppich's master's license was suspended for a year. He never sailed again.

To compensate for the *Cedarville*'s loss, Bradley acquired the Pittsburgh Fleet's *George A. Sloan* in 1966 and sent it to Fraser Shipyards in Superior for conversion. Built in 1943 and 620 feet long, the *Sloan* immediately became the second-largest and second-youngest ship in the Bradley Fleet after the *John G. Munson*.[24]

With consolidation of the Pittsburgh and Bradley fleets, the former Tin Stackers were once again back in the fold. The merger brought eight self-unloaders into the Great Lakes Fleet: *W. F. White, Irvin L. Clymer, Rogers City, T. W. Robinson, John G. Munson, Myron C. Taylor,*

George A. Sloan, and *Calcite II.* They joined forty-three red-hulled vessels from the old Pittsburgh fleet. The combination of ore carriers and self-unloaders carried long-term implications for Great Lakes Fleet that would slowly emerge as lake carriers grappled with even more changes in the coming years.

10

Into the Future

Along with acquiring new ships and a new name, Great Lakes Fleet faced two other momentous changes in 1967. It began building the first of a new generation of giant ore carriers, and it started carrying a product destined to change the face of the fleet and the Great Lakes shipping industry—taconite pellets.

Late in the navigation season, Great Lakes Fleet placed an order with American Ship Building Company of Lorain to construct the motor vessel *Roger Blough*. In announcing the project, Great Lakes Fleet sent a shock wave through the shipping industry. Not only would the *Blough* be the first new American ship built on the Great Lakes since 1960, but it would be, by far, the biggest vessel ever to sail the lakes. The Lake Carriers' Association called the *Blough*'s construction contract the industry's "most important and newsworthy event" of the year.[1]

The association's enthusiasm was appropriate because the *Blough* was nothing short of revolutionary. With an overall length of 858 feet and a beam of 105 feet, it would be 128 feet longer and 30 feet wider than the biggest vessels on the lakes. These dimensions would enable it to carry 43,900 gross tons of cargo—more than twice that of the biggest existing lakers. It would be the first ship built to take advantage of the giant new Poe Lock under construction at the Soo, and the first designed to handle taconite pellets. Most significantly, the *Blough* was the first ship Great Lakes Fleet would build as a self-unloader. To casual observers the ship would appear to be a straight-decker because it lacked the typical self-unloader's deck boom that could be swung out to deposit cargo on a dock. Instead, the ship would use a newly designed "shuttle boom" that could be extended from either side of the hull at the stern. This fifty-four-foot boom could discharge taconite pellets into a dockside hopper at the rate of 10,000 tons an hour. From the hopper the pellets would fall onto a conveyor system to be whisked off to the steel mill or to storage. Hulett unloading machines would never touch the *Blough*.

The decisions to make the *Blough* a self-unloader and to use a shuttle boom were part of a complex series of calculations involving the cargo it was intended to carry—taconite pellets—and the destinations to which it would carry that cargo—U.S. Steel docks in Gary or Conneaut.

Up to this point, self-unloaders were not considered feasible in the ore trade. Natural iron ore varied widely in texture, moisture, and size. It could come as powdery "fines," as large chunks, or as any size in between. It also tended to contain a significant amount of moisture, which made it prone to clumping in the summer and freezing during fall and winter. These characteristics made natural ore a difficult cargo for self-unloaders to handle. Fines flew off the unloading belts like dust while clumps of ore or large chunks jammed the hopper gates in the cargo hold. Taconite pellets did not share these drawbacks. Pellets were small, and uniform in size and consistency. They contained little moisture, so they would flow freely regardless of the temperature. Because the *Blough* was designed to exclusively carry taconite pellets, it made sense to build the ship as a self-unloader.

The *Blough*'s size also was a factor in making it a self-unloader. To take advantage of the new Poe Lock, the vessel would be built 105 feet wide, whereas Hulett unloaders were not effective in any ship more than 75 feet wide. In addition, the aging machines were increasingly expensive to operate and maintain. Steel mills and dock companies would soon have to decide whether to replace the machines or find an alternative. By making the *Blough* a self-unloader, the fleet avoided all these problems with dockside gear while reducing the amount of time the ship would spend in port.

Because it was the first of a new generation of ships, the *Blough* had design idiosyncrasies that would not appear in later vessels. Shipmasters insisted the vessel have its pilothouse on the bow so they could properly navigate narrow, winding channels. This was done even though experience in building seagoing freighters showed that a pilothouse on the stern worked just as well and cost less to build. The *Blough* also was designed so its spar deck would be low enough to load at the older ore docks, which relied on gravity to dump ore into ships. Finally, naval architects gave the ship a shuttle boom even though that would limit its use to docks with compatible hopper systems. The *Blough* would be dedicated to serving U.S. Steel docks, so that limitation was acceptable to the fleet.[2]

The broad-beamed *Roger Blough* seems to fill the Duluth ship canal as the vessel enters port for ore. The ship's shuttle unloader can extend from either of the openings visible on the stern. Ken Thro photo, courtesy Lake Superior Maritime Visitors Center.

The interest generated by Great Lakes Fleet's announcement made it clear that more giant ships would be built after the *Blough*. "In my judgment, the placement of this vessel order signals initiation of a new cycle in lake vessel construction," said James Hirshfield, president of the Lake Carriers' Association. "It provides a graphic illustration of a long-recognized fact that Great Lakes vessel operators invariably undertake to utilize to the fullest degree the capacity of the navigation locks and improved sections of the connecting channels."[3]

Hirshfield obviously wanted to show Congress and the Army Corps of Engineers that the new Poe Lock was worth the $40 million it cost to build. In reality, Great Lakes Fleet was not quite ready to "utilize to the fullest degree" all the navigation improvements made in recent years. When the *Blough* was being planned a year earlier, the fleet's original intention was to make it 1,000 feet long. The new lock certainly could accommodate a vessel of that length. But then concerns arose about whether the navigation channels connecting the Great Lakes were wide enough to handle such a long ship. Fleet officials were especially worried that a 1,000-foot vessel would be unable to make the tight turns in the St. Marys River. After much debate, they decided to proceed cautiously and reduce the ship's length. It would prove to be a costly decision.[4]

While Great Lakes Fleet was focusing its attention on the *Blough*, U.S. Steel's sprawling Minntac plant started producing taconite pellets. The first production line was fired up during the week of October 12.

On October 27 the steamer *August Ziesing* tied up on the west side of ore dock Number 6 in Duluth. At 8 A.M. the dock chutes came down and, with a rumbling hiss, taconite pellets began sliding into the *Ziesing*'s hold. By 11:45 A.M. the steamer held 12,594 gross tons of pellets. A short time later it backed away from the dock and began its voyage to Conneaut with pellets bound for mills in U.S. Steel's Pittsburgh District.[5]

Trainloads of Minntac pellets began streaming into the Duluth dock. *A. H. Ferbert* loaded the second pellet cargo on October 28, and four days later *Leon Fraser* picked up the third. Oliver Iron Mining Company, the Duluth, Missabe and Iron Range Railway, and Great Lakes Fleet quickly took advantage of taconite's resistance to freezing. The ore docks shipped their last natural ore cargo on November 12, but they continued shipping taconite pellets until *Cason J. Callaway* took the final load on December 28. Two days later *Callaway* was the last ship of the season to pass through the Soo Locks before they were shut down for the winter. It was the latest closing ever at the Soo.[6]

The lock that was to accommodate the *Blough* and other superships was essentially completed in 1968. Work had been delayed by labor strikes and by a decision to redesign the lock after the Corps of Engineers chose to expand it to 1,200 feet long by 110 feet wide.[7] Great Lakes Fleet's flagship *Philip R. Clarke* became the first ship to transit the new lock on October 30 as part of a test to determine how well it operated. The lock remained closed the rest of the season as workers continued testing and fine-tuning the equipment. The Poe Lock officially opened June 26, 1969, when the upbound *Clarke* entered the chamber for the ceremonial first passage. In command was Captain George Ashby, who had started sailing for the Pittsburgh fleet in 1936. Joining him in the pilothouse for the celebration was ninety-five-year-old Captain Thomas Small, who had been wheelsman aboard the steamer *Colgate Hoyt* when the ore-laden whaleback carried the first cargo through the original Poe Lock on August 4, 1896.[8]

After Great Lakes Fleet announced it would build the *Blough,* only one other company unveiled plans to build its own giant freighter. During the 1968 season Bethlehem Steel said it would construct the *Stewart J. Cort*—the first ship to be 1,000 feet long. The situation started to change, however, when President Richard Nixon signed the Merchant Marine Act of 1970. The law included Great Lakes fleets in establishing statutory construction reserve funds. This financial incentive prompted several companies, including U.S. Steel, to begin thinking about building their own 1,000-foot ore carriers.[9]

Another law affecting Great Lakes shipping was the Water Quality Improvement Act of 1970. In response to the nation's growing concern with environmental pollution, Great Lakes Fleet and other shipowners now had to grapple with the problem of sewage disposal aboard their vessels. Handling shipboard sewage was a relatively new problem. For decades all Great Lakes ships dealt with sewage and wash water by simply pumping it overboard. Under the new environmental law, ship operators had two options: keep wastewater on board until it could be pumped out at a treatment plant, or adequately treat wastewater before pumping it overboard. Great Lakes Fleet chose the latter.

The fleet's three newest ships—*Arthur M. Anderson, Callaway,* and *Clarke*—were built with equipment on board to treat their own sewage. The process of testing systems that could be installed in older ships began in 1971 with *T. W. Robinson* of the old Bradley fleet. After considerable experimentation and expense, the fleet settled on a biological system that treated effluent to the same level as that discharged by many municipal sewage treatment plants.

The fleet's other major environmental problem was smoke produced by its coal-burning ships. Triple expansion steam engines used only 14 percent of the energy available in coal, and they produced a lot of smoke if firemen did not properly handle their boiler fires. The thick, black plumes that sometimes billowed from ships' smokestacks had raised public ire for decades, prompting many cities to adopt ordinances limiting the amount of smoke ships could produce while in port. New federal and state environmental laws cracked down on this kind of pollution, and the fleet tried unsuccessfully to fine-tune its ships' engines and boilers to minimize smoke. "We went through great efforts to improve efficiency in firing the boiler, but it was just impossible," said Ralph Bertz, who had come to the fleet's engineering department from U.S. Steel's Michigan Limestone operation. "So we started a program to convert from coal to oil."

The first ships to be converted to oil fuel were the vessels with turbine engines, such as *Benjamin F. Fairless* and *Irving S. Olds*. Work began in the early 1970s, and the results were positive. The new oil-burners produced far less pollution than the coal-burners had. A substantial bonus was the savings that resulted from having fewer firemen and coal passers on the crew rosters. "The conversions were economically justified," Bertz said, "but the impetus was the environmental regulations that were promulgated at that time."

The 1971 season was supposed to bring the long-anticipated debut of the *Blough*. That expectation came to a tragic end June 24 when fire broke out in the ship's engine room just ten days before the vessel was scheduled to be completed. About one hundred men were working aboard the ship just after 10 A.M. when oil leaking from a gasket in an oil line was ignited by a high-intensity lightbulb. The smoky blaze spread quickly, forcing the shipyard workers to abandon the vessel. Four men working in the ship's bottom tanks were trapped and died as the blaze raged for much of the day. By the time the flames were extinguished, the aft deckhouse was wrecked and the engine room badly damaged. Repairs eventually would cost $13 million. Bertz "inherited" the *Blough* project right after the fire. "We had to regroup," he said, recalling the devastation he found inside the ship. "We had to rebuild a complete engine room and rebuild the engines." The *Blough* had been scheduled to enter regular operation that fall, but the fire pushed back its scheduled completion to the middle of 1972.[10]

Bertz was well qualified to take on the *Blough* project. He joined U.S. Steel in 1949 and spent twelve years working on engineering projects on the Mesabi Range, including the experimental taconite plants. He then was transferred to the Michigan Limestone operation in Rogers City. It was his first contact with marine matters, and he ended up in charge of the Bradley fleet's modest shipyard when it was placed under Michigan Limestone's management. In 1971 he was promoted to head of Great Lakes Fleet's engineering office and transferred to Cleveland.

Work on the *Blough* had begun when the after section's keel was laid on September 3, 1968. A little more than a year later, shipyard crews put down the keel for the forward section. The two sections were joined in July 1970. It was the last time a Great Lakes ship would be built from the keel up in the traditional style of shipbuilding. After the engine room was rebuilt following the fire, the *Blough*'s maiden voyage was scheduled for June 14, 1972. The responsibility and honor of commanding the fleet's biggest ship was assigned to Joseph Neil Rolfson, Jr., of Bay Village, Ohio. Captain Rolfson joined the Pittsburgh fleet in 1937, rising to second mate by 1940 and to first mate in 1948. Eleven years later he climbed aboard *Eugene J. Buffington* as master. Rolfson's selection to bring out the *Blough* was appropriate, not only because of his experience but because of his heritage. His father had sailed for the fleet for forty-two years and had brought out his own new ships—the *Myron C. Taylor, William A. Irvin,* and *Leon Fraser*. A little less than a year after its fire, the *Blough* departed the Lorain shipyard bound for Two Harbors to load 41,608 gross tons of pellets. The delay had cost Great Lakes Fleet the honor of bringing out the first of the so-called Poe-class ships—vessels that could transit the Soo only through the Poe Lock. Bethlehem Steel's *Cort*—the first 1,000-footer—had sailed the previous month.[11]

Once out of the shipyard, the *Blough*'s problems continued. Early trips were marred by severe vibration that caused a relentless hammering noise and sent dishes dancing off the messroom tables. A firm in the Netherlands was hired to run sophisticated water tank tests on a model of the *Blough*. The tests showed that the vibration resulted from two types of cavitation caused by the ship's propeller. Cavitation is the rapid formation and collapse of vapor pockets in a flowing liquid in regions of extremely low pressure. This typically occurs at the tips of a ship's rapidly turning propeller. The *Blough*'s propeller blades were producing sheet cavitation as they turned and vortex cavitation as they passed close to the hull. The tank tests indicated that attaching a "tunnel" structure to the hull above the propeller would reduce the cavitation. The tunnel worked and the vibration problem was solved.

Considering the vast differences between the *Blough* and other ships, it was not surprising that a major problem arose. Engines on the fleet's older ships produced a maximum of 7,000 horsepower, while the *Blough*'s twin diesels linked to a single propeller shaft produced 15,000 horsepower. "It was new ground for us," Bertz said. In the years to come the *Blough*'s original propeller would be replaced with a more efficient "highly skewed" propeller, giving the ship the smooth ride it enjoys today. Despite the improvements, not everyone was enamored of the big ship. Captain Jessie Cooper, who served for a time as the *Blough*'s master, once called it "the hardest ship I've ever run in my life."[12]

Along with the *Blough* fire, the 1971 season brought other changes to the fleet. Among them was the decision to move the Great Lakes Fleet office from Cleveland to Duluth, where the Pittsburgh Steamship Company had first set up headquarters seventy years earlier. The fleet even moved back into the same building—the old Wolvin Building, which by now U.S. Steel had purchased and renamed the Missabe Building. U.S. Steel simply said it wanted the fleet based closer to the source of its iron ore. So starting on November 29, fleet employees began the monthlong task of moving the traffic, industrial engineering, personnel, accounting, and other administrative offices to Duluth. Overseeing the move was William Ransom, who had taken the role of general manager in April after Captain Joseph Parilla retired. He reported to Christian Beukema, who had become U.S. Steel's vice president of ore, limestone, and lake shipping on December 1, 1968, after

Charles Khoury relinquished command of the fleet. Khoury would go on to become president and chief executive officer of Great Lakes Towing Company several years after leaving the fleet.[13]

Beukema was a strong and often outspoken proponent of extending the Great Lakes navigation season beyond its traditional April-to-December limits. Using his position as head of both mining and shipping operations, he pushed hard to persuade the federal government to keep the Soo Locks open late into the winter and to provide icebreakers to keep crucial waters navigable. His goal was not to operate ships year-round, although that certainly was a tantalizing possibility. Rather, he wanted to run his ships to the point each winter at which costs caused by delays and damage made it no longer worthwhile. U.S. Steel's interest in extending the season lay in its desire to better compete against increasingly aggressive and often government-subsidized foreign competitors. With taconite providing an opportunity to ship ore in cold weather, Beukema wanted to get the maximum return from the corporation's investment in the ships of Great Lakes Fleet and the costly Minntac facility. Because the fleet was self-insured, he could operate its vessels into the winter months without risking the loss of insurance coverage or facing extraordinarily high premiums.[14]

Great Lakes Fleet's initial experiment with extending the season came in 1967 with the availability of Minntac's first taconite pellets. That year five Tin Stackers ran late into the winter carrying pellets. The downbound *Callaway* locked through the Soo on December 30, bringing to an end the locks' longest season on record. The winter navigation experiment was considered successful, hampered only by an insufficient supply of pellets.

Because of late-season demand for iron ore, the Corps of Engineers agreed to a request by the Lake Carriers' Association to extend the 1968 season by keeping the Soo open until mid-January 1969. Great Lakes Fleet operated six ships into the first week of January, learning the hard way that it needed a better way to supply fuel for coal-burners on the winter run and better heat for the crews' quarters in older vessels. The season ended when *Enders M. Voorhees* made the final Soo transit at 11:59 P.M. on January 4, 1969, with taconite pellets destined for Lorain. It was another new record for the locks' latest closing.

With experience behind them, the men working on U.S. Steel's ships, railroads, and ore docks began to really push the limits of the navigation season. Six ships stretched the 1969 season into mid-January 1970. When the *Olds* loaded taconite pellets at Two Harbors on January 9, it marked only the second time on the upper Great Lakes

The *Arthur M. Anderson* slogs along in a well-maintained "track" through the ice. The AAA-class vessels were workhorses of extended-season operations. The Howard Weis Collection, Copyright Doris Sampson.

that ore had been loaded aboard a ship after December 31. The *Olds* locked through on January 11, taking the title as latest vessel to transit the locks. Accompanied by the *Clarke,* it set course for Lorain escorted by the Coast Guard's heavy icebreaker *Mackinaw.* Meanwhile, the *Fairless* loaded at Escanaba on December 24, setting a new record there for late operations.

After three years of experimentation, the Lake Carriers' Association decided to make its big push in winter navigation during the 1970 season. It asked the Army Corps of Engineers to keep one lock open through January 1971. As it had been the previous three years, Great Lakes Fleet was the strongest supporter of an extended season.

Crewmen began reporting aboard the *Clarke* and *Olds* on March 14, 1970—just eight weeks after the vessels had laid up for the winter. "We never had such a short winter layup," Captain George Ashby, master of the *Clarke,* commented as the new season began.[15] Although the nation's economy was slowing, lake tonnage remained high throughout the season. So when winter arrived, Great Lakes Fleet was ready with the largest contingent of ships on the winter run: *Anderson, Callaway, Clarke, Voorhees, Olds, Fairless,* and *Fraser.* During the bitterly cold days of January, the seven ships made a total of thirty-eight trips carrying Minntac pellets from Two Harbors to Gary and South Chicago. *Clarke* was the last downbound ship through the Soo that season, locking through on January 29, 1971. The passage marked the 304th day of continuous operation for both the *Clarke* and the Soo. Again, it was a new operating record for the locks.[16]

Winter navigation in 1970–71 offered several surprising advantages. Ships encountered ice during only 17 percent of the 1,620-mile journey. The *Callaway* recorded two of its fastest trips of the season in January. With few other vessels operating, there seldom were delays at the docks or at the Soo. In evaluating the season, Beukema termed it a clear success: "First of all, we lifted 602,000 tons of taconite pellets from Minnesota to the lower lake ports after everyone else quit. Importantly, we handled this tonnage at less cost than if we had left it to be handled by older ships of our fleet during the normal 1971 season." He noted that his ships had sustained no appreciable damage from ice during four years of extended-season operations, while they had received extensive ice damage when resuming navigation each spring. This was a clear indication that insurance rates for ships operating late in the season should not be affected by past experiences of spring navigation. Beukema also claimed winter navigation would work even better if more ships were sailing. More traffic in the St. Marys River and Straits of Mackinac would help keep the vessel tracks open through the ice and reduce the need for Coast Guard assistance.[17]

Winter navigation did not always go as smoothly as it did during the 1970–71 season. The weather was much more severe the following winter, allowing Great Lakes Fleet's seven-ship task force to carry only 461,000 gross tons of pellets during January 1972. Toward the end of the month, the *Clarke* encountered a blizzard that left it stuck in the ice of Whitefish Bay. Captain Ashby took the setback philosophically. "I found out that the *Clarke* is unable to push all the ice in Whitefish Bay to make headway," he told reporters. "We were just plain stuck." Riding along on that trip was Beukema, who got a good look at the rigors of extended-season navigation. "We learned a lot of lessons about ice operations and saw first-hand what a bind a ship can get into when it's operating under conditions which dump a lot of snow on the ice. This mixture produces a nearly impenetrable mass and is something we'll have to learn to contend with in the future," he said.[18]

Beukema continued his push for efficiency, and the extended season played a big role in the effort. During the 1972 season, Fraser Shipyards in Superior installed a bubbler system on the *Fraser*. The system used compressed air to send a stream of bubbles through openings

Stuck fast in ice, the *Leon Fraser* gets up a head of steam while the Coast Guard icebreaker *Mackinaw* cuts a path around the vessel during extended-season operations in February 1972. U.S. Coast Guard.

along the ship's keel. The bubbles would, in effect, lubricate the hull to help it slip through ice. Also that season, the new *Blough* entered service and immediately took its place in the winter fleet. The *Blough* was ideally suited for winter work. It was powerful, and its 105-foot beam cut a big swath through the ice for other ships to follow.

During the 1972 and 1973 seasons, Great Lakes Fleet operated ships into the first week of February. Federal funding for the experiment in winter navigation was supposed to expire in 1974, but was extended two and a half years because so many questions remained about the project's feasibility. With limited time to make their case, the Lake Carriers' Association, Coast Guard, Corps of Engineers, and Great Lakes Fleet agreed it was time to extend the season all the way to year-round operation. The Corps of Engineers agreed to keep the Poe and MacArthur locks open throughout the winter. The Coast Guard beefed up its ice-breaking capabilities by bringing the polar icebreaker *Westwind* into the lakes to aid the *Mackinaw*.

Going into the historic year-round season, the Great Lakes Fleet winter task force would look significantly different than it had the past few years. U.S. Steel approved a project to lengthen the three AAA-class ships—*Anderson, Callaway,* and *Clarke.* "They are still performing very well, and we felt that efficiency and productivity could be increased in a shorter period of time by lengthening rather than by building new vessels," a fleet spokesman said.

First into Fraser Shipyards in Superior was the *Callaway.* Shipyard workers put the vessel into dry dock on April 13, 1974, then cut through the hull just aft of amidship. The stern section was floated out of dry dock and a newly built 120-foot mid-body section was floated in to take its place. After workers painstakingly lined up the forward and mid-body sections, they floated the stern back into the dry dock and aligned it with the other sections through careful ballasting. Workers then welded the sections together. A new 16.5-ton rudder was added to handle the ship's increased size. By May 26 the newly lengthened ship was ready to emerge from dry dock.

Returning to service just a few weeks after leaving the shipyard, the 767-foot *Callaway* steamed into the Duluth, Missabe and Iron Range Railway ore dock at Duluth and loaded 26,634 gross tons of ore bound for Gary. The *Clarke* entered Fraser in October to undergo the same procedure. The *Anderson* underwent the same process in the spring of 1975. The fleet also announced that it planned to add a 128-foot mid-body to the *John G. Munson,* bringing its length to 768 feet.[19]

A newly lengthened *Cason J. Callaway* awaits the Hulett unloaders at U.S. Steel's Gary Works. *Callaway* was the first of the fleet's three AAA-class ships to be lengthened by 120 feet. U.S. Steel photo.

Led by the *Blough* and fortified by the newly lengthened *Callaway* and *Clarke,* Great Lakes Fleet plowed into the winter of 1974-75 with eight ships. November was mild, signaling a favorable start to the endeavor. Between December 16, 1974, and March 15, 1975, ships from Great Lakes Fleet and other companies carried 7 million tons of cargo. Great Lakes Fleet hauled taconite between Two Harbors and U.S. Steel's lower lake ports. The *Munson* added a new dimension to the winter run by carrying twenty coal cargoes from Conneaut to Duluth. In mid-March the Two Harbors ore dock closed briefly for maintenance as the Duluth ore dock opened for the new season. While the historic season was deemed successful, participants agreed that problems remained in shoreline erosion and ice control. Better radar also was needed to guide ships during the frequent periods of reduced visibility.

Navigation continued without pause into 1975. Tonnage lagged, however, as the aftershocks of the 1974 Arab oil embargo helped push the nation's economy into its worst recession since World War II. That fall, as the fleets prepared for another test of winter navigation, the Great Lakes shipping industry suffered one of its most shocking losses in decades.

Shortly before 3 P.M. on November 9, 1975, Columbia Steamship Company's *Edmund Fitzgerald* steamed away from the Burlington Northern ore docks in Superior with taconite pellets scheduled for delivery to Detroit's Zug Island. The day was mild and pleasant, but foul weather was predicted for later that night and the next day. It was expected to be a typical autumn storm. Captain Ernest McSorley and the twenty-eight men aboard the *Fitzgerald* had no reason to worry.

Not long after its departure, the 729-foot *Fitzgerald* fell into line behind the *Anderson*, which was just leaving Two Harbors and also downbound with taconite pellets. Captain Jesse "Bernie" Cooper, master of the *Anderson*, exchanged information with McSorley by radio. As the vessels proceeded down Lake Superior, the mild autumn day ebbed and conditions began to worsen. The weather deteriorated quickly that evening, and by 9 P.M. the National Weather Service had posted gale warnings. At 1 A.M. the *Fitzgerald* transmitted a routine weather report that it was encountering intermittent heavy rain, snow flurries, and heavy seas driven by sixty-mile-per-hour winds.[20]

Cooper and McSorley steered their vessels on a northerly route, skirting the south shore of Isle Royale to avoid the brunt of the wind. The National Weather Service's morning report showed the storm was intensifying, but not to an unusual degree. The noon report, however, was much different. Winds up to ninety-two miles per hour were predicted for the height of the storm. Using information broadcast by the Weather Service, Captain Cooper followed his standard practice of drawing his own weather map. He did not like the results. "It was a small, intense storm," he recalled years later. "On the weather map it had the appearances of a hurricane, with a very tight cloud curl and small center. It was very intense. It was very nasty if you were in the middle, but if you got 150 to 200 miles away from it, there wasn't much there."[21]

By the afternoon of November 10 both ships had reached the eastern end of Lake Superior. They remained in contact, but the faster *Fitzgerald* had gradually overtaken the *Anderson* and opened up a seventeen-mile lead on the Tin Stacker. Radio reports from the *Fitzgerald* indicated it was encountering rain and snow that sometimes reduced visibility to zero. The seas continued to grow, sending waves over the *Anderson*'s deck from both sides. Despite the rough weather, everything seemed normal.

Then, about 3:20 P.M., Captain Cooper looked at his radarscope and noticed the *Fitzgerald* was passing a treacherous area known as Six Fathom Shoal, which extends north from Caribou Island. To the master's practiced eye it appeared the *Fitzgerald* was passing much too close to the shoal, which, as its name implies, is a rocky outcropping covered by just thirty to forty feet of water. Cooper was concerned but made no effort to contact the *Fitzgerald,* since offering unsolicited advice would have violated one of the unwritten rules of courtesy among shipmasters. Yet, as he later told officials of Great Lakes Fleet, he was certain that McSorley and the *Fitzgerald* had strayed into the dangerous waters of Six Fathom Shoal. "I swear he went in there," Captain Cooper said. "In fact, we were talking about it. We were concerned that he was in too close, that he was going to hit that shoal off Caribou."[22] At 3:30 P.M. the *Fitzgerald* radioed to the *Anderson* that it had suffered minor damage to some deck gear. The *Fitzgerald* also reported that it was listing—an indication the ship was taking on more water than its powerful pumps could discharge.

The two vessels pushed on as the storm reached its furious peak. At one point McSorley radioed that *Fitzgerald* had lost its radar. He asked Cooper for assistance in tracking the *Fitzgerald*'s progress toward the shelter of Whitefish Bay. Captain Cooper readily agreed to help, and for the next few hours the *Anderson*'s pilothouse crew kept the *Fitzgerald* apprised of its position.

At 5 P.M. the *Anderson* radioed the *Fitzgerald* that it was about thirty-five miles north of Whitefish Bay. The winds by now had risen to eighty miles per hour. Tremendous sixteen- to twenty-five-foot waves were overtaking the *Anderson* from astern, sending green water pounding over its deck. In the ship's darkened pilothouse, Cooper and other officers watched the glowing radar screen that depicted the *Fitzgerald*'s progress.

By nightfall the *Fitzgerald* was only fourteen miles from Whitefish Bay. McSorley again radioed the *Anderson*. This time he asked Cooper to stay close in case the *Fitzgerald* encountered difficulty. Other than that, McSorley gave little indication that his ship was in trouble. "There was no panic or anything," Cooper said later.[23]

About 7 P.M., Captain Cooper momentarily left the *Anderson*'s pilothouse to retrieve a pipe from his cabin. While he was gone, First Mate Morgan Clark took radar bearings on an approaching ship that was leaving Whitefish Bay. At 7:10 P.M. he called the *Fitzgerald* to advise McSorley of the oncoming vessel. As they completed their brief conversation, Clark asked: "By the way, how are you making out with your problems?" McSorley answered: "We are holding our own."[24]

In the next few minutes a furious snow squall engulfed the *Anderson*. Clark watched the radarscope as the *Fitzgerald*'s image moved into "a white blob" caused by radar waves bouncing off waves and snow. Ten minutes later the snow squall passed. The radar screen cleared. To Clark's consternation, the *Fitzgerald*'s image was gone.

Captain Cooper, now back in the pilothouse, conferred with Clark about the *Fitzgerald*'s disappearance from the radarscope. They began calling the vessel on the radio and scanning the horizon for its lights or its silhouette. More radio calls followed to vessels anchored in Whitefish Bay and to the Coast Guard station at the Soo. Despite radio messages from the *Anderson*, the Coast Guard did not begin to seriously consider the *Fitzgerald*'s disappearance until about 9 P.M. Even then, no large Coast Guard ships were available anywhere near Whitefish Bay. Search aircraft could be launched from Traverse City, Michigan, but any rescue work on the water would have to be done by commercial vessels already in the area. The Coast Guard asked several ships, including the *Anderson*, to leave the shelter of Whitefish Bay to search the seething waters of Lake Superior for the *Fitzgerald* or its survivors.

In the dark, the odds were slim that anyone in the water could be found. In the violent conditions existing that night, the odds were even slimmer that anyone could survive. With considerable trepidation, the *Anderson*, the *William Clay Ford*, and other ships risked the wind and waves off Whitefish Point to search for the *Fitzgerald* or men who might have escaped as the vessel sank. About 5:30 A.M. the *Anderson* steamed through a field of floating debris—life jackets, life rafts, oil-covered ring buoys, gas canisters, oars. The wreckage was from the *Fitzgerald*. More debris turned up in the days to come, but no bodies were ever found. Out in the blackness that night, battered by the wind and huge waves, the *Fitzgerald* plunged beneath the surface so quickly that no one in the pilothouse had time to radio even a brief cry for help.

In the days and months and years to come, many theories were put forth in an attempt to explain the *Fitzgerald*'s sinking. A major dispute arose among the Coast Guard, the National Transportation Safety Board, and the Lake Carriers' Association. The two government agencies maintained that improperly secured or collapsing hatch covers allowed the *Fitzgerald*'s hold to fill with water until the vessel lost buoyancy and plunged to the bottom. The Lake Carriers' Association countered that the one-piece hatch covers used on modern ships like the *Fitzgerald* had never failed. It said the steamer damaged its hull when it passed near Six Fathom Shoal about 3:20 P.M. on November 10.

Captain Cooper agreed with the Lake Carriers' Association. The captain, of course, was no stranger to the bitter weather brewed over the Great Lakes. He was aboard the *Peter A. B. Widener* during the Armistice Day Storm of 1940, and he had endured a lot of foul weather in the thirty-five years that followed. The storm that sank the *Fitzgerald*, he said, was not as bad as the Armistice Day Storm, and it was not severe enough to sink a modern ship—unless it was damaged. He kept thinking back to the *Fitzgerald*'s first report of trouble just minutes after passing over Six Fathom Shoal. "I know one thing," Captain Cooper said. "At 3:20 in the afternoon that ship received a mortal wound. She either bottomed out or suffered a stress fracture. I think she bottomed out."[25]

In the years following the disaster, Captain Cooper never cared to speculate on the reasons for McSorley's actions or why he did not call for help after the *Fitzgerald* was damaged. "I honestly believe they knew they were in trouble, but Whitefish Bay was only fourteen miles away [in the end] and he [McSorley] thought he could make that," Cooper said.[26]

A few weeks after its harrowing experience on Lake Superior, the *Anderson* began the second winter of the year-round navigation experiment. Great Lakes Fleet had planned to run nine ships during the winter, but the nation's sluggish economy prompted the fleet to cut that number to six. This second winter was even more successful than the first. American and Canadian fleets, led by the Tin Stackers, carried 6 million tons of cargo through the Soo Locks between December 16, 1975, and March 10, 1976, when the 1976 season officially began. The locks recorded 480 vessel passages during that period.

Despite its success, the winter run was grueling for ships and men. Water vapor condensing in frigid air produced surface fog that sometimes cut visibility to zero. Snow blowing across pack ice could do the same. Temperatures on Lake Superior could plunge to thirty-five degrees below zero in the depths of winter, with wind chill readings as low as eighty below. These arctic conditions were particularly hard on the deckhands, who spent hours on deck handling lines and using hoses to clear ice from hatch covers, winches, and other gear. Sailors for Great Lakes Fleet—all volunteers during the winter run—took solace in wages that were up to 10 percent higher than their normal pay. Nonetheless, many shipmasters, officers, and unlicensed crewmen openly opposed the extended-season operations.

The worst part of the extended season generally came each year in December and January, when storms were frequent and rapidly forming ice choked loading slips and other protected waters. After mid-January, pack ice built up in Whitefish Bay, the St. Marys River, and the Straits of Mackinac, often forcing ships to travel in convoys or to seek assistance from Coast Guard icebreakers. By February much of the ice was formed and stabilized. Icebreakers could cut a track through the ice and maintain it with relative ease until the restless spring winds of March put the ice in motion once again.

Battling the ice was part of every trip during the winter. Excerpts from the log of the *Clarke* for the first weeks of 1976 describe some of the problems that the ship and Captain Joseph Murphy encountered during winter navigation on Lake Michigan and Lake Superior:

January 20:
"1130 hours . . . Beset 16 miles west of Mackinac Bridge. Awaiting cutter assistance. Very hard, heavy ice."
"1230 . . . Under way with cutter *Westwind*."

January 23:
"0435 . . . Encountered hard, heavy ice a few miles east of Whitehall [Michigan] Light. Difficult progress between Garden Island and two miles west of Lansing Shoal. Cutter *Westwind* was assisting steamers *Ferbert* and *Fraser* eastbound."

January 30:
"0440 . . . Johnson's Point [in the St. Marys River] . . . Snow . . ."
"0612 . . . Beset—Assistance required."
"1040 . . . Cutter *Woodrush* arrived."
"1225 . . . Freed—Under way."

The *Clarke*'s ice problems in January were quickly resolved with Coast Guard assistance. But on its fifty-first trip of the season, the ship encountered the worst conditions of the winter. What was normally a two-and-a-half-day trip stretched into six days as the ship battled ice jams near the Soo.

February 2:
"1016 . . . Depart Gary."

February 3:
"1034 . . . Mackinaw Bridge."
"1330 . . . Detour Light"
"1627 . . . Off Johnson's Point . . . Temperature 17 degrees . . . Snow . . ."
"1630 . . . Beset abeam Johnson's Point Light."
"1805 . . . Under way."
"1957 . . . Ordered to stop by Coast Guard. Congested Soo Harbor."

February 4:
Stopped in ice all day.

February 5:
"1149 . . . Tried to break loose."
"1244 . . . Under way—freed by cutter *Mackinaw.*"
"1345 . . . Arrive Soo."
"1500 . . . Depart Soo."
"1726 . . . Parisienne Island . . . [under way] in ice . . . temperature 8 degrees"
"1850 . . . Stuck in ice."

At this point, the *Clarke*'s chief engineer began trying to free the ship by pumping out ballast water according to a series of pre-arranged plans.

"1900 . . . Pumping out [ballast water] to Plan 3"
"1925 . . . Beset 2 miles north of Whitefish Point."
"1940 . . . Pumps secured."
"2045 . . . Called Coast Guard Control for assistance."

February 6:
"2244 . . . Under way. Maneuvering in ice."
"2345 . . . Stopped in ice—7 miles north northwest of Whitefish Point . . . temperature 5 degrees . . . snow."

February 7:
"0250 . . . Working engine to get free."
"0445 . . . Freed—Under way."
"1453 . . . Off Copper Harbor . . . Temperature 17 degrees . . . Wind west by south at 40 miles an hour . . . occasional snow."
"1502 . . . Reduced speed 10 rpms."
"1510 . . . Reduced speed 5 more rpms."

February 8:
"0049 . . . Off Grand Marais, Minn. . . . Wind north northwest at 40 miles an hour."
"0624 . . . Arrive Two Harbors."[27]

A severe winter brought the 1976–77 season to an early close when the downbound *Blough, Clarke,* and *Callaway* locked through the Soo on January 23, 1977. Year-round navigation resumed during the winter of 1977-78 when Great Lakes Fleet ran seven ships between Two Harbors and lower lake ports. It marked the third time the Soo Locks had remained open all year, but this time it was a struggle. Steel mills were desperate for ore following a five-and-a-half-month strike. Another bitter winter clogged the waters with ice. At one point, eighteen ships were stuck in ice in the St. Marys River. Food had to be airlifted to some ships, and sick crewmen were taken off by helicopter. The *Olds* was plowing through an ice track when unexpectedly heavy ice forced it to a sudden stop. Columbia Steamship Company's *Armco,* following close behind, rammed the *Olds,* causing $250,000 in damage. Ships in other fleets were damaged in other mishaps. Even the icebreaker *Westwind* had to be taken out of service for major repairs.

As the experiment in extending the season continued, it faced growing opposition from environmentalists who feared that keeping the rivers open would harm animal and fish habitats. Many people who lived in the remote areas along the St. Marys River also opposed winter navigation because it disrupted ice fishing and made travel difficult for island residents accustomed to driving their cars across solid ice during the coldest months.

By early 1979, as government funding for the extended season neared its end, Christian Beukema sometimes was portrayed as a prophet trying to lead unwilling companies, unions, insurers and government agencies into the promised land of a longer navigation season. In an article in *Industry Week,* he complained that other fleets were unwilling to join U.S. Steel in a solid commitment to making winter navigation work. Beukema offered to share with other fleets what his men had learned through experience over the past decade. He invited representatives from other fleets to ride the Tin Stackers during the winter run to see how it could be done. He found few takers. The government-funded demonstration program ended after the 1978-79 season. After that, the target date for closing the Soo Locks was set in mid- to late January each year.[28]

Looking back at ten years of effort, Beukema declared the extended-season experiment a success. Great Lakes Fleet was the shipping industry's leader throughout the program. It always ran the most ships and carried the most tonnage after the traditional end of the season. Beukema maintained that his unique position as head of both raw materials and shipping for U.S. Steel enabled him to see the overall potential cost savings offered by the longer season. His other advantage was that the cost of running the fleet was built into the cost of making the Steel Corporation's products. Winter navigation was prohibitively expensive for independent fleets that sold their services to other companies at a profit, but Great Lakes Fleet's role as a cost company enabled the Steel Corporation to wring the greatest financial advantage from a longer season. "Other companies could have done it all along. They had taconite long before we had it," Beukema said. "I was the single fellow in the total steel industry—of all the companies in the steel industry—who had the total responsibility from the mines through to the inventories at the furnaces. I had the transportation all the way through. I knew what could be done in that respect. Whereas, the [other] ore companies might be independent of the steamboat company, the steamboat company is interested in its profit and sold to a steel company at a price. I operated as an integrated cost."[29]

While it was battling the elements to extend the season, Great Lakes Fleet also was fighting a serious shortage of sailors—a problem felt throughout the shipping industry. The fleet even published a brochure in the mid-1970s advising young men about the job opportunities and working conditions aboard its forty-two vessels. The focus was on attracting men to entry-level jobs such as deckhand, wiper, and porter, which paid $850 a month and provided free meals and living quarters for much of the year. New employees also received one week of paid vacation during each of their first two years on the job.

As Great Lakes Fleet pushed for more efficiency through a longer season, it also joined other lake fleets in seeking the ultimate in cost-effectiveness—the 1,000-foot ship. Playing it safe with the length of *Blough* turned out to be a mistake. The *Cort* experienced no trouble negotiating the St. Marys River or other connecting waterways. Other fleets quickly stepped forward to order 1,000-footers, filling the building berths at the shrinking number of shipyards around the lakes. Officials at Great Lakes Fleet were considering building two or three 1,000-footers, but by the time they were ready to act the fleet faced a four-year backlog of shipyard orders.[30]

Unwilling to wait, Beukema and Great Lakes Fleet in 1975 signed a twenty-five-year charter with Litton Great Lakes Corporation for *Presque Isle,* a tug and barge integrated to function as a single 1,000-foot vessel. Litton built the $35 million *Presque Isle* as part of an unsuccessful foray into Great Lakes shipping and shipbuilding. The tug was 141 feet long and the barge 974 feet long. When the tug slid into the notch at the barge's stern, the two vessels became essentially a single ship with an overall length of 1,000 feet. Litton chose to build the tug and barge unit to take advantage of Coast Guard manning rules that allowed tugs with barges to function with a smaller crew than a conventional self-propelled ship. As built, the vessel could function with a crew of fifteen—about half that of a conventional ship.[31]

Presque Isle originally was intended for the Wilson Marine Transit Company, which Litton had purchased several years earlier. By the time the vessel was completed in 1973, however, Litton was ready to sell the Wilson fleet. The *Presque Isle* sailed as an "orphan" for two years before Great Lakes Fleet chartered it. It was a good move for both parties. Litton found a long-term home for its costly mistake, and Great Lakes Fleet got its first 1,000-footer right away at a price that allowed it to operate the ship at a profit. The vessel also was a good fit for the fleet's operations. It could carry 52,000 gross tons of ore, stone, or coal and was powerful enough to run in ice during extended-season operations. Most importantly, its deck structures were low enough to allow it to load at existing ore docks and at U.S. Steel's limestone plant in Rogers City. As part of its charter, Great Lakes Fleet chose to operate *Presque Isle* with a full crew as if it were a conventional ship.[32]

While negotiations proceeded with Litton over *Presque Isle*, Great Lakes Fleet also was proceeding with plans to build a 1,000-footer. In the spring of 1974, Ralph Bertz and William Buhrmann, the fleet's head of operations, went to the Soo and boarded one of the fleet's ships on which Christian Beukema was a guest. "We spent the rest of the trip going over plans for a 1,000-footer," Bertz recalled. "He [Beukema] was laying down certain requirements. We had orders to build the safest ship possible. The criteria laid down for building the big ones was that they would be devoted exclusively to the pellet trade and would haul their maximum capacity in pellets."[33]

Beukema's edict to build a safe ship was especially pointed. "In '58 we lost the *Bradley*," Bertz said. "After that, U.S. Steel was the first to issue loading and ballasting manuals to aid in safe loading of the ships. When I came to the fleet, the byword was 'By God, we're going to operate and build safe ships.'" As a key safety feature, the new ship would be designed differently amidships than other 1,000-footers. Other ships were being built with large interior "hopper" holds extending out to the shell plating on the hull. This enabled them to profitably carry cargoes such as coal, which requires relatively large cubic volume. The new 1,000-footer for Great Lakes Fleet would have its interior hopper nestled entirely between big side tanks. These tanks could be used for ballast water and also serve to reduce the chances that a collision or grounding could breach the hopper and allow it to flood. Using bigger side tanks reduced the cubic capacity available for cargo. However, that would not matter because the ship would carry only taconite pellets—a heavy, dense cargo that required considerably less cubic capacity than coal or stone.[34]

Decisions involving some parts of the ship's design had to be made in tandem with other divisions of U.S. Steel. Unlike the *Presque Isle* and *Blough*, the new 1,000-footer would be too big to load under the chutes at the ore docks in Two Harbors and Duluth. At this point, officials for the Duluth, Missabe and Iron Range Railway were brought into the design process. They agreed to modify the railroad's ore docks by adding shiploaders that used conveyor belts on long booms to reach the ship's hatches.[35]

Another critical decision focused on the ship's unloading system. U.S. Steel's docks in Gary and Conneaut were phasing out their Hulett unloaders and bridge cranes because of the rising cost of operating and maintaining them. Both docks had built hopper systems to receive ore from short unloading shuttles like those used on the *Blough*. Inside Great Lakes Fleet, a debate began over whether the new ship should be built with a shuttle boom like the *Blough* or with a long deck-mounted boom like those used on other self-unloaders. The key point of the argument was that a longer boom would add versatility to the ship, preventing it from being a captive of a few ports with the hopper system.

The debate was finally decided by the fleet's contract with its labor unions. When the Bradley fleet was merged into the Pittsburgh fleet, sailors in each fleet maintained allegiance to separate unions. Which union got to man a particular ship was spelled out in their contracts: the union representing former Bradley sailors manned ships with deck-mounted self-unloading booms, while the union representing the Pittsburgh sailors manned ships without such booms. This was a simple way to solve union jurisdiction in 1967 because the former Bradley vessels were the only ones with self-unloading gear. As new ships were built as self-unloaders, however, it threatened to cause labor problems. To avoid conflict, the fleet decided to build its 1,000-footer with a shuttle boom. The boom-versus-shuttle debate did not completely disappear, however. Anticipating that the ship's unloading shuttle someday might be swapped for a deck-mounted boom, Bertz made sure the vessel was designed for easy conversion from one system to the other.[36]

In April 1974, Great Lakes Fleet contracted with Bay Shipbuilding Company in Sturgeon Bay, Wisconsin, to build a 1,000-foot ship to be named *Edwin H. Gott*, in honor of the former chief executive officer of U.S. Steel Corporation. Before work began, models of the new ship were built and tested in water tanks to avoid the vibration problems that had plagued the *Blough*. The tests predicted the new ship's operating characteristics and also helped engineers determine which of five possible bow shapes would be most efficient.

Because of the backlog of orders for new ships, the *Gott*'s hull was not begun until November 9, 1977. Work progressed quickly and the hull was floated out of dry dock the following July. On October 31, 1978, Mary Gott, wife of Edwin Gott, smashed a bottle of champagne across the new ship's rounded bow as the shipyard erupted in the bellows and shrieks of ships' horns. U.S. Steel President David Roderick told assembled spectators that "this new vessel represents a new era in Great Lakes shipping for United States Steel and an affirmation of our belief in the future for our domestic steelmaking."[37]

The object of Roderick's praise represented the peak of Great Lakes shipbuilding. At 1,004 feet in overall length and 105 feet in width, the *Gott* overshadowed the *Blough* and simply dwarfed every other vessel in the fleet. The gigantic ship could carry 61,000 gross tons of taconite pellets in a single trip. With one crew it could haul as much cargo as five of the fleet's 600-footers and discharge that cargo in the time it took shoreside gear to unload one or two of the smaller ships. Below decks, twin diesel engines generating more than 19,000 horsepower enabled the ship to cruise at sixteen and one-half miles per hour. Two variable-pitch propellers allowed the *Gott*'s master to drive the ship astern without reversing the engines. The engine room was controlled from a soundproof booth dominated by a long panel of gauges and more than 150 warning lights to alert engineers to everything from "oil in bilge water" to "port engine bearing temperature high."[38]

The *Gott*'s appearance also was strikingly different from that of its fleetmates. A full, rounded bow and square stern allowed more space inside the hull for cargo and machinery. Relatively small hatches ran down the center of the deck. All crews quarters were located in a large deckhouse on the stern. Inside, unlicensed crewmen were quartered in double rooms. The large, well-equipped galley and spacious dining rooms looked like a cafeteria normally found in office buildings ashore. The pilothouse, also on the stern, spanned almost the entire width of the ship, stretching ninety-six feet from end to end. It towered a full nine stories above the lowest level of the engine room. A small elevator was even installed to give the men easier access to the ship's towering superstructure.

The *Gott* was taken to Milwaukee to complete its fitout. Its maiden voyage began on February 19, 1979—an unusually late date for a ship's first trip. While on Lake Superior bound for Two Harbors, the ship encountered heavy ice. Traveling behind the Coast Guard's *Mackinaw*, the *Gott* was crunching through windrows of ice up to fifteen feet high. Anticipating the *Gott*'s role in winter navigation, Great Lakes Fleet had ordered it built with a reinforced bow. Special high-strength steel—much of it made by U.S. Steel in Gary and South Chicago—was used in the bow's structural members and the hull plates, and impact-resistant steel was used as reinforcement of the bow and forward "shoulder" areas. Unfortunately, the ice was so heavily windrowed that it pushed above the reinforced portion of the bow and punctured the hull. The first sign of trouble came when Ralph Bertz, riding along with Christian Beukema, noticed a gauge that indicated water in the forepeak ballast tank. After alerting the master, Bertz and Beukema went below and found the forepeak flooding. Though not serious, it certainly was not what they wanted to see on a maiden voyage.

When the *Gott* arrived in Two Harbors, Beukema ordered the tug *Edna G.* to circle the ship to inspect the hull. The tug's crewmen not only reported on the damage to the bow, but also delivered an even more startling piece of news. One of the *Gott*'s two massive rudders was missing. Somewhere on Lake Superior the rudder had simply fallen off. The ship was so big and handled so well that the wheelsmen had not noticed the difference.[39]

The *Gott* did not resume its maiden voyage until April 21, when it departed Two Harbors with 59,375 gross tons of taconite pellets. Repairs had been made during the winter, but the dispute over responsibility for the missing rudder was just beginning. Great Lakes Fleet insisted that a construction defect caused it to fall off, while the shipyard maintained that the ice claimed it. Extensive tests finally revealed a defect where the rudder was welded to a steel member. The fleet and the shipyard agreed to use a different kind of steel in this link to avoid future problems. The change was made to the *Gott*'s rudders and incorporated into the second 1,000-footer Great Lakes Fleet was building. With its rudder problems out of the way, the *Gott* quickly proved to be a success. In July it set a record at Two Harbors by loading 61,333 gross tons of taconite pellets. The next month it topped that with a cargo of 61,572 gross tons.[40]

While the *Gott* was under construction in 1976, Great Lakes Fleet had ordered a second 1,000-footer to be named *Edgar B. Speer.* American Ship Building Company started construction in 1977 and was supposed to deliver it in 1979. A strike at the shipyard delayed its completion, however, and the *Speer* was not floated out of dry dock until May

Shipyard workers put the finishing touches on the *Edgar B. Speer*, the second 1,000-footer built for Great Lakes Fleet. Although nearly identical to the *Edwin H. Gott*, the *Speer* underwent too many changes to be considered a true sister ship to the *Gott*. U.S. Steel photo.

1980. Lessons learned in building the *Gott* and other big ships were incorporated into construction of the *Speer*. The new 1,000-footer was a close cousin to the *Gott*, but because of the changes the two were different enough that they could not be called sister ships. The *Speer* was christened on June 4. Captain William Simonds of Ashtabula, Ohio, was in command and Chief Engineer John Wilson of Sault Ste. Marie, Michigan, was overseeing the engine room when the ship made its maiden voyage a short time later. Its first cargo was 55,000 tons of taconite pellets loaded at Two Harbors and consigned for the Gary works.

Commanding one of the new giants was a challenge for shipmasters who had spent decades sailing 600-footers or even a few years aboard the lengthened AAA-class ships. "I'd say it takes a year to get used to the changes; so you're as good on the big boat as you were on a boat with the pilothouse forward. After you get used to it [the pilothouse aft], you like it better," said Captain Bill Wilson, who closed out his career commanding the *Speer*. Masters and mates had to get used to new points of reference and the blind spot within 600 feet of the ship's bow. They also had to master the seemingly impossible task of sliding a 105-foot-wide ship into the 110-foot-wide Poe Lock. "The ship is enormous, your pivot point for turning is different, all your marks for mak-

Numbered shuttle loaders extend across the deck of the *Edwin H. Gott* as it loads taconite pellets in Two Harbors in 1984. The shuttles were added to the Two Harbors and Duluth ore docks to serve the 105-foot-wide vessels. Each shuttle arm houses a conveyor belt that delivers a stream of pellets from the dock into the hold. Author's photo.

ing turns in the rivers had to change, the weight's enormous," Wilson said. "You've got to have more precision navigating in the rivers because you don't have room to make a mistake." Although many considered the big ships to be ungainly and ugly, those who sailed them for Great Lakes Fleet developed a genuine affection for the 1,000-footers. "It's pretty big for the harbors," Wilson said of the *Speer,* "but out on the lake, if it's handled right, it will take any storm you can dish out."[41]

As the *Blough, Gott,* and *Speer* entered service, Great Lakes Fleet sold off many of the 600-foot straight-deckers that had served so well for so many decades. In 1978 it sold *Richard V. Lindabury* to the Kinsman Transit Company and sold for scrap the *William P. Palmer, Percival Roberts Jr., William B. Schiller,* and *Richard Trimble.* The following year Kinsman bought *William A. McGonagle* while *James A. Farrell,* the former flagship, went to the boneyard. The *Speer*'s arrival in 1980 sent seven more ships to the scrapyard: *Eugene J. Buffington, D. M. Clemson, Thomas F. Cole, Alva C. Dinkey, D. G. Kerr, Governor Miller,* and *J. P. Morgan Jr.* The same year *Peter A. B. Widener* was sold for use as a storage barge.[42]

The Poe-class ships arrived on the eve of a critical period for Great Lakes shipping and for Great Lakes Fleet. Overall, the industry was undergoing a widespread and relatively rapid conversion from straight-deck vessels to self-unloaders. This trend resulted from a complicated combination of factors involving cargo, vessel size, and rising costs. Steel mills were abandoning the use of raw ore and moving toward exclusive use of taconite pellets to provide iron ore for their furnaces. This reduced the amount of tonnage for the fleets to carry. At the same time, hauling taconite allowed fleets to operate longer each season. Combined with the growing size of ships, these factors meant fleets needed fewer vessels to carry the tonnage required by the steel mills.

Fewer ships, however, presented a dilemma to the mills and dock companies. To operate at peak economic efficiency, dockside unloading gear needed a steady stream of ships to service. As fleets shrank, use of shoreside rigs dropped. Yet their costs remained constant or increased because of escalating employee wages and rising maintenance costs for the aging equipment.

Taking all the factors into account, it began to make more sense from an economic standpoint to put unloading gear on each ship and to do away with shoreside equipment. Converting ships to self-unloaders saved the steel mills money because they could get rid of their Hulett unloaders and bridge cranes. The fleets would benefit because their ships would no longer be delayed by a backlog of vessels waiting to be unloaded. This would enable the smaller fleets to operate even more efficiently.[43]

While the ships were undergoing the revolutionary change to self-unloaders, the steel industry was sliding into a trap that had grown since the end of World War II. Throughout much of the 1950s and 1960s, America's steel mills had roared along with little foreign competition. By the late 1970s, however, steel made by government-subsidized mills in other countries was pouring into the United States. American steel exports fell because aging U.S. mills could no longer compete with more modern and more efficient mills located overseas. In the early 1980s, the steel industry tumbled into a painful period of massive retrenchment. From this crucible of disaster emerged the Great Lakes Fleet that sails today.

11

Modern Times

The 1980 navigation season opened on a low note, overshadowed by the shaky state of the nation's economy. Fleets fitted out most of their ships based on cargo commitments already made or on expectations that iron ore tonnage would reach the spectacular level of 92 million gross tons hauled the previous season. Great Lakes Fleet began the season operating twenty-four of its thirty-one ships. Less than a month after the season began, however, the economy worsened. Steel mills slashed their orders for taconite pellets, and mines started cutting production. By summer 40 percent of American ships on the Great Lakes were idle. A slight surge in demand for taconite late in the season pushed overall tonnage that year to 72.9 million gross tons. Nonetheless, when *Cason J. Callaway* made the last passage through the Soo on December 31, it marked the locks' earliest closing since 1967.[1]

The nation's economic troubles continued in 1981. Carmakers stalled and the construction industry sagged. The steel industry vigorously pursued retrenchment, and company after company began closing mills and laying off employees. These troubled industries were prime customers of Great Lakes shipping, and their woes were felt by all the fleets. Prospects for the navigation season seemed to be about the same as for the previous, troubled year.

As part of its response to the growing crisis in the steel industry, U.S. Steel began closing mills and divesting itself of parts of its operations. The vertical integration that so enamored the company's founders now was seen as an archaic notion that kept the company from realizing its full profit potential. In one of those divestitures, the company announced on June 5 the single greatest change in its Great Lakes fleet since the Steel Trust was formed in 1901. The steelmaker revealed it was turning Great Lakes Fleet into a wholly owned subsidiary company. The fleet was renamed USS Great Lakes Fleet and designated a common carrier. For the first time in eighty years, it would compete with other fleets on the open market and seek customers other than United States Steel Corporation.

"The move will enable us to fully develop profitable Great Lakes shipping opportunities with outside parties as has been done so successfully in our ocean shipping business with the Navios group of companies," said Thomas Marshall, vice president of resource development for U.S. Steel. USS Great Lakes Fleet would continue carrying cargoes to U.S. Steel's mills, but it also would pursue contracts to haul cargo for other companies of all sizes. In addition, a second subsidiary, named USS Great Lakes Fleet Services, would use the fleet's management organization to seek contracts to manage the ships of other companies.

Christian Beukema, who had undergone heart surgery several years earlier, had taken early retirement in 1980, so the Steel Corporation's executives elected William Buhrmann, the fleet's general manager, to lead the new companies.[2] Like his predecessors, Buhrmann had a lengthy history with U.S. Steel. He joined the company right out of college in 1952 and assumed a low-level supervisory job in one of the company's mills. Progressing through the ranks of management over the years, Buhrmann eventually became assistant to the general manager of the famed Homestead Works. He was transferred to Duluth in 1973 to become manager of fleet operations for Great Lakes Fleet. Three years later he became general manager.[3]

For the first time ever, the management staff of Great Lakes Fleet had to draw up a marketing plan and develop a strategy for grabbing a piece of the shrinking market for lake shipping. At the time, the fleet's main cargoes were taconite pellets and stone carried for the Steel Corporation, with occasional grain cargoes carried down the St. Lawrence Seaway to keep ships busy during slow times. Now it needed to court other steelmakers as well as the myriad smaller users of lake shipping—utilities needing coal for power plants, salt companies shipping their product to commercial users and cities, and construction companies needing stone. Backed by the U.S. Steel name and the long history of the Pittsburgh fleet, the fleet's salesmen had an advantage when calling upon prospective customers. "You had a reputation for being reliable," Buhrmann said. "When you mentioned the name, no one asked you, 'What was that name again?'" The drawback was that they had to convince would-be customers that when the steel industry recovered, the fleet would stick with its new customers and not abandon them to rush back to hauling only U.S. Steel cargoes.

To Buhrmann, these were exciting times. Two key factors came into play to make USS Great Lakes Fleet successful. First was Great Lakes Fleet's long-standing culture of knowing and controlling costs. In its early years, the fleet was a cost company. As part of the production chain for U.S. Steel, its costs were factored into the cost of making steel. Its emphasis was on controlling costs rather than making a profit. Indeed, the fleet's profits and losses were a closely held secret, known only by a few managers. U.S. Steel did not want the knowledge of profits to affect decisions made to control costs. When the fleet changed from being a cost company to being a common carrier, it had to begin making a profit. This was achieved because the fleet was knowledgeable about even the smallest of its costs. All those years of Harry Coulby, A. F. Harvey, and Donald Potts harping about costs at the annual meetings began to pay their biggest dividends. Now the fleet was able to increase profitability and compete against other fleets by controlling its costs rather than by raising its rates—just as Andrew Carnegie had done in his steel mills during the previous century.

The second factor in the fleet's success was its wide array of ships. The big ships—*Edwin H. Gott, Edgar B. Speer,* and *Roger Blough*—could continue efficiently meeting U.S. Steel's taconite needs and eventually serve other steelmakers. Now, however, the 600-foot self-unloaders like *Calcite II* and *Myron C. Taylor* no longer were liabilities. They could be used to reach new customers located in the small harbors and on the narrow, winding rivers that dot the shoreline of the Great Lakes. "We had to look at how to make their size a positive rather than a negative," Buhrmann said. "We had to maximize our opportunity to 'sell' our smaller vessels." In the years to come, skillful masters would take these smaller ships into ports rarely visited by Tin Stackers. They began crawling through narrow rivers to reach ports like Manistee and Ontonagon, Michigan. They carried coal to Dunkirk, New York, and Ashland, Wisconsin, They loaded stone at Drummond Island and Alabaster, Michigan. They began handling niche cargoes such as slag, petroleum coke, gypsum, potash, foundry sand, and salt. Because they were self-unloaders, the ships could serve any dock within reach of their booms. The fact that they were diesel-powered meant they were economical enough to operate while the coal-burners went to the scrapyard.[4]

Great Lakes Fleet's new status added urgency to its drive for greater efficiency. To operate most economically, all its ships had to be self-unloaders. So as the 1981 shipping season neared its end, the fleet sent *Arthur M. Anderson, Philip R. Clarke,* and *Cason J. Callaway* into Fraser

Shipyards in Superior to be converted to self-unloaders. All three ships had been launched in 1952 at 647 feet in length and 19,700 gross tons capacity. They were lengthened in 1974 to 767 feet and 26,500 gross tons capacity. Converting them to self-unloaders would take away some of their cargo capacity, but that loss of productivity would be greatly outweighed by reducing their average unloading time from seventeen hours to just five or six. In addition, the ships would be able to serve far more potential customers that lacked shoreside unloading gear.[5] Another key factor also played into the decision to convert the ships. U.S. Steel's dock operation in Lorain was getting rid of its unloading equipment, and the corporation's giant complex at Gary wanted to do the same. To ensure that the fleet had enough vessels to serve these ports, the corporation's steelmaking arm agreed to pay part of the cost of the conversions.[6]

By the time the *Callaway* arrived at Fraser Shipyards in late August, work on the conversion project was already well under way. Since April shipyard workers had been building sixty-nine sections of hopper bottom that would comprise the interior of the ships' cargo holds. Each section was prefabricated with the conveyor structure, piping, and electrical wiring in place. First to enter dry dock was the *Callaway*. It was followed that fall by *Clarke* and then *Anderson* on November 1. Much of the work had to be done in dry dock to maintain the structural integrity of the ships. After each ship entered the dock, workers removed the bottom of its cargo hold. When that was completed, they began the painstaking task of lowering twenty-three sections of hopper bottom through the hatches of each vessel. The sections were then aligned inside the hold and welded together to form a continuous hopper bottom. The sides of this new hold were sloped to permit cargo to flow down to seventy-four hydraulically operated gates, which opened onto a seventy-eight-inch-wide loop conveyor belt running beneath the hold.

The after end of each ship also required extensive work to accommodate the loop belt conveyor that lifted cargo from the hold to the deck. Unlike the fleet's older self-unloaders, these three ships would have their 250-foot-long booms installed just in front of the after deckhouse. The booms could be raised up to eighteen degrees above the deck and rotate ninety degrees to either side. A substantial increase in each ship's electrical capacity was required to operate the boom, gates, and other additional equipment. All told, the project added about 1,500 tons of steel to each ship.

All three ships were ready to return to service at the start of the 1982 navigation season. First out of the yard was the *Callaway*. The

The *Philip R. Clarke* and *Cason J. Callaway* sit in Fraser Shipyards on March 1, 1982, before beginning their first season of operation as self-unloaders. The two ships, along with the *Arthur M. Anderson,* were converted during the winter of 1981–82. Wes Harkins photo.

ship steamed across Duluth-Superior harbor on April 20 to the Duluth, Missabe and Iron Range Railway ore docks. The men on the ore dock dumped several pockets of taconite pellets into the ship so its crew could test the vessel's new conveyor system and belts. When everything checked out, they went ahead and loaded the rest of the vessel. The next day *Clarke* went through the same procedure. Finally, on April 29, *Anderson* went under the chutes to check its equipment and then load its first cargo with its new profile. With their new gear the ships could carry 25,300 gross tons of taconite pellets and unload them at the rate of about 6,000 tons an hour.[7]

While Fraser was converting the AAA-class ships, the Duluth, Missabe and Iron Range Railway began a project to add a shiploader system to its Dock 6 in Duluth. Workers first added seventeen feet to part of the dock's concrete foundation to keep 1,000-footers from striking the dock superstructure. Then they removed the old loading chutes, gates, and operating equipment on the portion of the dock closest to shore. They replaced the old equipment with conveyor shuttles that could be lowered across the deck of a Poe-class ship to load taconite pellets the full width of its hold. A portion of the slip alongside the dock was dredged to thirty feet to allow 1,000-footers to load at the new facility as well as to enable vessels to pass each other when entering or leaving the dock. When completed in 1983, the new shiploader gave the railroad's Duluth dock the same ability to load 1,000-footers as its Two Harbors dock possessed. In the years to come, however, Minntac's pellets would be shipped out of Two Harbors while the Duluth dock generally handled pellets from other customers.

That meant Great Lakes Fleet's big ships would load most of their cargos in Two Harbors.

Great Lakes Fleet's introduction of its three new self-unloaders in 1982 was overshadowed by the nation's deepening economic recession. U.S. Steel's Minntac plant had canceled plans to expand, prompting the fleet to shelve plans to build a third 1,000-footer. Steel mills and ore mines throughout the Midwest operated at reduced capacity and laid off workers. Many closed permanently. That meant greatly reduced demand for ships, and Great Lakes Fleet was among the companies feeling the crunch. The *Blough,* laid up on September 12, 1981, did not fit out in 1982. Indeed, it would remain idle for five more years, largely a victim of its peculiar size. It was too small to run in the pellet trade as efficiently as the *Gott* and *Speer,* and it was too big to handle the smaller cargoes and docks like the AAA-class vessels.[8] Many of the fleet's thirty-one ships remained in layup along with the *Blough,* signaling a painful retrenchment for the organization. The fleet ran a few of its remaining straight-deckers, such as *Robert C. Stanley,* in 1981, but the only straight-decker to fit out in 1982 was *Benjamin F. Fairless,* which loaded grain in Duluth on its first trip on April 9. It steamed into Duluth on May 19 scheduled to load taconite pellets but instead was sent into layup. It was the last straight-decker to sail as a Tin Stacker, ending an era that spanned eighty-one years.

As the season continued, *Speer* and *Anderson* were laid up in midsummer, joining the idle *Callaway.* Many of the fleet's men were laid off, most with no realistic hope of ever returning to work as sailors. For them the recession meant the loss of good jobs which just a few years earlier had seemed as solid as the steel decks of the ships they sailed. Unlicensed men now were earning $18,000 to $25,000 a year, while officers made $35,000 to $45,000.[9] Only the most senior men were able to keep their jobs during these lean years. Positions were so scarce that in some cases men with twenty-five years of seniority were able to work only as temporary replacements for men on vacation. Making the decisions that resulted in these layoffs was "very traumatic," Buhrmann recalled. "The personnel decisions of people being let go were the hardest to make," he said. "We had a thirty-year work force that would be laid off. There was a lot of soul-searching. It was a crisis."[10]

The 1982 season turned out dismally. Iron ore tonnage fell from nearly 75 million gross tons the previous season to just 38.5 million. The 1983 season was markedly better as ore tonnage rebounded to 52 million gross tons, yet many ships remained in layup that year. Acknowledging that its needs were changing and seeking to cut costs,

Great Lakes Fleet sent five more idle straight-deckers to the scrapyard. This group included *B. F. Affleck, Joshua A. Hatfield, John Hulst, Horace Johnson,* and *August Ziesing.* More followed in 1984 when *Eugene W. Pargny, Eugene P. Thomas,* and *Homer D. Williams* were sold for scrap. Even though these three were the more efficient diesel-powered vessels, they lacked unloading gear and were just too old to be candidates for conversion. The fleet would keep a few straight-deckers in reserve as late as 1988, when it finally sold *Irving S. Olds* and *Benjamin F. Fairless* for scrap. The only straight-deckers to escape the cutting torches were *William A. Irvin* and *Leon Fraser.* The *Irvin* was sold to Duluth's convention center board and opened in 1986 as a museum. The *Fraser,* a workhorse of the extended season, was sold in 1985 for conversion to a cement carrier and renamed *Alpena.*[11]

As U.S. Steel continued its battle to survive and prosper in more competitive domestic and world markets, the corporation decided to sell its transportation companies. This included not only Great Lakes Fleet but also railroads such as the Duluth, Missabe and Iron Range and the Erie, Joliet and Eastern. During the next several years, Bill Buhrmann would make about ten sales presentations to banking groups and other potential buyers. Driving the decision to sell the fleet was a need for "better utilization of assets," Buhrmann said. "The company no longer had a need to be sole owner; it was entrepreneurship."[12]

Even as Great Lakes Fleet sold off its old straight-deckers, it continued investing in the efficiency of its self-unloaders. In the early 1980s the Caterpillar Company approached the fleet with a proposal to share the cost of repowering one of its ships. As part of the arrangement, the ship would use a big new diesel engine that Caterpillar wanted to test before starting up a production line. At this point only two ships remained in the fleet that burned coal and were powered by triple expansion engines: *George A. Sloan* and *Irvin L. Clymer.* Because the *Sloan* was bigger and newer, it was chosen to receive the new engine.

Late in 1984, tugboats nudged the 620-foot *Sloan* into Fraser Shipyards. Over the winter workers removed the ship's 2,500-horsepower steam engine and replaced it with a pre-production model of the Caterpillar 3612—a twelve-cylinder engine capable of producing 4,500 horsepower. It was a complicated project that required careful planning to ensure everything fit into the ship's engine room. The *Sloan* had to retain its boilers to provide steam for winches, ballast pumps, and the steering engine. Workers added fuel tanks and a soundproof control booth in the engine room. In addition, the ship received a new rudder and a variable-pitch propeller. Engine controls were installed in the pilothouse so the master could directly control engine speed and direction. When it emerged from the shipyard, the *Sloan* initially had some minor problems in the link between the engine and propeller shaft, but those were soon solved. The new diesel engine gave the *Sloan* more power, increased its speed, and lowered its fuel costs. Without it, the vessel undoubtedly would have been scrapped.[13]

As part of its continuing efforts to make diesel power even cheaper, Great Lakes Fleet became increasingly involved in experimenting with inexpensive, low-grade fuel oil to power its ships. This heavy fuel was so thick that it could not flow at ambient temperatures. It had to be heated to about 120 degrees before it could be pumped from the fuel tank into a "day tank." There it was heated to about 200 degrees so it could flow into a centrifuge, which removed sludge and water from the fuel before it was burned. The fleet's engineers even experimented with burning the accumulated sludge in other boilers. "We tried to use everything we could in the fuel," said Ralph Bertz, head of engineering for the fleet. "We probably became one of the world's leaders in using heavy fuel in diesel engines."[14]

As the 1980s progressed, the steel industry slowly recovered as it became leaner and more efficient. The industry's improving fortunes, along with a more robust national economy, brought better days for Great Lakes shipping. Ore tonnage climbed to 58.4 million in 1985. Tonnage slipped the following season largely because of a nationwide labor dispute involving 22,100 members of the United Steelworkers of America and USX, as the Steel Corporation was now known. The dispute—the two sides disagreed on whether it was a strike or a lockout—shut down USX mills around the country, including the Minntac plant, on August 1. Great Lakes Fleet's four active ore carriers—*Speer, Gott, Callaway,* and *Anderson*—proceeded to ports around the lakes, where union members disembarked and the remaining licensed crewmen laid up the ships. Later in the season the *John G. Munson, Sloan, Calcite II,* and *Taylor* were idled when negotiations failed to produce a new contract for 138 sailors aboard those vessels who were members of the Steelworkers union.[15]

Although the steel industry was improving, USX remained determined to raise cash and cut its financial obligations by selling portions of the corporation. Great Lakes Fleet was a marketable commodity because it was profitable and comprised about 20 percent of the "lift" or tonnage capacity on the Great Lakes. In June 1988, USX announced that a majority stake in Great Lakes Fleet, the Duluth, Missabe and Iron Range Railway, and other USX transportation properties would be sold to Blackstone Capital Partners for $500 million. USX would retain a 49 percent interest in the operations. "The outlook for steel is solid," said Blackstone spokesman Jim Mossman. "We believe there is a solid future for these transportation companies."[16]

Blackstone Capital Partners was a Wall Street investment firm formed in 1985 by Peter Peterson, former U.S. secretary of commerce; Stephen Schwarsman, formerly of the Shearson, Lehman Brothers investment firm; and David Stockman, former director of the Office of Management and Budget under President Ronald Reagan. Blackstone immediately announced it would honor all contracts and labor agreements. All the properties were grouped under a company named Transtar Inc. in Monroeville, Pennsylvania, and all accounting, purchasing, and employee-relation operations were consolidated there. Beyond those changes, however, Blackstone was content to let the management of Great Lakes Fleet run its own operation without interference. Bill Buhrmann, who remained president, said the new owners recognized that the fleet's managers had the talent needed to operate the Tin Stackers as an independent company. "It worked out very well," he recalled nine years after the sale. "They [Blackstone] are good in not being hands-on in the operation. They let us run the place."[17]

As Great Lakes Fleet entered the 1990s, it took its place among the scores of Rust Belt companies that had reinvented themselves not just to survive but to again become competitive and prosperous. The tumult of the previous decade could easily have proven fatal, yet the fleet's managers and employees successfully adapted to a new era in Great Lakes shipping. While the 1980s brought pain and hardship to many, they also brought opportunity. The fleet was slimmed down and inefficient vessels sold. The Missabe Building in downtown Duluth was charged with excitement as the fleet sought new customers, served new ports, and carried new cargoes. "We really felt that we were in charge of our own destiny," Buhrmann said.[18]

USS Great Lakes Fleet now was made up of twelve ships. Because of their design and unloading gear, *Gott, Speer,* and *Blough* remained dedicated to carrying taconite pellets. *Presque Isle* continued to carry taconite pellets but also began making more frequent calls at Rogers City to load stone. *Anderson, Clarke, Callaway,* and *Munson* assumed more versatile roles, carrying taconite pellets along with increasing quantities of stone and coal. *Taylor, Sloan,* and *Calcite II* were largely dedicated to carrying stone and coal to a growing array of customers

The *John G. Munson* steams toward Duluth bearing the livery adopted by USS Great Lakes Fleet at the beginning of the 1990 navigation season. Former Bradley vessels, which had retained their traditional grey hulls, were repainted in red to match the other vessels in Great Lakes Fleet. All vessels were given new diagonal gray and black stripes on their bows. Author's photo.

located in small ports. Decisions made years or decades earlier to repower these smaller ships enabled them to survive while giving the fleet the vessels it needed to efficiently serve niche markets. "I think our company has changed with the times," Buhrmann told a trade publication in 1993. "We have pared down to recognize the 1990s as defined for us by our customers."[19]

To emphasize their new status as a truly independent fleet, all the ships were given the same paint scheme in 1990. The former Bradley ships gave up their gray hulls for the Pittsburgh fleet's traditional red. To acknowledge the Bradley heritage, all the ships' bows were painted with a distinctive diagonal gray stripe and a narrower black stripe. All ships retained the silver-and-black smokestacks that long ago earned their predecessors the nickname Tin Stackers.

The only question mark in the fleet now was the 552-foot *Clymer.* The "Swervin' Irvin," as it was fondly nicknamed by sailors, spent much of the 1970s laid up in Alpena, Michigan. Once the pride of the Bradley fleet, it now was just a short, old coal-burner. Its only saving grace was the steam turbine engine that had been installed in 1954. Although its future looked bleak, the ship got a reprieve in 1980 when it was towed to Fraser Shipyards to be refitted and given a bow thruster taken from the wrecked Cleveland-Cliffs steamer *Frontenac*. The *Clymer* returned to service on the lower lakes in 1981 carrying limestone, coal, and salt, a cargo often reserved for older ships because of the mortal corrosion it causes in the cargo hold. The *Clymer* operated on and off through the 1980s as one of the last coal-burning ships on the lakes. Ralph Bertz told the vessel's engineer and crew that if they did not carefully control the *Clymer*'s smoke, the ship would have to be taken out of service. The men kept tight rein on emissions, and the vessel continued to operate without causing pollution problems. The *Clymer*'s end finally came in 1990 when it was taken to Fraser Shipyards and laid up. It was scrapped four years later in Duluth.[20]

Buhrmann's watch at the helm of Great Lakes Fleet ended in 1993, when he retired after spending forty years with the Steel Corporation, its subsidiaries, and, finally, the independent fleet. Reflecting on his career, Buhrmann felt he was fortunate to have been part of the fleet during some of its most critical years. Great Lakes shipping seems to run in cycles, he said. Construction of bigger locks at the Soo or the deepening of navigation channels enables the industry to make significant changes, then it remains static for the next several decades. His tenure at Great Lakes Fleet occurred during one of these periods of great change as the company moved toward bigger ships, a smaller fleet, and independent status. "I felt very lucky as far as my time with the fleet," he said. "The years I was there will not be duplicated for a long, long time. The building program, the switch from common carrier to an entrepreneur—I'm glad I was a part of that."[21]

Replacing Buhrmann as president in July 1993 was R. Neil Stalker, another U.S. Steel veteran who had cast his lot with Blackstone. Unlike the fleet's previous presidents, Stalker came from a financial background rather than from the ranks of ship managers or mill supervisors. He held several positions at the U.S. Steel's Quebec Cartier Mining Company in Port Cartier, Quebec, before joining Great Lakes Fleet in 1979 as general accountant. In 1985 he was elected the fleet's comptroller. When Blackstone purchased the fleet three years later, Stalker was again elected to the role of its chief financial officer and added assistant treasurer to his title. He also became chief financial officer of the Duluth, Missabe and Iron Range Railway in 1992 before Transtar's board of directors elected him to the presidency.[22]

The nation's economy heated up during the 1990s, prompting a strong resurgence of Great Lakes shipping. By 1994 taconite tonnage was back up to 58.5 million. It was a far cry from the 80 to 90 million tons carried before the recession of the early 1980s, but still a substantial amount considering how much America's steel industry shrank during those years. Although much smaller—it now had about 370 employees and eleven ships—USS Great Lakes Fleet remained a powerful force on the lakes. With a combined cargo capacity of 391,450 gross tons, its ships served customers from Duluth to Ogdensburg, New York. USX Corporation remained the fleet's biggest customer as well as a significant shareholder. In 1995 the fleet carried 23.5 million net tons of cargo. Of that amount, nearly 65 percent was taconite pellets, limestone, and coal destined for USX facilities in Gary, Lorain, and Conneaut. The remainder included coal, stone, slag, petroleum coke, gypsum, potash, foundry sand, and salt shipped to customers around the Great Lakes. As the decade wore on, the fleet even began moving taconite pellets from non-USX ports such as Silver Bay and Superior as well as delivering pellets and coal to customers in Canada.[23]

Some traditional trade routes changed during the mid-1990s as the steel and iron ore industries continued efforts to streamline their operations. One of the most significant changes came as Mesabi Range taconite plants began producing "flux pellets." In the past, steel mills added limestone to their blast furnaces to remove impurities or flux from molten steel. As part of an effort to add value to their product, taconite producers figured out a way to incorporate limestone into taconite pellets. As more steel mills adapted their operations to this new type of pellet, demand grew for shipment of limestone to Duluth, where it could be hauled by rail to the Minnesota taconite plants.

On July 25, 1995, *Presque Isle* nudged against the inner end of the Duluth, Missabe and Iron Range Railway ore dock in Duluth, where a new hopper and conveyor system had been built to handle the unloading and storage of limestone. *Presque Isle* positioned the end of its boom over the hopper and began discharging stone, becoming the first vessel to use the new device. Gleaming white mounds of limestone soon lay alongside the dull gray piles of taconite in the dock's storage yard, and a growing number of Great Lakes Fleet's ships—particularly the *Presque Isle* and the AAA-class vessels—were arriving in Duluth to unload stone.[24]

The dream of year-round navigation on the Great Lakes faded during the 1990s, largely because of reduced demands by the steel industry and concerns that environmental damage could result from vessel traffic in the St. Marys River during the winter. Nonetheless, the navigation season now typically is ten months long because the slimmed-down fleets need that much time to meet their tonnage commitments. Using experience gained in its pioneering work during the 1970s, Great Lakes Fleet continues to operate several ships on an extended season. During the 1994 navigation season, the *Callaway* operated from April 8, 1994, to February 12, 1995. During those 310 days of service it carried nearly 2 million tons of cargo. After the Soo Locks closed on January 15, the ship continued taking cargoes from Escanaba, Michigan, earning the honor of carrying the last taconite cargo of the season for shipment to Gary. The following season the *Callaway* and *Anderson* loaded cargos out of Escanaba until mid-February, finally laying up on February 16, 1996.[25]

As an independent operation, Great Lakes Fleet now devotes considerable energy to marketing. Its sales force continuously travels around the lakes calling on potential customers and maintaining relationships with existing ones. Much emphasis is placed on meeting the growing expectations of customers. "We realize the expectations of our customers include not only efficient vessels and responsive home office and vessel personnel, but also an attitude of 'partnership' with them toward finding long-term answers to specific cargo-handling needs," the company says in its sales brochure. "We embrace this initiative and have demonstrated it for years through site visits by both fleet management and vessel personnel; attention to customers' cyclic delivery patterns; recommendations for delivery of multiple grades or commodities in one shipment; coordination with rail transporters for minimal demurrage; and by taking special efforts to ensure cargoes are delivered to the exact location and in the manner specified by the customer."[26]

A key to the fleet's marketing efforts is promotion of competitive pricing and, in particular, its efficient vessel dispatching. The *Anderson, Clarke, Callaway,* and *Munson* are billed as "interchangeable" ships that are constantly moving about the lakes and can be dispatched to meet customers' schedules. *Calcite II* and *Taylor* are described the same way in their ability to handle smaller cargoes. "This is a unique aspect of our Fleet—especially in the smaller vessel sizes. We can save customers considerable time and demurrage expense through substitution of one vessel for another of the same size if the initially nominated vessel is delayed by weather, encounters a dock availability conflict, or otherwise could not maintain a desired delivery schedule."[27]

The fleet also touts its vessels' versatility. As self-unloaders, they need minimal dockside facilities. The cargo holds of the four AAA-class vessels are divided into seven compartments, giving each ship the abil-

The *George A. Sloan* squirms up the narrow Manistee River carrying coal bound for a customer's dock in Manistee, Michigan. Although 65 percent of the tonnage carried by Great Lakes Fleet is consigned to the mills of USX Corporation, the fleet uses vessels like the *Sloan* to serve smaller customers in ports around the Great Lakes. Author's photo.

ity to carry different types or grades of commodities to a single location or to several sites. The smaller ships, with their bow thrusters, are capable of reaching virtually any dock with sufficient depth to handle a ship.

To better meet its customers' needs, Great Lakes Fleet continued to modify its vessels. In the mid-1990s it lengthened the unloading booms on the *Callaway, Clarke, Anderson,* and *Sloan.* "This allows for better unload[ing] of cargoes at draft-restricted docks and enables higher stockpiling to make maximum use of some customers' limited storage areas."[28] The fleet also fitted the *Sloan* with new hydraulic steering gear in 1996. The equipment enabled the ship's rudder to turn up to seventy degrees for increased maneuverability.

The biggest change, however, was the addition of a 280-foot unloading boom—the longest on the Great Lakes and possibly in the world—to the *Gott.* When the vessel was built, it was designed so it could easily be converted from the short shuttle unloader to a deck-mounted boom. The altering of the *Gott*'s familiar profile was performed by Bay Shipbuilding Company in Sturgeon Bay following the 1995 navigation season. Adding a boom to the *Gott* enabled it to overcome its greatest limitation—the short shuttle unloader that made it a captive of USX docks. With its new boom, the *Gott* could do a better job of unloading at USX docks and also make forays into the pellet-hauling business for steelmakers that lacked dockside conveyor systems required by the shuttle unloaders. The ship's increased versatility quickly became apparent. On July 8, 1996, it loaded for the first time at Taconite Harbor, the loading facility of Erie Mining Company. Later that year and again in 1997 it took on occasional cargoes of taconite pellets in Superior. The vessel began making deliveries to new ports, particularly to the steel mill in Nanticoke, Ontario.

Changes also were made in the pilothouse during these years. In the early 1990s, Great Lakes Fleet took the lead in adopting "electronic chart" technology for most of its ships. This revolutionary technology uses signals from Global Positioning System satellites to pinpoint a ship's location. This information is then used to depict the ship's location on an electronic navigation chart displayed on a computer screen. The master or mate can see the ship's actual position on the chart in relationship to channels, navigation buoys, and submerged obstructions. Images from the ship's radar can be added to the picture, enabling officers to see other vessels and determine when buoys have been moved off station.

The leadership of Great Lakes Fleet changed again in 1997 when R. Neil Stalker was appointed treasurer of Transtar Inc. Robert S. Rosati, president and chief executive officer of Transtar, was elected president of the fleet. Named as the fleet's general manager was Adolph Ojard, a man with a strong heritage in Great Lakes shipping as well as an extensive background in the handling of taconite pellets and other commodities for U.S. Steel Corporation.[29]

The *John G. Munson*'s deck boom hovers over the hopper on the Duluth, Missabe and Iron Range Railway ore dock in Duluth as the vessel unloads limestone on May 31, 1998. The railroad added the hopper and a conveyor system to the dock in 1996 to receive stone bound for taconite plants on the Mesabi Range. The stone is piled in a storage yard adjacent to the dock until it can be loaded into empty ore cars for the journey north. Author's photo.

Ojard's father, Adolph, Sr., spent twenty-five years as master of the tug *Edna G.* in Two Harbors. Both Ojard and the distinctive yellow-and-maroon steam tug became something of a legend as they helped Pittsburgh ships in and out of the ore docks. The family connection gave the younger Ojard firsthand experience with U.S. Steel's ships and its Two Harbors ore docks.

Ojard graduated from the University of Minnesota and then returned to Two Harbors in 1970 to work for the Duluth, Missabe and Iron Range Railway. His experience over the next decade varied widely. He loaded ships, served as union president, and in 1980 was promoted to manager of the port. Three years later he moved on to U.S. Steel's Bessemer and Lake Erie Railroad in Conneaut. Transtar transferred him back to the Duluth, Missabe and Iron Range Railway in 1991 to assume the role of superintendent of transportation. That was followed in 1994 by a three-year stint in Mobile, Alabama, as president of Transtar's Gulf and Warrior Navigation Company, a tug-barge operation, and Mobile River Terminal, a bulk cargo dock that serves oceangoing vessels.[30]

A few months after Rosati and Ojard began their new jobs, Great Lakes Fleet announced on November 1, 1997, that it had purchased all capital stock of Litton Great Lakes Corporation, owner of the *Presque Isle*. The purchase gave the fleet control of the integrated tug-barge it had managed since the mid-1970s. Litton Great Lakes Corporation changed its name to GLF Great Lakes Corporation and moved its headquarters from Erie, Pennsylvania, to Transtar's home in Monroeville, Pennsylvania. The purchase of the management company cleared the way for Great Lakes Fleet to acquire actual ownership of *Presque Isle* in 1998.[31]

As USS Great Lakes Fleet prepares for a new century, it faces serious challenges—just as it has throughout its history. The most apparent uncertainty now concerns the age of its ships. The fleet owns just two ships that are less than twenty years old—*Gott* and *Speer*. The *Taylor* and *Calcite II* are approaching seventy years of age, and *Sloan* was built during World War II. The *Munson, Anderson, Callaway,* and *Clarke*—workhorses of the fleet for decades—will soon be fifty years old. Even with top-quality maintenance, the steam turbines powering these latter four ships cannot last forever.

The rising costs of building and maintaining ships will be a substantial hurdle for Great Lakes Fleet. The *Speer* and *Gott* were designed to last fifty years and built strong enough to be lengthened to 1,300 feet.[32] Yet the astronomical cost of building an additional large lock at the Soo makes it possible that these ships will outlive their usefulness before they ever have a chance to be lengthened. Constructing even a medium-size vessel in the late 1990s or early in the next century would be prohibitively expensive without some sort of government tax incentives or construction subsidy. Maintenance costs are equally troubling. Simply putting a ship into dry dock for Coast Guard inspection costs upward of $500,000.[33] These costs raise questions about how long the fleet will continue seeking the certification needed to keep its oldest vessels in operation.

As part of its ever-changing nature, the Great Lakes shipping industry in recent years has moved toward converting self-unloading ships to unpowered barges pushed by tugs. This allows sound hulls to continue operating without the considerable expenses of large crews and installation of new engines. This trend raises the question of whether Great Lakes Fleet's middle-aged vessels might eventually operate as self-unloading barges. For its part, the fleet's management has been cryptic regarding the age of its ships or their futures: "Plans are to maintain these vessels to meet the high internal standards set throughout the years by our Fleet as well as meeting all requirements of outside governing agencies. Economic needs will shape future Company plans."[34]

Given its heritage of innovation, it seems likely that USS Great Lakes Fleet will evolve to meet its latest challenges. The Midwest Rust Belt, once written off by pundits as doomed, is healthy and still the nation's center for producing automobiles and steel. It also is well stocked with the minerals essential to an industrial economy. The need for waterborne transportation is there. No doubt the accumulated will of men like Augustus Wolvin, Harry Coulby, and thousands of now-departed sailors is still there, too. "We will be a progressive, profitable company twenty years from now," a spokesman for Great Lakes Fleet said, "continuing to provide a large share of the bulk transportation on the Great Lakes due to running an efficient yet profitable company which serves its customers well."[35]

As the sun approached its zenith over the Great Lakes on November 1, 1997, a small, efficient fleet of ships and men toiled in America's heartland. Spread across the Great Lakes from Buffalo to Duluth, these eleven vessels bore the distinctive silver smokestacks of USS Great Lakes Fleet.

Downbound on Lake Superior was *Edwin H. Gott,* laden with Minntac taconite pellets loaded at Two Harbors and bound for Conneaut. Under the loading shuttles at Two Harbors was *Edgar B. Speer.*

The ship's first mate watched carefully as millions of marble-size taconite pellets were whisked into the hold in a hissing blur. Within a few days this cargo of more than 60,000 tons of pellets would be dockside in Conneaut. As part of the fleet's taconite shuttle, *Presque Isle* was entering Lake Michigan bound for the sprawling USX steel complex in Gary. Not far behind was *Roger Blough*, also carrying pellets for Gary.

Scattered across the lakes were the versatile ships of the fleet's AAA class. *Arthur M. Anderson* loaded coal in South Chicago for delivery to nearby Gary while *Philip R. Clarke* discharged cargo in the tiny port of Huron, Ohio. *Cason J. Callaway* was tied up alongside the Duluth, Missabe and Iron Range Railway's ore dock in Duluth while its unloading gear dumped thousands of tons of limestone into the dock's stone hopper. When this task was finished, its crew would winch the ship into position to begin loading taconite pellets at the same dock.

Many years after the demise of the Bradley Transportation Company, four of its former ships continued carrying cargoes for Great Lakes Fleet. *John G. Munson* stood under the loading spouts in Cedarville, Michigan, taking on limestone. The *Munson* was scheduled to load only part of its cargo there before sailing the short distance to Calcite, Michigan, to complete its load bound for Duluth. *George A. Sloan* sidled up to a dock in Menominee, Michigan, to unload coal while *Myron C. Taylor* motored toward Fairport, Ohio, to discharge limestone. The venerable *Calcite II* was bound for Buffalo, New York, to unload its cargo before steaming into Port Colborne, Ontario, to load stone.

Today, Great Lakes Fleet can claim but few of the superlatives that once made it famous. It no longer is the largest fleet on the lakes, nor does it have the newest or biggest ships. Yet it can claim one title with pride: survivor. When change was vital, changes were made. When innovation was essential, innovations were made. When other fleets went out of business, the men and women of this fleet kept it going. Today, Great Lakes Fleet is no longer a link in the chain for the United States Steel Corporation. Instead, it has become a link in the chain for scores of companies that depend upon the low-cost transportation of iron ore, coal, stone, and other commodities. Nearly a century after its formation, the Pittsburgh Steamship Company sails on.

Appendix 1

Vessels Built, Purchased, and Acquired

PITTSBURGH STEAMSHIP COMPANY

Year	Built	Purchased/Acquired
1904		*Murphy*
1905	*Corey, Frick, Gary, Perkins*	*Shaw*
1906	*Morgan, Ream, Rogers, Widener*	
1907	*Baker, Cole, Lynch, Phipps*	
1909	*Buffington, Dinkey, Morgan Jr.*	
1910	*Dickson, Olcott, Palmer, Schiller*	
1911		*Crawford, Filbert, House*
1913	*Farrell, Roberts Jr., Trimble*	
1916	*Kerr, McGonagle*	*Collins, Ferbert* (renamed *Strom*), *McLean, Mitchell, Richardson, Shiras, Frontier* (supply boat)
1917	*Clemson, Pargny, Williams, Ziesing*	
1923	*Hatfield, Lindabury*	
1927	*Affleck, Harvey*	
1929	*Clyde, Johnson, Taylor*	
1930	*Lamont, Thomas*	
1938	*Irvin, Miller, Watson, Hulst*	
1940	*Sandy* (supply boat)	
1942	*Fairless, Ferbert, Fraser, Olds, Voorhees*	
1943	*Avery, Sloan, Stanley*	
1947	*Ojibway* (supply boat)	

Pittsburgh Steamship Company (con't)

Year	Built	Purchased/Acquired
1952	*Anderson, Callaway, Clarke*	
1964		*Hood* (transferred from American Steel & Wire Co.)
1972	*Blough*	
1979	*Gott*	
1980	*Speer*	
1997		*Presque Isle*

BRADLEY TRANSPORTATION COMPANY

Year	Built	Purchased/Acquired
1912	*Calcite*	
1915	*White*	
1917	*Bradley* (renamed *Munson* in 1927, *Clymer* in 1951)	
1923	*Taylor* (renamed *Rogers City* in 1957)	
1925	*Robinson*	
1927	*Bradley*	
1952	*Munson*	
1956		*Taylor, Harvey* (renamed *Cedarville*)
1960		*Clyde* (renamed *Calcite II*)
1966		*Sloan*

Source: *United States Steel Corporation Great Lakes Fleet Bulk and Self-Unloaders Vessel Register,* 1968, 1971, 1984.

Appendix 2

Vessels Sold, Scrapped, and Lost

PITTSBURGH STEAMSHIP COMPANY

Year	Sold	Scrapped or sold for scrapping	Lost
1902			Wilson, 129[*]
1905	Thompson, Bartlett, Colby Hoyt, **Whitworth, Russell**, 105, 107, 109, 110, 111, 116, 126, 127, 201, 202		LaFayette, Madeira
1906			Grecian
1909	Wolvin, Palmer, 117, 118, 130, 131, 132, 133, 134		Trevor
1911	Cambria		Joliet
1912	Marina, Masaba, Corona, Maruba, Corsica, **Malta**		
1913	Gilbert		Matoa
1914	Griffin, LaSalle, Wawatam, Mariska, Briton		
1915	Colgate		
1917	German		
1918	Manola, Saxon, Roman, Van Hise		
1920			Superior City
1922	Mather		
1923	**Marcia**		
1924	**Maida, Carrington**		

Pittsburgh Steamship Company (con't)

Year	Sold	Scrapped or sold for scrapping	Lost
1926	Eads, Ericsson, **Holley, 137**		
1927	Cort, Crescent City, Coralia Empire City, Rockefeller, Neilson		
1928	Bessemer, Siemens, Watt, **Manda**		
1929	Maritana, Mariposa, **Thomas**		
1936	McDougall, Fairbairn, Maricopa, **Maia, Bell, Nasmyth, Jenney, Corliss, Krupp, Martha, Magna**		
1939	Stephenson, **Marsala**		
1940	Linn, Shaw, Murphy, Richardson, **Bryn Mawr, Smeaton, Manila**		
1943	McLean		
1944	Collins, Shiras, Rensselaer, Black, Strom, Mitchell, Fulton, Queen City, Zenith City		
1945	Princeton, Houghton, Maunaloa, **Fritz, Roebling**		
1946	Mataafa		
1947	Frontier (supply boat)		
1948	Harvard		
1952	Sandy (supply boat)		
1954	Poe, Morse, Cornell, Bunsen, Malietoa		
1956	Taylor, Harvey (transferred to Bradley fleet)		
1960	Clyde (transferred to Bradley fleet)		
1961		Ellwood, Gates	
1962	Edenborn, Hill		
1963	Corey, Gary		
1964	Frick, Perkins		
1965	Baker, Lynch, Morgan, Ream		

Pittsburgh Steamship Company (con't)

Year	Sold	Scrapped or sold for scrapping	Lost
1966	House, Sloan (transferred to Bradley fleet)		
1968	Dickson, Olcott		
1974		Crawford, Hood	
1975		Rogers	
1976		Filbert, Phipps	
1978	Lindabury	Palmer, Roberts Jr., Schiller, Trimble	
1979	McGonagle	Farrell	
1980	Widener	Buffington, Clemson, Cole, Dinkey, Kerr, Miller, Morgan Jr.	
1983		Affleck, Hatfield, Hulst, Johnson, Ziesing	
1984		Pargny, Williams, Thomas	
1985	Irvin		
1986		Avery, Watson	
1987		Lamont, Stanley, Voorhees, Ferbert	
1988		Fairless, Olds	
1991	Fraser		

BRADLEY TRANSPORTATION COMPANY

Year	Sold	Scrapped or sold for scrapping	Lost
1961		Calcite	
1958			Bradley
1965			Cedarville
1976	White		
1987		Rogers City, Robinson	
1994		Clymer	

* Barges indicated in bold.

APPENDIX 3

Vessels of the Pittsburgh Steamship Company
Vessels are listed under the final name they held while part of the Pittsburgh Steamship Company or Bradley Transportation Company.

Dimensions given generally are those used when the vessel was built or acquired by the fleet. Dimensions are rounded to the nearest whole number.

The year given for a vessel's construction is that which is most widely listed in company records or other accepted sources. Bear in mind some ships were built in one year but were not formally delivered to the fleet until the following year.

VESSEL: *B. F. Affleck*
REGISTRY: US 226895
YEAR BUILT: 1927
SHIPYARD: Toledo Ship Building Co., Toledo. Hull No. 178
ENGINE: Triple expansion steam engine
GROSS TONNAGE: 7,992
LENGTH—KEEL: 580 feet
LENGTH—OVERALL: 604 feet
BEAM: 60 feet
DEPTH: 32 feet
HISTORY: Last sailed in 1979. Sold for scrap in 1983. Scrapped in 1988 in Port Colborne, Ontario.

VESSEL: *Arthur M. Anderson*
REGISTRY: US 264207
YEAR BUILT: 1952
SHIPYARD: American Ship Building Co., Lorain. Hull No. 868
ENGINE: Steam turbine
GROSS TONNAGE: 11,623
LENGTH—KEEL: 620
LENGTH—OVERALL: 647 feet

BEAM: 70 feet
DEPTH: 36 feet
HISTORY: Lengthened 120 feet in 1975, converted to self-unloader 1981-82. Operational in 1998.

VESSEL: *Sewell Avery*
REGISTRY: US 243483
YEAR BUILT: 1943
SHIPYARD: American Ship Building Co., Lorain. Hull No. 827
ENGINE: Double compound steam engine
GROSS TONNAGE: 8,758
LENGTH—KEEL: 595 feet
LENGTH—OVERALL: 620 feet
BEAM: 60 feet
DEPTH: 35 feet
HISTORY: Briefly named *Lancashire* before turned over to Pittsburgh Steamship Co. Laid up in 1981 and sold in 1986 for use as a dock in Sault Ste. Marie, Ontario.

VESSEL: *George F. Baker*
REGISTRY: US 204225
YEAR BUILT: 1907
SHIPYARD: Superior Shipbuilding Co., Superior. Hull No. 517
ENGINE: Triple expansion
GROSS TONNAGE: 7,240
LENGTH—KEEL: 587 feet
LENGTH—OVERALL: 601 feet
BEAM: 58 feet
DEPTH: 28 feet
HISTORY: Sold to the Kinsman fleet in 1965 and renamed *Henry Steinbrenner*. Scrapped in 1979 in Ashtabula, Ohio.

VESSEL: *E. B. Bartlett*
REGISTRY: US 136205
YEAR BUILT: 1891
SHIPYARD: American Steel Barge Co., Superior. Hull No. 113
ENGINE: Triple expansion
GROSS TONNAGE: 1,400
LENGTH—KEEL: 265 feet
BEAM: 38 feet
DEPTH: 24 feet
HISTORY: Sold saltwater in 1905. Sank in 1916 in the Cape Cod Canal. Raised and scuttled in deep water.

VESSEL: *Sir Isaac Lothian Bell*
REGISTRY: US 116741

YEAR BUILT: 1896
SHIPYARD: F.W. Wheeler Co., West Bay City. Hull No. 118
ENGINE: Barge
GROSS TONNAGE: 3,418
LENGTH—KEEL: 365 feet
BEAM: 45 feet
DEPTH: 23 feet
HISTORY: Sold Canadian in 1936 and renamed *Blanch H.* Renamed *Black River* in 1949. Rebuilt as diesel-driven ship in 1952. Renamed *Tuxpancliffe* in 1979. Scrapped in 1983 in Corpus Christi, Texas.

VESSEL: *Sir Henry Bessemer*
REGISTRY: US 116723
YEAR BUILT: 1896
SHIPYARD: Globe Iron Works, Cleveland. Hull No. 66
ENGINE: Triple expansion
GROSS TONNAGE: 4,321
LENGTH—KEEL: 413 feet
BEAM: 48 feet
DEPTH: 24 feet
HISTORY: Sold in 1928 and renamed *Michael J. Bartelme.* Renamed *Wolverine* in 1944. Sank in 1955; later raised and laid up. Scrapped in 1971 at Green Bay.

NAME: *Clarence A. Black*
REGISTRY: 127300
YEAR BUILT: 1898
SHIPYARD: Cleveland Shipbuilding Co., Lorain. Hull No. 31
ENGINE: Triple expansion
GROSS TONNAGE: 4,150
LENGTH—KEEL: 413 feet
LENGTH—OVERALL: 434 feet
BEAM: 50 feet
DEPTH: 28 feet
HISTORY: Sold in 1944 as part of wartime tonnage exchange. Scrapped in 1947 in Hamilton, Ontario.

NAME: *Roger Blough*
REGISTRY: US 533062
YEAR BUILT: 1972
SHIPYARD: American Ship Building Co., Lorain. Hull No. 900
ENGINE: Twin diesels
GROSS TONNAGE: 22,041
LENGTH—KEEL: 833 feet
LENGTH—OVERALL: 858 feet
BEAM: 105 feet

DEPTH: 39 feet
HISTORY: Operational in 1998.

VESSEL: *Carl D. Bradley*
REGISTRY: US 226776
YEAR BUILT: 1927
SHIPYARD: American Ship Building Co., Lorain. Hull No. 797
ENGINE: Steam turbine, electric motor
GROSS TONNAGE: 10,028
LENGTH—KEEL: 615 feet
LENGTH—OVERALL: 638 feet
BEAM: 65 feet
DEPTH: 33 feet
HISTORY: Sank on Nov. 18, 1958, off Gull Island, Lake Michigan.

VESSEL: *Briton*
REGISTRY: US 3493
YEAR BUILT: 1890
SHIPYARD: Globe Iron Works, Cleveland. Hull No. 39
ENGINE: Triple expansion
GROSS TONNAGE: 2,348
LENGTH—KEEL: 296 feet
BEAM: 40 feet
DEPTH: 20 feet
HISTORY: Sold in 1914. Sent to saltwater in 1917 and returned to the lakes in 1923. Wrecked on Nov. 13, 1929, on Point Abino, Lake Erie.

VESSEL: *Bryn Mawr*
REGISTRY: US 3845
YEAR BUILT: 1900
SHIPYARD: Chicago Shipbuilding Co., South Chicago. Hull No. 41
ENGINE: Barge
GROSS TONNAGE: 4,294
LENGTH—KEEL: 400 feet
LENGTH—OVERALL: 412 feet
BEAM: 50 feet
DEPTH: 24 feet
HISTORY: Sold Canadian in 1940 and renamed *Bryn Barge*. Used as breakwall in 1968.

NAME: *Eugene J. Buffington*
REGISTRY: US 206147
YEAR BUILT: 1909
SHIPYARD: American Ship Building Co., Lorain. Hull No. 366
ENGINE: Triple expansion
NET TONNAGE: 5,857

LENGTH—KEEL: 580 feet
LENGTH—OVERALL: 601 feet
BEAM: 58 feet
DEPTH: 32 feet
HISTORY: Laid up after 1974. Sold for scrap in 1980 and scrapped in Spain.

VESSEL: *Robert W. E. Bunsen*
REGISTRY: US 111294
YEAR BUILT: 1900
SHIPYARD: Chicago Shipbuilding Co., Chicago. Hull No. 40
ENGINE: Quadruple expansion
GROSS TONNAGE: 5,181
LENGTH—KEEL: 439 feet
BEAM: 50 feet
DEPTH: 25 feet
HISTORY: Sold in 1954. Converted to a crane barge and renamed *Marquis Roen*. Sold for off-lakes use in 1973.

VESSEL: *Calcite*
REGISTRY: US 209763
YEAR BUILT: 1912
SHIPYARD: Detroit Ship Building Co., Wyandotte. Hull No. 188
ENGINE: Quadruple expansion
GROSS TONNAGE: 3,996
LENGTH—KEEL: 416 feet
LENGTH—OVERALL: 436 feet
BEAM: 54 feet
DEPTH: 29 feet
HISTORY: First ship of the Bradley Transportation Co. Remained in fleet until scrapped in 1961.

VESSEL: *Calcite II*
REGISTRY: US 228886
YEAR BUILT: 1929
SHIPYARD: American Ship Building Co., Lorain. Hull No. 804
ENGINE: Triple expansion
GROSS TONNAGE: 8,188
LENGTH—KEEL: 588 feet
LENGTH—OVERALL: 604 feet
BEAM: 60 feet
DEPTH: 32 feet
HISTORY: Built for Pittsburgh Steamship Co. as *William G. Clyde*. Transferred to Bradley Division in 1960. Converted to self-unloader and given present name in 1961. Repowered with diesel engine in 1964. Operational in 1998.

VESSEL: *Cason J. Callaway*
REGISTRY: US 264349
YEAR BUILT: 1952
SHIPYARD: Great Lakes Engineering Works, Ecorse. Hull No. 297
ENGINE: Steam turbine
GROSS TONNAGE: 11,591
LENGTH—KEEL: 620 feet
LENGTH—OVERALL: 647 feet
BEAM: 70 feet
DEPTH: 36 feet
HISTORY: Lengthened 120 feet in 1974, converted to self-unloader 1981-82. Operational in 1998.

VESSEL: *Cambria*
REGISTRY: US 126420
YEAR BUILT: 1887
SHIPYARD: Globe Iron Works, Cleveland. Hull No. 12
ENGINE: Triple expansion
GROSS TONNAGE: 1,878
LENGTH—KEEL: 280 feet
LENGTH—OVERALL:
BEAM: 40 feet
DEPTH: 20 feet
HISTORY: Sold in 1910 to Port Huron and Duluth Steamship Co. Converted to passenger steamer and renamed *Lakeland*. Sank in 1924 near Sturgeon Bay, Wis.

VESSEL: *Carrington*
REGISTRY: US 127172
YEAR BUILT: 1897
SHIPYARD: Chicago Shipbuilding Co., South Chicago.
ENGINE: Barge
GROSS TONNAGE: 3,180
LENGTH—KEEL: 352 feet
LENGTH—OVERALL: 366 feet
BEAM: 44 feet
DEPTH: 22 feet
HISTORY: Sold in 1924. Renamed *Cordova* in 1927 and *Delkote* in 1939. Scrapped in 1967.

VESSEL: *Cedarville*
REGISTRY: US 226492
YEAR BUILT: 1927
SHIPYARD: Great Lakes Engineering Works, River Rouge. Hull No. 255
ENGINE: Triple expansion

GROSS TONNAGE: 7,973
LENGTH—KEEL: 588 feet
LENGTH—OVERALL: 603 feet
BEAM: 60 feet
DEPTH: 32 feet
HISTORY: Built for Pittsburgh Steamship Co. as *A. F. Harvey*. Transferred to Bradley Transportation Co. in 1956, converted to a self-unloader, and renamed *Cedarville*. Sank on May 7, 1965, with the loss of 10 men after colliding with Norwegian freighter *Topdalsfjord* two miles east of Mackinac Bridge.

VESSEL: *Philip R. Clarke*
REGISTRY: US 263699
YEAR BUILT: 1952
SHIPYARD: American Ship Building Co., Lorain. Hull No. 867
ENGINE: Steam turbine
GROSS TONNAGE: 11,623
LENGTH—KEEL: 620 feet
LENGTH—OVERALL: 647 feet
BEAM: 70 feet
DEPTH: 36 feet
HISTORY: Lengthened 120 feet in 1974 and converted to self-unloader in 1981-82. Operational in 1998.

VESSEL: *D. M. Clemson*
REGISTRY: US 214599
YEAR BUILT: 1917
SHIPYARD: American Ship Building Co., Lorain. Hull No. 716
ENGINE: Triple expansion
GROSS TONNAGE: 7,377
LENGTH—KEEL: 580 feet
LENGTH—OVERALL: 600 feet
BEAM: 60 feet
DEPTH: 32 feet
HISTORY: Sold for scrap in 1980. Scrapped in 1981 in Thunder Bay, Ontario.

VESSEL: *Irvin L. Clymer*
REGISTRY: US 215018
YEAR BUILT: 1917
SHIPYARD: American Ship Building Co., Lorain. Hull No. 718
ENGINE: Triple expansion
GROSS TONNAGE: 7,203
LENGTH—KEEL: 530 feet
LENGTH—OVERALL: 552 feet
BEAM: 60 feet

DEPTH: 32 feet
HISTORY: Built as *Carl D. Bradley.* Renamed *John G. Munson* in 1927. Given final name in 1952. Repowered with steam turbine in 1954. Operated as one of the last coal-fired steamers on the Great Lakes. Laid up in 1990 and scrapped in 1994 in Duluth.

VESSEL: *Joseph L. Colby*
REGISTRY: US 76933
YEAR BUILT: 1890
SHIPYARD: American Steel Barge Co., Superior. Hull No. 108
ENGINE: Two-cylinder steam engine
GROSS TONNAGE: 1,245
LENGTH—KEEL: 265 feet
LENGTH—OVERALL: 280 feet
BEAM: 36 feet
DEPTH: 22 feet
HISTORY: Sold saltwater in 1905 and renamed *Bay State.* Returned to the lakes in 1922 and converted to sand dredge. Scrapped in 1935 at Chicago.

NAME: *Thomas F. Cole*
REGISTRY: US 203891
YEAR BUILT: 1907
SHIPYARD: Great Lakes Engineering Works, Ecorse. Hull No. 27
ENGINE: Triple expansion
GROSS TONNAGE: 7,739
LENGTH—KEEL: 580 feet
LENGTH—OVERALL: 605 feet
BEAM: 58 feet
DEPTH: 32 feet
HISTORY: Sold for scrap in 1980 and scrapped in 1981 in Thunder Bay, Ontario.

VESSEL: *James B. Colgate*
REGISTRY: US 77019
YEAR BUILT: 1892
SHIPYARD: American Steel Barge Co., Superior. Hull No. 121
ENGINE: Triple expansion
GROSS TONNAGE: 1,713
LENGTH—KEEL: 308 feet
BEAM: 38 feet
DEPTH: 24 feet
HISTORY: Sold in 1915. Sank on Oct. 20, 1918, in Lake Erie.

VESSEL: *E. C. Collins*
REGISTRY: US 200666
YEAR BUILT: 1904
SHIPYARD: American Shipbuilding Co., Lorain. Hull No. 329

ENGINE: Triple expansion
GROSS TONNAGE: 4,787
LENGTH—KEEL: 420 feet
LENGTH—OVERALL: 440 feet
BEAM: 50 feet
DEPTH: 28 feet
HISTORY: Built as *Edwin F. Holmes*. Purchased by Pittsburgh Steamship Co. in 1916 and renamed *E. C. Collins*. Sold in 1944. Eventually renamed *J. B. Ford*. Still in service in 1998 as a cement storage barge for Inland Lakes Transportation Co.

VESSEL: *Coralia*
REGISTRY: US 127129
YEAR BUILT: 1896
SHIPYARD: Globe Iron Works, Cleveland. Hull No. 64
ENGINE: Triple expansion
GROSS TONNAGE: 4,330
LENGTH—KEEL: 413 feet
LENGTH—OVERALL: 418 feet
BEAM: 48 feet
DEPTH: 24 feet
HISTORY: Sold to Nicholson-Universal Steamship Co. in 1927. Renamed *T. H. Browning* in 1950 and *L. D. Browning* in 1952. Used for grain storage after 1955. Scrapped in 1965 at Hamilton, Ontario.

NAME: *William E. Corey*
REGISTRY: US 202296
YEAR BUILT: 1905
SHIPYARD: Chicago Shipbuilding Co., Chicago. Hull No. 67
ENGINE: Triple expansion
GROSS TONNAGE: 6,485
LENGTH—KEEL: 557 feet
LENGTH—OVERALL: 569 feet
BEAM: 56 feet
DEPTH: 31 feet
HISTORY: Sold in 1963 to Upper Lakes Shipping Ltd. and renamed *Ridgetown*. Sunk as breakwall in 1970 at Nanticoke, Ontario. Later raised and sunk as breakwall at Port Credit, Ontario.

VESSEL: *George H. Corliss*
REGISTRY: US 86363
YEAR BUILT: 1896
SHIPYARD: Chicago Shipbuilding Co., Chicago. Hull No. 23
ENGINE: Barge
GROSS TONNAGE: 3,259
LENGTH—KEEL: 352 feet

LENGTH—OVERALL: 378 feet
BEAM: 44 feet
DEPTH: 22 feet
HISTORY: Sold in 1936 and renamed *Ethel* in 1937, *Ethel J.* in 1939, *Portadoc* in 1945, and *Kenordoc* in 1961. Sold for scrap in 1973.

VESSEL: *Cornell*
REGISTRY: US 127451
YEAR BUILT: 1900
SHIPYARD: Chicago Shipbuilding Co., Chicago. Hull No. 42
ENGINE: Quadruple expansion
GROSS TONNAGE: 5,082
LENGTH—KEEL: 454 feet
LENGTH—OVERALL: 474 feet
BEAM: 50 feet
DEPTH: 28 feet
HISTORY: Sold in 1954 and renamed *Neptune*. Renamed *Cornell* in 1958. Sold in 1961 and scrapped in Yugoslavia.

VESSEL: *Corona*
REGISTRY: US 126505
YEAR BUILT: 1888
SHIPYARD: Globe Iron Works, Cleveland. Hull No. 18
ENGINE: Triple expansion
GROSS TONNAGE: 2,408
LENGTH—KEEL: 299 feet
BEAM: 41 feet
DEPTH: 21 feet
HISTORY: Sold in 1912 and renamed *Paipoonge*. Eventually sold saltwater. Renamed *Dorte Jensen* in 1924 and *Kaupo* in 1927. Sunk in 1940 in an attempt by British forces to block the harbor mouth at Dieppe, France. Cut down and abandoned in 1947.

VESSEL: *Corsica*
REGISTRY: US 126474
YEAR BUILT: 1888
SHIPYARD: Globe Iron Works, Cleveland. Hull No. 15
ENGINE: Triple expansion
GROSS TONNAGE: 2,364
LENGTH—KEEL: 299 feet
BEAM: 41 feet
DEPTH: 22 feet
HISTORY: Sold in 1912. Taken to the East Coast in 1917 and scrapped in 1926.

VESSEL: *Henry Cort*
REGISTRY: US 150587

YEAR BUILT: 1892
SHIPYARD: American Steel Barge Co., Superior. Hull No. 125
ENGINE: Triple expansion
GROSS TONNAGE: 2,234
LENGTH—KEEL: 320 feet
LENGTH—OVERALL: 335 feet
BEAM: 42 feet
DEPTH: 25 feet
HISTORY: Originally built as package freighter *Pillsbury*. Acquired by Bessemer Steamship Co. and renamed *Henry Cort*. Sold by Pittsburgh Steamship Co. in 1927. Wrecked on Dec. 1, 1934, off pierhead at Muskegon, Mich.

VESSEL: *George G. Crawford*
REGISTRY: US 204251
YEAR BUILT: 1907
SHIPYARD: American Ship Building Co., Lorain. Hull No. 347
ENGINE: Triple expansion
GROSS TONNAGE: 7,699
LENGTH—KEEL: 585 feet
LENGTH—OVERALL: 605 feet
BEAM: 60 feet
DEPTH: 32 feet
HISTORY: Built as *LeGrand S. DeGraff*. Acquired by Pittsburgh Steamship Co. in 1911 and renamed *George G. Crawford*. Sold for scrap in 1974 and scrapped the following year.

VESSEL: *Crescent City*
REGISTRY: US 127176
YEAR BUILT: 1897
SHIPYARD: Chicago Shipbuilding Co., Chicago. Hull No. 25
ENGINE: Quadruple expansion
GROSS TONNAGE: 4,213
LENGTH—KEEL: 406 feet
LENGTH—OVERALL: 420 feet
BEAM: 48 feet
DEPTH: 24 feet
HISTORY: Sold in 1927 and converted to carry automobiles. Renamed *Carl W. Meyers* in 1950. Scrapped in 1959 in Hamilton, Ontario.

VESSEL: *William B. Dickson*
REGISTRY: US 207981
YEAR BUILT: 1910
SHIPYARD: Great Lakes Engineering Works, Ecorse. Hull No. 75
ENGINE: Triple expansion
GROSS TONNAGE: 7,568

LENGTH—KEEL: 587 feet
LENGTH—OVERALL: 601 feet
BEAM: 58 feet
DEPTH: 28 feet
HISTORY: Sold in 1968 and renamed *Merle M. McCurdy*. Laid up in 1985 and scrapped in 1988 in Port Colborne, Ontario.

NAME: *Alva C. Dinkey*
REGISTRY: US 206090
YEAR BUILT: 1909
SHIPYARD: American Ship Building Co., Lorain. Hull No. 365
ENGINE: Triple expansion
GROSS TONNAGE: 7,702
LENGTH—KEEL: 580 feet
LENGTH—OVERALL: 601 feet
BEAM: 58 feet
DEPTH: 32 feet
HISTORY: Sold for scrap in 1980. Towed to Spain that year for scrapping.

VESSEL: *James B. Eads*
REGISTRY: US 86307
YEAR BUILT: 1893
SHIPYARD: Globe Iron Works, Cleveland. Hull No. 53
ENGINE: Triple expansion
GROSS TONNAGE: 2,995
LENGTH—KEEL: 330 feet
LENGTH—OVERALL: 346 feet
BEAM: 42 feet
DEPTH: 24 feet
HISTORY: Originally built as *Globe*. Lengthened 72 feet in 1899, renamed *James B. Eads* in 1900. Sold in 1926 to Nassau Ship and Dredge Co. Sold for scrap in 1963 and scrapped in 1967.

NAME: *William Edenborn*
REGISTRY: US 81702
YEAR BUILT: 1900
SHIPYARD: West Bay City Shipbuilding Co., West Bay City. Hull No. 40
ENGINE: Quadruple expansion
GROSS TONNAGE: 5,085
LENGTH—KEEL: 478 feet
LENGTH—OVERALL: 497 feet
BEAM: 52 feet
DEPTH: 30 feet
HISTORY: Sold to the city of Cleveland in 1962 and sunk as a breakwater. Later cut down and covered.

NAME: *Isaac L. Ellwood*
REGISTRY: US 100707
YEAR BUILT: 1900
SHIPYARD: West Bay City Shipbuilding Co., West Bay City. Hull No. 39
ENGINE: Quadruple expansion
GROSS TONNAGE: 5,143
LENGTH—KEEL: 478 feet
LENGTH—OVERALL: 497 feet
BEAM: 52 feet
DEPTH: 30 feet
HISTORY: Sold for scrap in 1961 and scrapped in Conneaut, Ohio. At the time it was one of the fleet's four remaining original vessels.

NAME: *Empire City*
REGISTRY: US 136623
YEAR BUILT: 1897
SHIPYARD: Cleveland Shipbuilding Co., Cleveland. Hull No. 28
ENGINE: Quadruple expansion
GROSS TONNAGE: 4,118
LENGTH—KEEL: 405 feet
LENGTH—OVERALL: 425 feet
BEAM: 48 feet
DEPTH: 24 feet
HISTORY: Sold to Empire Steamship Co. in 1929, converted to self-unloader and renamed *Sumatra*. Renamed *Dolomite* in 1962. Scrapped in 1968.

NAME: *John Ericsson*
REGISTRY: US 77226
YEAR BUILT: 1896
SHIPYARD: American Steel Barge Co., Superior. Hull No. 138
ENGINE: Triple expansion
GROSS TONNAGE: 3,200
LENGTH—KEEL: 390 feet
LENGTH—OVERALL: 405 feet
BEAM: 48 feet
DEPTH: 27 feet
HISTORY: Sold in 1926. Eventually donated to Hamilton, Ontario, for use as a museum. The effort failed and the vessel was scrapped in 1968.

VESSEL: *Sir William Fairbairn*
REGISTRY: US 116736
YEAR BUILT: 1896
SHIPYARD: Detroit Dry Dock Co., Wyandotte. Hull No. 124
ENGINE: Triple expansion
GROSS TONNAGE: 4,220

LENGTH—KEEL: 424 feet
LENGTH—OVERALL: 436 feet
BEAM: 46 feet
DEPTH: 28 feet
HISTORY: Sold in 1936. Operated for many years by Hutchinson Steamship Co. Used for grain storage in 1959. Scrapped in 1964.

VESSEL: *Benjamin F. Fairless*
REGISTRY: US 242260
YEAR BUILT: 1942
SHIPYARD: American Ship Building Co., Lorain. Hull No. 824
ENGINE: Steam turbine
GROSS TONNAGE: 10,294
LENGTH—KEEL: 614 feet
LENGTH—OVERALL: 639 feet
BEAM: 67 feet
DEPTH: 35 feet
HISTORY: Laid up in 1982 as the last straight-decker to sail for Great Lakes Fleet. Sold for scrap in 1988 and scrapped in Taiwan.

VESSEL: *James A. Farrell*
REGISTRY: US 210912
YEAR BUILT: 1913
SHIPYARD: American Ship Building Co., Lorain. Hull No. 397
ENGINE: Triple expansion
GROSS TONNAGE: 7,770
LENGTH—KEEL: 580 feet
LENGTH—OVERALL: 600 feet
BEAM: 58 feet
DEPTH: 32 feet
HISTORY: Idled in 1973, sold for scrap in 1979, scrapped in Duluth in 1980.

VESSEL: *A. H. Ferbert*
REGISTRY: US 242024
YEAR BUILT: 1942
SHIPYARD: Great Lakes Engineering Works, Ecorse. Hull No. 289
ENGINE: Steam turbine
GROSS TONNAGE: 10,294
LENGTH—KEEL: 614 feet
LENGTH—OVERALL: 639 feet
BEAM: 67 feet
DEPTH: 35 feet
HISTORY: Sold for scrap in 1987 and towed to Turkey for scrapping in 1988.

VESSEL: *William J. Filbert*
REGISTRY: US 204417

YEAR BUILT: 1907
SHIPYARD: American Ship Building Co., Lorain. Hull No. 348
ENGINE: Triple expansion
GROSS TONNAGE: 7,699
LENGTH—KEEL: 585 feet
LENGTH—OVERALL: 605 feet
BEAM: 60 feet
DEPTH: 32 feet
HISTORY: Built as *William M. Mills*. Purchased and renamed by Pittsburgh Steamship Co. in 1911. Sold for scrap in 1976 and scrapped the following year in Duluth.

VESSEL: *Leon Fraser*
REGISTRY: US 241856
YEAR BUILT: 1942
SHIPYARD: Great Lakes Engineering Works, Ecorse. Hull No. 287
ENGINE: Steam turbine
GROSS TONNAGE: 10,294
LENGTH—KEEL: 614 feet
LENGTH—OVERALL: 639 feet
BEAM: 67 feet
DEPTH: 35 feet
HISTORY: Sold in 1985. Shortened 120 feet, converted to a cement carrier and renamed *Alpena*. Still operational in 1998.

NAME: *Henry C. Frick*
REGISTRY: US 202296
YEAR BUILT: 1905
SHIPYARD: West Bay City Ship Building Co., West Bay City. Hull No. 615
ENGINE: Triple expansion
GROSS TONNAGE: 6,834
LENGTH—KEEL: 549 feet
LENGTH—OVERALL: 569 feet
BEAM: 56 feet
DEPTH: 31 feet
HISTORY: Sold Canadian in 1964 and renamed *Michipicoten*. Sold for scrap in 1972. Sank off Anticosti Island in the lower St. Lawrence River while being towed overseas for scrapping.

NAME: *John Fritz*
REGISTRY: US 77308
YEAR BUILT: 1898
ENGINE: Barge
SHIPYARD: West Bay City Shipbuilding Co., West Bay City. Hull No. 125
GROSS TONNAGE: 4,693
LENGTH—KEEL: 338 feet

LENGTH—OVERALL: 450 feet
BEAM: 50 feet
DEPTH: 28 feet
HISTORY: Sold Canadian in 1945. Sold for use as a breakwall in 1968.

VESSEL: *Frontier*
REGISTRY: US 207434
YEAR BUILT: 1910
SHIPYARD: Buffalo Dry Dock Co., Buffalo.
ENGINE: Triple expansion
LENGTH—KEEL: 120 feet
LENGTH—OVERALL: 140 feet
BEAM: 39 feet
DEPTH: 15 feet
HISTORY: Built as *Niagara Frontier* for the International Ferry Co. of Buffalo to serve as ferry between Buffalo and Fort Erie. Acquired by Pittsburgh Steamship Co. in 1917. Modified to serve as a supply boat and renamed *Frontier*. Affectionately nicknamed "The Gutwagon" by Pittsburgh sailors, the *Frontier* was sold in 1947.

NAME: *Robert Fulton*
REGISTRY: US 111129
YEAR BUILT: 1896
SHIPYARD: Detroit Dry Dock Co., Wyandotte. Hull No. 125
ENGINE: Triple expansion
GROSS TONNAGE: 4,219
LENGTH—KEEL: 424 feet
LENGTH—OVERALL: 434 feet
BEAM: 45 feet
DEPTH: 28 feet
HISTORY: Sold in 1944. Scrapped in 1948.

NAME: *Elbert H. Gary*
REGISTRY: US 202014
YEAR BUILT: 1905
SHIPYARD: Chicago Ship Building Co., Chicago. Hull No. 66
ENGINE: Triple expansion
GROSS TONNAGE: 6,584
LENGTH—KEEL: 549 feet
LENGTH—OVERALL: 569 feet
BEAM: 56 feet
DEPTH: 31 feet
HISTORY: Sold to Kinsman Transit Co. in 1963 and renamed *R. E. Webster*. Sold for scrap in 1972 and towed to Spain for scrapping in 1973.

NAME: *John W. Gates*
REGISTRY: US 77407

YEAR BUILT: 1900
SHIPYARD: American Shipbuilding Co., Lorain. Hull No. 37
ENGINE: Quadruple expansion
GROSS TONNAGE: 5,946
LENGTH—KEEL: 478 feet
LENGTH—OVERALL: 497 feet
BEAM: 52 feet
DEPTH: 30 feet
HISTORY: Sold for scrap in 1961 and scrapped in Conneaut, Ohio. At the time it was one of four remaining vessels from the original fleet.

VESSEL: *German*
REGISTRY: US 86122
YEAR BUILT: 1890
SHIPYARD: Globe Iron Works, Cleveland. Hull No. 38
ENGINE: Triple expansion
GROSS TONNAGE: 2,348
LENGTH—KEEL: 296 feet
BEAM: 40 feet
DEPTH: 21 feet
HISTORY: Sold in 1917 and renamed *Yankee*. Sank in the Atlantic Ocean off New York in 1919.

VESSEL: *W. H. Gilbert*
REGISTRY: US 81382
YEAR BUILT: 1892
SHIPYARD: Globe Iron Works, Cleveland. Hull No. 89
ENGINE: Triple expansion
GROSS TONNAGE: 2,820
LENGTH—KEEL: 328 feet
BEAM: 42 feet
DEPTH: 20 feet
HISTORY: Sold in 1913. Sank in 1914 after a collision off Thunder Bay Island in Lake Huron.

VESSEL: *Edwin H. Gott*
REGISTRY: US 600648
YEAR BUILT: 1979
SHIPYARD: Bay Shipbuilding Corp., Sturgeon Bay. Hull No. 718
ENGINE: Twin diesels
GROSS TONNAGE: 35,592
LENGTH—KEEL: 993 feet
LENGTH—OVERALL: 1,004 feet
BEAM: 105 feet
DEPTH: 56 feet
HISTORY: Shuttle boom self-unloader replaced with 280-foot conventional boom during winter of 1995-96. Operational in 1998.

VESSEL: *Governor Miller*
REGISTRY: US 237394
YEAR BUILT: 1938
SHIPYARD: American Ship Building Co., Lorain. Hull No. 810
ENGINE: Steam turbine
GROSS TONNAGE: 8,240
LENGTH—KEEL: 586 feet
LENGTH—OVERALL: 610 feet
BEAM: 60 feet
DEPTH: 32 feet
HISTORY: Sold for scrap in 1980. Towed to Spain that year for scrapping.

VESSEL: *Grecian*
REGISTRY: US 86136
YEAR BUILT: 1890
SHIPYARD: Globe Iron Works, Cleveland. Hull No. 40
ENGINE: Triple expansion
GROSS TONNAGE: 2,348
LENGTH—KEEL: 296 feet
LENGTH—OVERALL:
BEAM: 40 feet
DEPTH: 21 feet
HISTORY: Grounded on June 7, 1906, near Detour, Mich. Released but sank three days later off Thunder Bay Island, Lake Huron, while under tow to a shipyard.

VESSEL: *Griffin*
REGISTRY: US 86140
YEAR BUILT: 1891
SHIPYARD: Cleveland Shipbuilding Co., Cleveland. Hull No. 12
ENGINE: Triple expansion
GROSS TONNAGE: 1,856
LENGTH—KEEL: 266 feet
BEAM: 38 feet
DEPTH: 19 feet
HISTORY: Sold in 1914. Converted to a crane ship in 1918. Cranes removed 1929. Converted to a self-unloading sand dredge in 1938 and renamed *Joseph S. Scobel*. Sold for scrap in 1970 and scrapped in 1971 in Port Colborne, Ontario.

NAME: *Harvard*
REGISTRY: US 96507
YEAR BUILT: 1900
SHIPYARD: Detroit Shipbuilding Co., Wyandotte. Hull No. 134
ENGINE: Quadruple expansion
GROSS TONNAGE: 4,565

LENGTH—KEEL: 461 feet
LENGTH—OVERALL: 474 feet
BEAM: 50 feet
DEPTH: 29 feet
HISTORY: Sold in 1948. Scrapped in 1961 in Cleveland.

VESSEL: *Joshua A. Hatfield*
REGISTRY: US 222798
YEAR BUILT: 1923
SHIPYARD: American Ship Building Co., Lorain. Hull No. 782
ENGINE: Triple expansion
GROSS TONNAGE: 7,940
LENGTH—KEEL: 580 feet
LENGTH—OVERALL: 600 feet
BEAM: 60 feet
DEPTH: 32 feet
HISTORY: Idled in 1977. Sold for scrap in 1983. Scrapped 1988-89 in Duluth.

VESSEL: *James J. Hill*
REGISTRY: US 77409
YEAR BUILT: 1900
SHIPYARD: American Shipbuilding Co., Lorain. Hull No. 38
ENGINE: Quadruple expansion
GROSS TONNAGE: 5,946
LENGTH—KEEL: 478 feet
LENGTH—OVERALL: 497 feet
BEAM: 52 feet
DEPTH: 30 feet
HISTORY: Sold to city of Cleveland in 1962 and sunk as breakwater. Later cut down and covered.

VESSEL: **Alexander Holley**
REGISTRY: US 107237
YEAR BUILT: 1896
SHIPYARD: Superior Steel Barge Co., Superior. Hull No. 139
ENGINE: Barge
GROSS TONNAGE: 2,721
LENGTH—KEEL: 352 feet
LENGTH—OVERALL: 376 feet
BEAM: 46 feet
DEPTH: 26 feet
HISTORY: Sold Canadian in 1926. Sold for scrap in 1965.

NAME: *Clifford F. Hood*
REGISTRY: US 3925
YEAR BUILT: 1902

SHIPYARD: West Bay City Ship Building Co., West Bay City. Hull No. 605
ENGINE: Triple expansion
GROSS TONNAGE: 7,760
LENGTH—KEEL: 414 feet
LENGTH—OVERALL: 434 feet
BEAM: 50 feet
DEPTH: 32 feet
HISTORY: Launched as bulk freighter *Bransford.* Acquired by Pittsburgh Steamship Co. in 1915 and renamed *John H. McLean.* Acquired in 1943 by U.S. Steel subsidiary American Steel and Wire Co. and renamed *Clifford F. Hood.* Deck cranes added in 1944 to handle semi-finished steel products. Transferred back to Pittsburgh fleet in 1964. Sold for scrap in 1973 and towed overseas for scrapping the following year.

VESSEL: *Douglass Houghton*
REGISTRY: US 157552
YEAR BUILT: 1899
SHIPYARD: Globe Iron Works, Cleveland. Hull No. 78
ENGINE: Quadruple expansion
GROSS TONNAGE: 5,332
LENGTH—KEEL: 456 feet
LENGTH—OVERALL: 475 feet
BEAM: 50 feet
DEPTH: 24 feet
HISTORY: Originally one of three distinctive ships in the Pittsburgh Steamship Co. with two smokestacks. Reboilered in 1910 for better economy and one smokestack removed. Sold in 1945 to Upper Lakes and St. Lawrence Transportation Co. Sunk for a breakwater at Toronto in 1969.

VESSEL: *Francis E. House*
REGISTRY: US 203917
YEAR BUILT: 1906
SHIPYARD: Chicago Shipbuilding Co., Chicago. Hull No. 72
ENGINE: Triple expansion
GROSS TONNAGE: 7,769
LENGTH—KEEL: 592 feet
LENGTH—OVERALL: 605 feet
BEAM: 60 feet
DEPTH: 27 feet
HISTORY: Built as *William B. Kerr.* Acquired by Pittsburgh Steamship Co. in 1911 and renamed. Sold in 1966 to Kinsman Transit Co. and renamed *Kinsman Independent.* Sold for scrap in 1974 and towed to Spain that year.

VESSEL: *Colgate Hoyt*
REGISTRY: US 126642

YEAR BUILT: 1890
SHIPYARD: American Steel Barge Co., Superior. Hull No. 106
ENGINE: Triple expansion
GROSS TONNAGE: 1,253
LENGTH—KEEL: 276 feet
BEAM: 36 feet
DEPTH: 19 feet
HISTORY: Sold in 1905 and renamed *Bay City*. Taken to the East Coast in 1906 and renamed *Thurmond*. Wrecked off New Jersey in 1909.

VESSEL: *John Hulst*
REGISTRY: US 237380
YEAR BUILT: 1938
SHIPYARD: Great Lakes Engineering Works, Ecorse. Hull No. 286
ENGINE: Steam turbine
GROSS TONNAGE: 8,302
LENGTH—KEEL: 586 feet
LENGTH—OVERALL: 611 feet
BEAM: 60 feet
DEPTH: 32 feet
HISTORY: Sold for scrap in 1983. Scrapped in 1985 in Thunder Bay, Ontario.

VESSEL: *William A. Irvin*
REGISTRY: US 237395
YEAR BUILT: 1938
SHIPYARD: American Ship Building Co., Lorain. Hull No. 811
ENGINE: Steam turbine
GROSS TONNAGE: 8,240
LENGTH—KEEL: 586 feet
LENGTH—OVERALL: 610 feet
BEAM: 60 feet
DEPTH: 32 feet
HISTORY: Launched in late 1937, the vessel made its maiden voyage on June 25, 1938. Laid up in 1978. Sold to the Duluth Convention Center Board in 1986 for $110,000. Opened in 1986 as a museum in Duluth.

VESSEL: *W. LeBaron Jenney*
REGISTRY: US 81561
YEAR BUILT: 1897
SHIPYARD: F.W. Wheeler and Co., West Bay City. Hull No. 120
ENGINE: Barge
GROSS TONNAGE: 3,422
LENGTH—KEEL: 366 feet
LENGTH—OVERALL: 378 feet
BEAM: 44 feet
DEPTH: 26 feet

HISTORY: Sold in 1936 to Lakehead Transportation. Over the years renamed *Alfred, Alfred J., Collingdoc,* and *K. A. Powell.* Scrapped in 1973.

VESSEL: *Horace Johnson*
REGISTRY: US 228899
YEAR BUILT: 1929
SHIPYARD: American Ship Building Co., Lorain. Hull No. 805
ENGINE: Triple expansion
GROSS TONNAGE: 7,938
LENGTH—KEEL: 580 feet
LENGTH—OVERALL: 604 feet
BEAM: 60 feet
DEPTH: 32 feet
HISTORY: Sold for scrap in 1983. Scrapped in 1984 in Thunder Bay, Ontario.

VESSEL: *Joliet*
REGISTRY: 76873
YEAR BUILT: 1890
SHIPYARD: Cleveland Shipbuilding Co., Cleveland. Hull No. 7
ENGINE: Triple expansion
GROSS TONNAGE: 1,921
LENGTH—KEEL: 266 feet
LENGTH—OVERALL:
BEAM: 38 feet
DEPTH: 20 feet
HISTORY: Collided with fleetmate *Henry Phipps* on Sept. 22, 1911, and sank opposite Port Huron, Mich. Hull removed in 1963.

VESSEL: *D. G. Kerr*
REGISTRY: US 214147
YEAR BUILT: 1916
SHIPYARD: American Ship Building Co., Lorain. Hull No. 714
ENGINE: Triple expansion
GROSS TONNAGE: 8,017
LENGTH—KEEL: 580 feet
LENGTH—OVERALL: 600 feet
BEAM: 60 feet
DEPTH: 32 feet
HISTORY: Sold for scrap in 1980. Sank off the Azores while under tow to Spain.

VESSEL: *Alfred Krupp*
REGISTRY: US 107242
YEAR BUILT: 1896
SHIPYARD: Chicago Shipbuilding Co., Chicago. Hull No. 24
ENGINE: Barge
GROSS TONNAGE: 3,259

LENGTH—KEEL: 352 feet
LENGTH—OVERALL: 365 feet
BEAM: 45 feet
DEPTH: 27 feet
HISTORY: Sold in 1936. Scrapped in 1973.

NAME: *Lafayette*
REGISTRY: US 141857
YEAR BUILT: 1900
SHIPYARD: American Shipbuilding Co., Lorain. Hull No. 301
ENGINE: Quadruple expansion
GROSS TONNAGE: 5,113
LENGTH—KEEL: 454 feet
BEAM: 50 feet
DEPTH: 28 feet
HISTORY: Wrecked on Nov. 28, 1905, about six miles northeast of Two Harbors, Minn. Stern eventually salvaged and brought to Superior, where the engine was removed and installed in the *J. S. Ashley*.

VESSEL: *Thomas W. Lamont*
REGISTRY: US 229753
YEAR BUILT: 1930
SHIPYARD: Toledo Ship Building Co., Toledo. Hull No. 184
ENGINE: Triple expansion
GROSS TONNAGE: 7,964
LENGTH—KEEL: 580 feet
LENGTH—OVERALL: 604 feet
BEAM: 60 feet
DEPTH: 32 feet
HISTORY: Repowered with diesel engine in 1965. Scrapped in 1987 in Turkey.

VESSEL: *LaSalle*
REGISTRY: US 141050
YEAR BUILT: 1889
SHIPYARD: Cleveland Shipbuilding Co., Cleveland. Hull No. 6
ENGINE: Triple expansion
GROSS TONNAGE: 1,921
LENGTH—KEEL: 266 feet
LENGTH—OVERALL: 285 feet
BEAM: 38 feet
DEPTH: 19 feet
HISTORY: Sold in 1914. Renamed *Eastrich* in 1928, *Howard Hindman* in 1943, and *Forestdale* in 1952. Scrapped in 1961.

VESSEL: *Richard V. Lindabury*
REGISTRY: US 222849

YEAR BUILT: 1923
SHIPYARD: American Ship Building Co., Lorain. Hull No. 783
ENGINE: Triple expansion
GROSS TONNAGE: 7,940
LENGTH—KEEL: 586 feet
LENGTH—OVERALL: 600 feet
BEAM: 60 feet
DEPTH: 32 feet
HISTORY: Sold to Kinsman Transit Co. in 1978 and renamed *Kinsman Independent*. Idled in 1987 and scrapped in 1989 in Turkey.

VESSEL: *William R. Linn*
REGISTRY: US 81597
YEAR BUILT: 1898
SHIPYARD: Chicago Shipbuilding Co., Chicago. Hull No. 32
ENGINE: Quadruple expansion
GROSS TONNAGE: 4,328
LENGTH—KEEL: 400 feet
LENGTH—OVERALL: 420 feet
BEAM: 48 feet
DEPTH: 24 feet
HISTORY: Sold in 1940. Converted to a tanker and renamed *L. S. Wescoat*. Scrapped in 1965 in Germany.

VESSEL: *Thomas Lynch*
REGISTRY: US 204212
YEAR BUILT: 1907
SHIPYARD: Chicago Shipbuilding Co., Chicago. Hull No. 73
ENGINE: Triple expansion
GROSS TONNAGE: 7,240
LENGTH—KEEL: 586 feet
LENGTH—OVERALL: 601 feet
BEAM: 58 feet
DEPTH: 27 feet
HISTORY: Sold Canadian in 1965 and renamed *Wiarton*. Sunk for use as a dock in 1973 in Hamilton, Ontario.

VESSEL: *Madeira*
REGISTRY: US 93020
YEAR BUILT: 1900
SHIPYARD: Chicago Shipbuilding Co., Chicago. Hull No. 38
ENGINE: Barge
GROSS TONNAGE: 5,039
LENGTH—KEEL: 436 feet
BEAM: 50 feet
DEPTH: 24 feet
HISTORY: Wrecked on Nov. 28, 1905, at Split Rock, Minn.

NAME: *Magna*
REGISTRY: US 92740
YEAR BUILT: 1896
SHIPYARD: Chicago Shipbuilding Co., Chicago. Hull No. 22
ENGINE: Barge
GROSS TONNAGE: 3,259
LENGTH—KEEL: 352 feet
LENGTH—OVERALL: 366 feet
BEAM: 44 feet
DEPTH: 22 feet
HISTORY: Sold in 1936 to John T. Hutchinson. Sold for use as a breakwater in 1956. Raised and scrapped in 1962 at Ashtabula, Ohio.

VESSEL: *Maia*
REGISTRY: US 92894
YEAR BUILT: 1898
SHIPYARD: Chicago Shipbuilding Co., Chicago. Hull No. 33
ENGINE: Barge
GROSS TONNAGE: 3,804
LENGTH—KEEL: 376 feet
LENGTH—OVERALL: 390 feet
BEAM: 48 feet
DEPTH: 22 feet
HISTORY: Sold by Pittsburgh Steamship Co. in 1936. Sold in 1956 for use as a breakwater. Raised and scrapped in 1962 in Ashtabula, Ohio.

VESSEL: *Maida*
REGISTRY: US 92844
YEAR BUILT: 1898
SHIPYARD: Superior Shipbuilding Co., Superior. Hull No. 142
ENGINE: Barge
GROSS TONNAGE: 3,474
LENGTH—KEEL: 375 feet
LENGTH—OVERALL: 390 feet
BEAM: 46 feet
DEPTH: 22 feet
HISTORY: Sold in 1924. Converted to a self-unloading barge in 1929. Sold for scrap in 1968.

VESSEL: *Malietoa*
REGISTRY: US 92973
YEAR BUILT: 1899
SHIPYARD: Cleveland Shipbuilding Co., Lorain. Hull No. 36
ENGINE: Quadruple expansion
GROSS TONNAGE: 5,229
LENGTH—KEEL: 454 feet
LENGTH—OVERALL: 474 feet

BEAM: 50 feet
DEPTH: 24 feet
HISTORY: Begun as *Texas* but sold to Minnesota Steamship Co. and renamed *Malietoa* while still under construction. Sold in 1954. Scrapped in 1964 in Ashtabula, Ohio.

NAME: *Malta*
REGISTRY: US 92637
YEAR BUILT: 1895
SHIPYARD: Chicago Shipbuilding Co., Chicago. Hull No. 13
ENGINE: Barge
GROSS TONNAGE: 2,237
LENGTH—KEEL: 302 feet
BEAM: 40 feet
DEPTH: 25 feet
HISTORY: Sold in 1912 to Western Dry Dock Co. and renamed *Thunder Bay*. Shortened 54 feet in 1921. Renamed *Pinebranch* in 1937. Converted to a tanker in 1940. Renamed *Empire Stickleback* in 1941. Renamed *Pinebranch* in 1946. Sunk in 1960 as breakwater in Mulgrave, Nova Scotia.

VESSEL: *Manda*
REGISTRY: US 92696
YEAR BUILT: 1896
SHIPYARD: Chicago Shipbuilding Co., Chicago. Hull No. 17
ENGINE: Barge
GROSS TONNAGE: 3,256
LENGTH—KEEL: 352 feet
LENGTH—OVERALL: 366 feet
BEAM: 44 feet
DEPTH: 22 feet
HISTORY: Sold in 1928. Converted to a wrecking lighter by Great Lakes Towing Co. and renamed *Resolute*. Converted to a barge in 1942. Sold for scrap in 1972.

VESSEL: *Manila*
REGISTRY: US 92950
YEAR BUILT: 1899
SHIPYARD: Chicago Shipbuilding Co., Chicago. Hull No. 36
ENGINE: Barge
GROSS TONNAGE: 5,039
LENGTH—KEEL: 436 feet
LENGTH—OVERALL: 450 feet
BEAM: 50 feet
DEPTH: 24 feet
HISTORY: Sold by Pittsburgh Steamship Co. in 1940. Sold off the lakes in 1956.

NAME: *Manola*
REGISTRY: US 92170
YEAR BUILT: 1890
SHIPYARD: Globe Iron Works, Cleveland. Hull No. 30
ENGINE: Triple expansion
GROSS TONNAGE: 2,326
LENGTH—KEEL: 282 feet
BEAM: 40 feet
DEPTH: 21 feet
HISTORY: Sold in 1918 to Emergency Fleet Corp. of the U.S. Government. Vessel was cut in two for transfer to the East Coast. Forward half sank on Dec. 3, 1918, near Duck Island in Lake Ontario. Stern used in building steamer *Mapledawn*. That vessel wrecked on Nov. 30, 1924, near Midland, Ontario.

VESSEL: *Marcia*
REGISTRY: US 92638
YEAR BUILT: 1895
SHIPYARD: Chicago Shipbuilding Co., Chicago. Hull No. 12
ENGINE: Barge
GROSS TONNAGE: 2,237
LENGTH—KEEL: 302 feet
LENGTH—OVERALL: 318 feet
BEAM: 40 feet
DEPTH: 20 feet
HISTORY: Sold in 1923. Renamed *Mitschfibre* in 1924. Sold for scrap in 1966.

VESSEL: *Maricopa*
REGISTRY: US 92736
YEAR BUILT: 1896
SHIPYARD: Chicago Shipbuilding Co., Chicago. Hull No. 21
ENGINE: Triple expansion
GROSS TONNAGE: 4,223
LENGTH—KEEL: 406 feet
LENGTH—OVERALL: 425 feet
BEAM: 48 feet
DEPTH: 23 feet
HISTORY: Sold in 1936 to Geistman Transportation Co. and renamed *John P. Geistman* in 1937. Renamed *E. E. Johnson* in 1941, *Altadoc* in 1945, and *D. B. Weldon* in 1962. Used for grain storage at Goderich, Ontario, in 1962. Scrapped in 1975 at Thunder Bay, Ontario.

VESSEL: *Marina*
REGISTRY: US 92282
YEAR BUILT: 1891
SHIPYARD: Chicago Shipbuilding Co., Chicago. Hull No. 1

ENGINE: Triple expansion
GROSS TONNAGE: 2,432
LENGTH—KEEL: 292 feet
BEAM: 40 feet
DEPTH: 21 feet
HISTORY: Sold in 1912 and renamed *George A. Graham*. Wrecked on Oct. 7, 1917, on Manitoulin Island. Salvaged for scrap in 1937.

VESSEL: *Mariposa*
REGISTRY: US 92448
YEAR BUILT: 1892
SHIPYARD: Globe Iron Works, Cleveland. Hull No. 48
ENGINE: Triple expansion
GROSS TONNAGE: 2,831
LENGTH—KEEL: 330 feet
BEAM: 45 feet
DEPTH: 20 feet
HISTORY: Sold in 1929. Laid up after 1944 and scrapped in 1947 in Hamilton, Ontario.

NAME: *Mariska*
REGISTRY: US 92189
YEAR BUILT: 1890
SHIPYARD: Globe Iron Works, Cleveland. Hull No. 31
ENGINE: Triple expansion
GROSS TONNAGE: 2,325
LENGTH—KEEL: 282 feet
BEAM: 40 feet
DEPTH: 22 feet
HISTORY: Sold in 1916 to Basset Steamship Co. Renamed *Kamaris* in 1923 and *Quedoc* in 1926. Converted to a barge and renamed *H.S. & G. No. IX* in 1959. Scrapped in 1961.

VESSEL: *Maritana*
REGISTRY: US 92459
YEAR BUILT: 1892
SHIPYARD: Chicago Shipbuilding Co., Chicago. Hull No. 4
ENGINE: Triple expansion
GROSS TONNAGE: 2,957
LENGTH—KEEL: 330 feet
LENGTH—OVERALL: 348 feet
BEAM: 45 feet
DEPTH: 20 feet
HISTORY: Sold in 1929. Scrapped in 1947 in Hamilton, Ontario.

VESSEL: *Marsala*
REGISTRY: US 93021

YEAR BUILT: 1900
SHIPYARD: Chicago Shipbuilding Co., Chicago. Hull No. 39
ENGINE: Barge
GROSS TONNAGE: 5,039
LENGTH—KEEL: 436 feet
LENGTH—OVERALL: 450 feet
BEAM: 50 feet
DEPTH: 24 feet
HISTORY: Sold in 1939. Sold for off-lakes use in 1956.

NAME: *Martha*
REGISTRY: US 92697
YEAR BUILT: 1896
SHIPYARD: Chicago Shipbuilding Co., Chicago. Hull No. 18
ENGINE: Barge
GROSS TONNAGE: 3,256
LENGTH—KEEL: 352 feet
BEAM: 44 feet
DEPTH: 22 feet
HISTORY: Sold in 1936 and renamed *Florence* the following year. Renamed *Maureen H.* in 1938, *Florence J.* in 1939, and *Owendoc* in 1949. Scrapped in 1977.

VESSEL: *Maruba*
REGISTRY: US 92191
YEAR BUILT: 1890
SHIPYARD: Globe Iron Works, Cleveland. Hull No. 34
ENGINE: Triple expansion
GROSS TONNAGE: 2,311
LENGTH—KEEL: 290 feet
LENGTH—OVERALL: 320 feet
BEAM: 40 feet
DEPTH: 21 feet
HISTORY: Sold by Pittsburgh Steamship Co. in 1912. Sold saltwater in 1917. Returned to the lakes in 1925 and scrapped 1934.

VESSEL: *Masaba*
REGISTRY: US 92298
YEAR BUILT: 1891
SHIPYARD: Chicago Shipbuilding Co., Chicago. Hull No. 2
ENGINE: Triple expansion
GROSS TONNAGE: 2.432
LENGTH—KEEL: 240 feet
BEAM: 40 feet
DEPTH: 21 feet
HISTORY: Sold in 1912 and shortened 52 feet. Sent saltwater in 1918. Renamed *General Henrys* in 1920. Out of registry and presumed scrapped in 1924.

NAME: *Mataafa*
REGISTRY: US 150810
YEAR BUILT: 1899
SHIPYARD: Cleveland Shipbuilding Co., Cleveland. Hull No. 33
ENGINE: Quadruple expansion
GROSS TONNAGE: 4,840
LENGTH—KEEL: 429 feet
BEAM: 50 feet
DEPTH: 25 feet
HISTORY: Begun as *Pennsylvania*. Acquired by the Minnesota Steamship Company while still under construction and renamed *Mataafa*. Wrecked on Nov. 25, 1905, at Duluth with loss of nine lives. Rebuilt and returned to service. Sold by Pittsburgh Steamship Co. in 1946. Scrapped in 1965 in Germany.

VESSEL: *Samuel Mather*
REGISTRY: US 116484
YEAR BUILT: 1892
SHIPYARD: American Steel Barge Co., Superior. Hull No. 120
ENGINE: Triple expansion
GROSS TONNAGE: 1,713
LENGTH—KEEL: 308 feet
LENGTH—OVERALL: 328 feet
BEAM: 38 feet
DEPTH: 24 feet
HISTORY: Sold in 1922 and renamed *Clifton*. Lost with all hands on Sept. 22, 1924, on Lake Huron.

VESSEL: *Matoa*
REGISTRY: US 92204
YEAR BUILT: 1890
SHIPYARD: Globe Iron Works, Cleveland. Hull No. 34
ENGINE: Triple expansion
GROSS TONNAGE: 2,311
LENGTH—KEEL: 290 feet
BEAM: 40 feet
DEPTH: 21 feet
HISTORY: Abandoned to insurance underwriters after stranding near Harbor Beach, Mich., during the Nov. 8-13 storm of 1913. Vessel was salvaged and taken to the East Coast in 1915. Sold Canadian and returned to the lakes in 1922. Renamed *Glenrig* in 1923 and *Huguenot* in 1926. Scrapped in 1937.

VESSEL: *Maunaloa*
REGISTRY: US 92974
YEAR BUILT: 1899

SHIPYARD: Chicago Shipbuilding Co., South Chicago. Hull No. 37
ENGINE: Quadruple expansion
GROSS TONNAGE: 4,951
LENGTH—KEEL: 430 feet
BEAM: 40 feet
DEPTH: 21 feet
HISTORY: Built as *Tennessee*. Acquired by Minnesota Steamship Co. while under construction and renamed *Maunaloa*. Sold Canadian in 1945 and renamed *Maunaloa II*. Scrapped in 1972 in Hamilton, Ontario.

VESSEL: *Alexander McDougall*
REGISTRY: US 107372
YEAR BUILT: 1898
SHIPYARD: American Steel Barge Co., Superior. Hull No. 141
ENGINE: Quadruple expansion
GROSS TONNAGE: 3,686
LENGTH—KEEL: 413 feet
LENGTH—OVERALL: 433 feet
BEAM: 50 feet
DEPTH: 22 feet
HISTORY: Sold in 1936 to Buckeye Steamship Co. Scrapped in 1947 at Hamilton, Ontario.

VESSEL: *William A. McGonagle*
REGISTRY: US 214234
YEAR BUILT: 1916
SHIPYARD: Great Lakes Engineering Works, Ecorse. Hull No. 154
ENGINE: Triple expansion
GROSS TONNAGE: 7,041
LENGTH—KEEL: 580 feet
LENGTH—OVERALL: 600 feet
BEAM: 60 feet
DEPTH: 32 feet
HISTORY: Sold to Kinsman Transit Co. in 1978. Renamed *Henry Steinbrenner* in 1986. Scrapped in 1994 in Port Maitland, Ontario.

NAME: *Pentecost Mitchell*
REGISTRY: US 96663
YEAR BUILT: 1903
SHIPYARD: American Ship Building Co., Lorain. Hull No. 321
ENGINE: Triple expansion
GROSS TONNAGE: 4,644
LENGTH—KEEL: 414 feet
LENGTH—OVERALL: 434 feet
BEAM: 50 feet
DEPTH: 28 feet

HISTORY: Launched as *H. B. Hawgood*. Purchased by Pittsburgh Steamship Co. in 1916 and renamed *Pentecost Mitchell*. Sold in 1944. Scrapped in 1948.

VESSEL: *J. Pierpont Morgan*
REGISTRY: US 203155
YEAR BUILT: 1906
SHIPYARD: Chicago Shipbuilding Co., Chicago. Hull No. 68
ENGINE: Triple expansion
GROSS TONNAGE: 7,161
LENGTH—KEEL: 586 feet
LENGTH—OVERALL: 601 feet
BEAM: 58 feet
DEPTH: 27 feet
HISTORY: Sold Canadian in 1965 and renamed *Heron Bay* in 1966. Renamed *Heron B.* in 1978. Scrapped in 1979.

NAME: *J. P. Morgan Jr.*
REGISTRY: US 207250
YEAR BUILT: 1909
SHIPYARD: American Ship Building Co., Lorain. Hull No. 373
ENGINE: Triple expansion
GROSS TONNAGE: 7,694
LENGTH—KEEL: 580 feet
LENGTH—OVERALL: 601 feet
BEAM: 58 feet
DEPTH: 32 feet
HISTORY: Sold for scrap in 1980.

NAME: *Samuel F. B. Morse*
REGISTRY: US 116841
YEAR BUILT: 1898
SHIPYARD: F. W. Wheeler and Co., West Bay City. Hull No. 124
ENGINE: Quadruple expansion
GROSS TONNAGE: 4,936
LENGTH—KEEL: 456 feet
LENGTH—OVERALL: 476 feet
BEAM: 50 feet
DEPTH: 29 feet
HISTORY: Sold in 1954. Converted to a barge and renamed *WY Chem 105*. Used as a construction barge until scrapped in 1982.

VESSEL: *John G. Munson*
REGISTRY: US 264136
YEAR BUILT: 1952
SHIPYARD: Manitowoc Ship Building Co., Manitowoc. Hull No. 415
ENGINE: Steam turbine

GROSS TONNAGE: 13,143
LENGTH—KEEL: 640 feet
LENGTH—OVERALL: 666 feet
BEAM: 72 feet
DEPTH: 36 feet
HISTORY: Lengthened to 768 feet in 1976. Operational in 1998.

VESSEL: *Simon J. Murphy*
REGISTRY: US 116987
YEAR BUILT: 1900
SHIPYARD: Detroit Shipbuilding Co., Wyandotte. Hull No. 135
ENGINE: Triple expansion
GROSS TONNAGE: 4,869
LENGTH—KEEL: 435 feet
LENGTH—OVERALL: 451 feet
BEAM: 51 feet
DEPTH: 23 feet
HISTORY: Purchased in 1902 by Donora Iron Mines Company, a U.S. Steel subsidiary, and operated by Pittsburgh Steamship Co. until the fleet acquired it in 1904. Sold in 1940. Scrapped in 1961.

VESSEL: *James Nasmyth*
REGISTRY: US 77231
YEAR BUILT: 1896
SHIPYARD: F.W. Wheeler Co., West Bay City. Hull No. 117
ENGINE: Barge
GROSS TONNAGE: 3,423
LENGTH—KEEL: 365 feet
LENGTH—OVERALL: 382 feet
BEAM: 45 feet
DEPTH: 23 feet
HISTORY: Sold Canadian in 1936 and renamed *Merle H.* Renamed *Pic River* in 1949. Rebuilt as diesel-driven ship in 1953. Renamed *Pic R.* in 1978. Scrapped in 1984 in Hamilton, Ontario.

VESSEL: *James B. Neilson*
REGISTRY: US 81373
YEAR BUILT: 1891
SHIPYARD: American Steel Barge Co., Superior. Hull No. 124
ENGINE: Triple expansion
GROSS TONNAGE: 2,234
LENGTH—KEEL: 320 feet
LENGTH—OVERALL: 335 feet
BEAM: 42 feet
DEPTH: 25 feet
HISTORY: Originally named *Washburn*. Acquired by Bessemer Steamship Co.

and renamed *James B. Neilson*. Sold in 1927 to Spokane Steamship Co. Scrapped in 1936.

NAME: *Ojibway*
REGISTRY: US 251149
YEAR BUILT: 1946
SHIPYARD: Great Lakes Engineering Works, Ashtabula.
ENGINE: Diesel
GROSS TONNAGE: 53
LENGTH—KEEL: 60 feet
LENGTH—OVERALL: 64 feet
BEAM: 22 feet
DEPTH: 8 feet
HISTORY: Built for Pittsburgh Steamship Co. in 1947 to supply vessels passing through the Soo Locks. Operational in 1998.

VESSEL: *William J. Olcott*
REGISTRY: US 207790
YEAR BUILT: 1910
SHIPYARD: Great Lakes Engineering Works, Ecorse. Hull No. 74
ENGINE: Triple expansion
GROSS TONNAGE: 7,568
LENGTH—KEEL: 588 feet
LENGTH—OVERALL: 601 feet
BEAM: 58 feet
DEPTH: 28 feet
HISTORY: Sold in 1968. Renamed *George E. Seedhouse* in 1970. Sold for scrap in 1975 but converted to a floating workshop at Sturgeon Bay, Wis., and renamed *Bld. 315*. Shortened to 300 feet in 1988.

VESSEL: *Irving S. Olds*
REGISTRY: US 242261
YEAR BUILT: 1942
SHIPYARD: American Ship Building Co., Lorain. Hull No. 825
ENGINE: Steam turbine
GROSS TONNAGE: 10,294
LENGTH—KEEL: 614 feet
LENGTH—OVERALL: 639 feet
BEAM: 67 feet
DEPTH: 35 feet
HISTORY: Laid up in 1981. Sold for scrap in 1988 and towed to Taiwan.

VESSEL: *William P. Palmer*
REGISTRY: US 81705
YEAR BUILT: 1900
SHIPYARD: American Shipbuilding Co., Cleveland. Hull No. 400

ENGINE: Triple expansion
GROSS TONNAGE: 2,293
LENGTH—KEEL: 242 feet
BEAM: 42 feet
DEPTH: 27 feet
HISTORY: Sold in 1909. Wrecked in 1936 at Fairport, Ohio. Salvaged and scrapped.

NAME: *William P. Palmer*
REGISTRY: US 208157
YEAR BUILT: 1910
SHIPYARD: Great Lakes Engineering Works, Ecorse. Hull No. 76
ENGINE: Triple expansion
GROSS TONNAGE: 7,602
LENGTH—KEEL: 580 feet
LENGTH—OVERALL: 601 feet
BEAM: 58 feet
DEPTH: 32 feet
HISTORY: Retired in 1971. Sold for scrap in 1978 and scrapped in 1979 in Duluth.

VESSEL: *Eugene W. Pargny*
REGISTRY: US 214747
YEAR BUILT: 1917
SHIPYARD: American Ship Building Co., Lorain. Hull No. 719
ENGINE: Triple expansion
GROSS TONNAGE: 7,724
LENGTH—KEEL: 580 feet
LENGTH—OVERALL: 600 feet
BEAM: 60 feet
DEPTH: 32 feet
HISTORY: Repowered with diesel engine in 1951. Sold for scrap in 1984 and scrapped in Duluth in 1988.

NAME: *George W. Perkins*
REGISTRY: US 202166
YEAR BUILT: 1905
SHIPYARD: Superior Shipbuilding Co., Superior. Hull No. 512
ENGINE: Triple expansion
GROSS TONNAGE: 6,553
LENGTH—KEEL: 558 feet
LENGTH—OVERALL: 569 feet
BEAM: 56 feet
DEPTH: 26 feet
HISTORY: Sold in 1964 to Redwood Enterprises and renamed *Westdale*. Renamed *H. C. Heimbecker* in 1977. Scrapped in 1981 in Ashtabula, Ohio.

NAME: *Henry Phipps*
REGISTRY: US 204136
YEAR BUILT: 1907
SHIPYARD: West Bay City Shipbuilding Co., West Bay City. Hull No. 623
ENGINE: Triple expansion
GROSS TONNAGE: 7,703
LENGTH—KEEL: 580 feet
LENGTH—OVERALL: 601 feet
BEAM: 58 feet
DEPTH: 32 feet
HISTORY: Sold for scrap in 1976 and scrapped in Duluth in 1978.

NAME: *General Orlando M. Poe*
REGISTRY: US 86501
YEAR BUILT: 1903
SHIPYARD: American Ship Building Co., Lorain. Hull No. 81
ENGINE: Quadruple expansion
GROSS TONNAGE: 4,835
LENGTH—KEEL: 470 feet
LENGTH—OVERALL: 490 feet
BEAM: 50 feet
DEPTH: 29 feet
HISTORY: Sold in 1954. Converted to a barge and renamed *WY Chem 104*. Scrapped in 1956 in Hamilton, Ontario.

VESSEL: *Presque Isle*
REGISTRY: US 553416 (tug); US 553417 (barge)
YEAR BUILT: 1973 (tug); 1972 (barge bow); 1973 (barge hull)
SHIPYARD: Halter Marine Services Inc., New Orleans, La. (tug); Erie Marine Inc., Erie, Pa. (barge hull); Defoe Shipbuilding Company, Bay City, Mich. (barge bow)
ENGINE: Twin diesels
GROSS TONNAGE: 1,578 (tug); 22,621 (barge)
LENGTH—OVERALL: 141 feet (tug); 974 feet (barge); 1,000 feet combined
BEAM: 105 feet
DEPTH: 46 feet
HISTORY: Leased by Great Lakes Fleet in 1975. Vessel's management company acquired by the fleet in 1997 and vessel itself purchased by the fleet in 1998. Operational in 1998.

VESSEL: *Princeton*
REGISTRY: US 150876
YEAR BUILT: 1900
SHIPYARD: American Shipbuilding Co., Lorain. Hull No. 302
ENGINE: Quadruple expansion

GROSS TONNAGE: 5,125
LENGTH—KEEL: 454 feet
LENGTH—OVERALL: 474 feet
BEAM: 50 feet
DEPTH: 28 feet
HISTORY: Sold in 1945. Used as a construction barge in 1953. Sunk as breakwall at Burns Harbor, Ind., and possibly later scrapped.

NAME: *Queen City*
REGISTRY: US 20612
YEAR BUILT: 1896
SHIPYARD: Cleveland Shipbuilding Co., Cleveland. Hull No. 24
ENGINE: Triple expansion
GROSS TONNAGE: 4,070
LENGTH—KEEL: 401 feet
LENGTH—OVERALL: 425 feet
BEAM: 48 feet
DEPTH: 28 feet
HISTORY: Sold in 1944. Scrapped in 1947 at Hamilton, Ontario.

VESSEL: *Norman B. Ream*
REGISTRY: US 203543
YEAR BUILT: 1906
SHIPYARD: Chicago Shipbuilding Co., Chicago. Hull No. 70
ENGINE: Triple expansion
GROSS TONNAGE: 7,053
LENGTH—KEEL: 587 feet
BEAM: 58 feet
DEPTH: 28 feet
HISTORY: Sold in 1965 to Kinsman Transit Co. Renamed *Kinsman Enterprise*. Renamed *Hull No. 1* in 1979. Scrapped in 1989 in Turkey.

NAME: *Rensselaer*
REGISTRY: US 111302
YEAR BUILT: 1900
SHIPYARD: American Shipbuilding Co., Cleveland. Hull No. 402
ENGINE: Quadruple expansion
GROSS TONNAGE: 5,124
LENGTH—KEEL: 454 feet
LENGTH—OVERALL: 474 feet
BEAM: 50 feet
DEPTH: 28 feet
HISTORY: Sold in 1944. Scrapped in 1947 in Hamilton, Ontario.

VESSEL: *R. R. Richardson*
REGISTRY: US 77532

YEAR BUILT: 1902
SHIPYARD: American Shipbuilding Co., Lorain. Hull No. 317
ENGINE: Triple expansion
GROSS TONNAGE: 4,644
LENGTH—KEEL: 414 feet
LENGTH—OVERALL: 434 feet
BEAM: 50 feet
DEPTH: 28 feet
HISTORY: Built as *J. M. Jenks*. Purchased by Pittsburgh Steamship Co. in 1916 and renamed *R. R. Richardson*. Sold in 1940. Renamed *Ralph S. Caulkins* in 1944 and *Poweraux Peter* in 1963. Scrapped in 1965 in Germany.

VESSEL: *Percival Roberts Jr.*
REGISTRY: US 211025
YEAR BUILT: 1913
SHIPYARD: American Ship Building Co., Lorain. Hull No. 398
ENGINE: Triple expansion
GROSS TONNAGE: 7,748
LENGTH—KEEL: 580 feet
LENGTH—OVERALL: 600 feet
BEAM: 58 feet
DEPTH: 32 feet
HISTORY: Laid up in 1971. Sold for scrap in 1978 and scrapped in Duluth in 1980.

VESSEL: *T. W. Robinson*
REGISTRY: US 224836
YEAR BUILT: 1925
SHIPYARD: American Ship Building Co., Lorain. Hull No. 790
ENGINE: Steam turbine, electric motor.
GROSS TONNAGE: 7,726
LENGTH—KEEL: 566 feet
LENGTH—OVERALL: 588 feet
BEAM: 60 feet
DEPTH: 32 feet
HISTORY: Idled in 1982. Scrapped in Brazil in 1987.

NAME: *Frank Rockefeller*
REGISTRY: US 121015
YEAR BUILT: 1896
SHIPYARD: American Steel Barge Co., Superior. Hull No. 136
ENGINE: Triple expansion
GROSS TONNAGE: 2,759
LENGTH—KEEL: 366 feet
LENGTH—OVERALL: 380 feet
BEAM: 45 feet

DEPTH: 26 feet
HISTORY: Sold in 1927 to Central Dredging Co. and renamed *South Park*. Wrecked on Nov. 15, 1942, at Manistique, Mich. Salvaged and converted to a tanker in 1943 and renamed *Meteor*. Converted to a museum in 1972 in Superior, Wis.

VESSEL: *John A. Roebling*
REGISTRY: US 77310
YEAR BUILT: 1898
SHIPYARD: F.W. Wheeler and Co., West Bay City. Hull No. 126
ENGINE: Barge
GROSS TONNAGE: 4,693
LENGTH—KEEL: 436 feet
LENGTH—OVERALL: 450 feet
BEAM: 50 feet
DEPTH: 24 feet
HISTORY: Sold in 1945. Later used as a grain storage barge before being sold for use as breakwall in 1968.

NAME: *Henry H. Rogers*
REGISTRY: US 203332
YEAR BUILT: 1906
SHIPYARD: Chicago Ship Building Co., Chicago. Hull No. 69
ENGINE: Triple expansion
GROSS TONNAGE: 7,718
LENGTH—KEEL: 580 feet
LENGTH—OVERALL: 601 feet
BEAM: 58 feet
DEPTH: 32 feet
HISTORY: Sold for scrap in 1974 and scrapped in Duluth in 1975.

VESSEL: *Rogers City*
REGISTRY: US 223352
YEAR BUILT: 1923
SHIPYARD: American Ship Building Co., Lorain. Hull No. 787
ENGINE: Triple expansion
GROSS TONNAGE: 7,318
LENGTH—KEEL: 530 feet
LENGTH—OVERALL: 552 feet
BEAM: 60 feet
DEPTH: 32 feet
HISTORY: Built as *B. H. Taylor*. Repowered with steam turbine in 1955. Renamed *Rogers City* in 1957. Sold for scrap in 1987.

VESSEL: *Roman*
REGISTRY: US 110932

YEAR BUILT: 1890
SHIPYARD: Globe Iron Works, Cleveland. Hull No. 41
ENGINE: Triple expansion
GROSS TONNAGE: 2,348
LENGTH—KEEL: 296 feet
BEAM: 40 feet
DEPTH: 21 feet
HISTORY: Sold saltwater in 1918 and renamed *Libertas* in 1919. Sank off New Jersey in 1919. Salvaged and scrapped in 1924.

VESSEL: *John Scott Russell*
REGISTRY: 53256
YEAR BUILT: 1889
SHIPYARD: American Steel Barge Co., Superior. Hull No. 103
ENGINE: Barge
GROSS TONNAGE: 1,192
LENGTH—KEEL: 253 feet
BEAM: 36 feet
DEPTH: 19 feet
HISTORY: Built as 103. Sold in 1905. Later renamed *Berkshire*. Sank off New Jersey in 1909.

VESSEL: *Sandy*
YEAR BUILT: 1940
ENGINE: Gas
GROSS TONNAGE: Not available
LENGTH—KEEL: 38 feet
BEAM: 11 feet
DEPTH: 6 feet
HISTORY: Acquired by Pittsburgh Steamship Co. in 1941 as a supply boat. Sold in 1952.

VESSEL: *Saxon*
REGISTRY: US 116376
YEAR BUILT: 1890
SHIPYARD: Globe Iron Works, Cleveland. Hull No. 37
ENGINE: Triple expansion
GROSS TONNAGE: 2,348
LENGTH—KEEL: 296 feet
BEAM: 40 feet
DEPTH: 21 feet
HISTORY: Sold saltwater in 1918. Renamed *Amasis* in 1922 and *Anne Jensen* in 1924. Scrapped in 1927 in Denmark.

NAME: *William B. Schiller*
REGISTRY: US 207252

YEAR BUILT: 1910
SHIPYARD: American Ship Building Co., Lorain. Hull No. 372
ENGINE: Triple expansion
GROSS TONNAGE: 7,739
LENGTH—KEEL: 580 feet
LENGTH—OVERALL: 601 feet
BEAM: 58 feet
DEPTH: 32 feet
HISTORY: Laid up in 1974. Sold for scrap in 1978. Scrapping begun in 1983 in Duluth and completed in 1988.

VESSEL: *Howard L. Shaw*
REGISTRY: US 96524
YEAR BUILT: 1900
SHIPYARD: Detroit Ship Building Co., Wyandotte. Hull No. 136
ENGINE: Triple expansion
GROSS TONNAGE: 4,241
LENGTH—KEEL: 428 feet
LENGTH—OVERALL: 436 feet
BEAM: 51 feet
DEPTH: 28 feet
HISTORY: Purchased in 1902 by Donora Iron Mines Company, a U.S. Steel subsidiary, and operated by Pittsburgh Steamship Co. until the fleet acquired it in 1905. Sold Canadian in 1940. Sunk as a breakwall at Toronto in 1969.

VESSEL: *MacGilvray Shiras*
REGISTRY: US 200747
YEAR BUILT: 1904
SHIPYARD: American Shipbuilding Co., Cleveland. Hull No. 420
ENGINE: Triple expansion
GROSS TONNAGE: 4,803
LENGTH—KEEL: 420 feet
LENGTH—OVERALL: 440 feet
BEAM: 50 feet
DEPTH: 28 feet
NAMES: Launched as *Umbria*. Purchased by Pittsburgh Steamship Co. in 1916 and renamed *MacGilvray Shiras*. Sold in 1944. Damaged in Buffalo in 1959 and scrapped that year.

VESSEL: *Sir William Siemens*
REGISTRY: US 116732
YEAR BUILT: 1896
SHIPYARD: Globe Iron Works, Cleveland. Hull No. 67
ENGINE: Triple expansion
GROSS TONNAGE: 4,344
LENGTH—KEEL: 413 feet

LENGTH—OVERALL: 432 feet
BEAM: 48 feet
DEPTH: 24 feet
HISTORY: Sold in 1928 and renamed *William B. Pilkey* in 1929. Renamed *Frank E. Vigor* in 1941. Sank on April 27, 1944, following a collision near Southeast Shoal Light in Lake Erie.

VESSEL: *George A. Sloan*
REGISTRY: US 243416
YEAR BUILT: 1943
SHIPYARD: Great Lakes Engineering Works, Ecorse. Hull No. 292
ENGINE: Triple expansion
GROSS TONNAGE: 9,706
LENGTH—KEEL: 595 feet
LENGTH—OVERALL: 620 feet
BEAM: 60 feet
DEPTH: 35 feet
HISTORY: Built for U.S. Maritime Commission as *Hill Annex*. Acquired by Pittsburgh Steamship Co. in 1943 and renamed *George A. Sloan*. Acquired by Bradley Division in 1966 and converted to a self-unloader in 1967. Repowered with diesel engine in 1984-85. Operational in 1998.

VESSEL: *John Smeaton*
REGISTRY: US 77358
YEAR BUILT: 1899
SHIPYARD: American Steel Barge Co., Superior. Hull No. 143
ENGINE: Barge
GROSS TONNAGE: 5,049
LENGTH—KEEL: 458 feet
LENGTH—OVERALL: 461 feet
BEAM: 50 feet
DEPTH: 25 feet
HISTORY: Sold by Pittsburgh Steamship Co. in 1940. Sold for off-lakes use in 1956.

VESSEL: *Edgar B. Speer*
REGISTRY: US 521104
YEAR BUILT: 1980
SHIPYARD: American Shipbuilding Co., Lorain. Hull No. 908
ENGINE: Twin diesels
GROSS TONNAGE: 34,620
LENGTH—KEEL: 993 feet
LENGTH—OVERALL: 1,004 feet
BEAM: 105 feet
DEPTH: 54 feet
HISTORY: Operational in 1998.

VESSEL: *Robert C. Stanley*
REGISTRY: US 243843
YEAR BUILT: 1943
SHIPYARD: Great Lakes Engineering Works, Ecorse. Hull No. 294
ENGINE: Triple expansion
GROSS TONNAGE: 9,057
LENGTH—KEEL: 595 feet
LENGTH—OVERALL: 621 feet
BEAM: 60 feet
DEPTH: 35 feet
HISTORY: Laid up in 1981 and sold for scrap in 1987.

VESSEL: *George Stephenson*
REGISTRY: US 86367
YEAR BUILT: 1896
SHIPYARD: F.W. Wheeler and Co., West Bay City. Hull No. 116
ENGINE: Triple expansion
GROSS TONNAGE: 4,564
LENGTH—KEEL: 407 feet
LENGTH—OVERALL: 432 feet
BEAM: 49 feet
DEPTH: 24 feet
HISTORY: Sold in 1939. Used for grain storage in 1959 and scrapped in 1963.

NAME: *Herman C. Strom*
REGISTRY: US 200333
YEAR BUILT: 1903
SHIPYARD: Superior Shipbuilding Co., Superior. Hull No. 511
ENGINE: Triple expansion
GROSS TONNAGE: 4,803
LENGTH—KEEL: 420 feet
LENGTH—OVERALL: 440 feet
BEAM: 50 feet
DEPTH: 28 feet
HISTORY: Built for Hawgood Brothers as *Wisconsin*. Purchased by Pittsburgh Steamship Co. in 1916 and renamed *A. F. Harvey*. Renamed *A. H. Ferbert* in 1927 and *Herman C. Strom* in 1942. Sold in 1944. Scrapped in 1946.

VESSEL: *Superior*
REGISTRY: US 116357
YEAR BUILT: 1890
SHIPYARD: Cleveland Ship Building Co., Cleveland. Hull No. 10
ENGINE: Fore and aft
GROSS TONNAGE: 251
LENGTH—KEEL: 98 feet
BEAM: 30 feet

DEPTH: 10 feet
HISTORY: Built as wooden ferry for S.T. Norrell of Superior, Wis. Acquired by Pittsburgh Steamship Co. in 1902 and converted into a supply boat at Sault Ste. Marie. Sold to Pringle Barge Line Co. in 1917 and converted to a tug. Burned at Stag Island in the St. Clair River on May 6, 1920.

NAME: *Superior City*
REGISTRY: US 116820
YEAR BUILT: 1898
SHIPYARD: American Shipbuilding Co., Lorain. Hull No. 29
ENGINE: Quadruple expansion
GROSS TONNAGE: 4,795
LENGTH—KEEL: 429 feet
LENGTH—OVERALL: 450 feet
BEAM: 50 feet
DEPTH: 28 feet
HISTORY: Collided with steamer *Willis L. King* on Aug. 20, 1920, about five miles southeast of Whitefish Point. Sank with loss of 29 lives.

VESSEL: *Myron C. Taylor*
REGISTRY: US 228960
YEAR BUILT: 1929
SHIPYARD: Great Lakes Engineering Works, Ecorse. Hull No. 269
ENGINE: Triple expansion
GROSS TONNAGE: 8,233
LENGTH—KEEL: 588 feet
LENGTH—OVERALL: 604 feet
BEAM: 60 feet
DEPTH: 32 feet
HISTORY: Built for Pittsburgh Steamship Co. Transferred to Bradley Division and converted to self-unloader in 1956. Repowered with diesel engine in 1968. Operational in 1998.

VESSEL: *Eugene P. Thomas*
REGISTRY: US 229306
YEAR BUILT: 1930
SHIPYARD: Great Lakes Engineering Works, Ecorse. Hull No. 274
ENGINE: Triple expansion
GROSS TONNAGE: 7,860
LENGTH—KEEL: 580 feet
LENGTH—OVERALL: 603 feet
BEAM: 60 feet
DEPTH: 32 feet
HISTORY: Repowered with diesel engine in 1963. Laid up in 1981. Scrapped in 1985.

VESSEL: *Sidney G. Thomas*
REGISTRY: US 116764
YEAR BUILT: 1897
SHIPYARD: Globe Iron Works, Cleveland. Hull No. 68
ENGINE: Barge
GROSS TONNAGE: 3,200
LENGTH—KEEL: 366 feet
LENGTH—OVERALL: 378 feet
BEAM: 44 feet
DEPTH: 26 feet
HISTORY: Sold in 1929. Renamed *Swederope* in 1940. Sold for scrap in 1967.

VESSEL: *A. D. Thompson*
REGISTRY: US 106834
YEAR BUILT: 1891
SHIPYARD: American Steel Barge Co., Superior. Hull No. 114
ENGINE: Triple expansion
GROSS TONNAGE: 1,300
LENGTH—KEEL: 265 feet
BEAM: 38 feet
DEPTH: 24 feet
HISTORY: Sold saltwater in 1905 and renamed *Bay View*. Returned to the lakes in 1922 and converted to a sand dredge. Scrapped in 1936 at Chicago.

VESSEL: *John B. Trevor*
REGISTRY: US 77173
YEAR BUILT: 1895
SHIPYARD: American Steel Barge Co., Superior. Hull No. 135
ENGINE: Triple expansion
GROSS TONNAGE: 1,713
LENGTH—KEEL: 308 feet
BEAM: 38 feet
DEPTH: 24 feet
HISTORY: Abandoned by Pittsburgh Steamship Co. to insurance underwriters after stranding at Isle Royale on Oct. 13, 1909. Eventually rebuilt and operated as *Atikokan*. Sent to the ocean in 1918. Scrapped in 1935 in Halifax, Nova Scotia.

VESSEL: *Richard Trimble*
REGISTRY: US 211287
YEAR BUILT: 1913
SHIPYARD: American Ship Building Co., Lorain. Hull No. 707
ENGINE: Triple expansion
GROSS TONNAGE: 7,745
LENGTH—KEEL: 580 feet
LENGTH—OVERALL: 600 feet
BEAM: 58 feet

DEPTH: 32 feet
HISTORY: Laid up in 1974. Sold for scrap in 1978 and scrapped in 1979 in Duluth.

VESSEL: *Charles R. Van Hise*
REGISTRY: 127426
YEAR BUILT: 1900
SHIPYARD: Superior Shipbuilding Co., Superior. Hull No. 144
ENGINE: Quadruple expansion
GROSS TONNAGE: 5,117
LENGTH—KEEL: 458 feet
BEAM: 50 feet
DEPTH: 25 feet
HISTORY: Carried the first cargo for U.S. Steel's Pittsburgh Steamship Co. Sold in 1918 for saltwater service. Cut in two and the forward half towed to Montreal. World War I ended before the stern could be towed, so the bow was returned to the lakes. Rebuilt and lengthened in 1920 and renamed *A. E. R. Schneider*. Renamed *S. B. Way* in 1931, *J. M. Oag* in 1936 and *Capt. C. D. Secord* in 1936. Sold for scrap in 1968 and towed to Spain.

VESSEL: *Enders M. Voorhees*
REGISTRY: US 242023
YEAR BUILT: 1942
SHIPYARD: Great Lakes Engineering Works, Ecorse. Hull No. 288
ENGINE: Steam turbine
GROSS TONNAGE: 10,294
LENGTH—KEEL: 614 feet
LENGTH—OVERALL: 639 feet
BEAM: 67 feet
DEPTH: 35 feet
HISTORY: Sold for scrap in 1987 and towed to Turkey.

VESSEL: *Ralph H. Watson*
REGISTRY: US 237854
YEAR BUILT: 1938
SHIPYARD: Great Lakes Engineering Works, Ecorse. Hull No. 285
ENGINE: Steam turbine
GROSS TONNAGE: 8,302
LENGTH—KEEL: 586 feet
LENGTH—OVERALL: 611 feet
BEAM: 60 feet
DEPTH: 32 feet
HISTORY: Laid up after 1980. Sold for scrap in 1986 and towed to Turkey in 1989 for scrapping.

VESSEL: *James Watt*
REGISTRY: US 77236

YEAR BUILT: 1896
SHIPYARD: Cleveland Shipbuilding Co., Cleveland. Hull No. 26
ENGINE: Triple expansion
GROSS TONNAGE: 4,090
LENGTH—KEEL: 405 feet
LENGTH—OVERALL: 427 feet
BEAM: 48 feet
DEPTH: 28 feet
HISTORY: Sold to Jenkins Steamship Co. in 1928. Scrapped in 1961 in Spain.

VESSEL: *Wawatam*
REGISTRY: US 81324
YEAR BUILT: 1890
SHIPYARD: Cleveland Shipbuilding Co., Cleveland. Hull No. 11
ENGINE: Triple expansion
GROSS TONNAGE: 1,856
LENGTH—KEEL: 266 feet
BEAM: 38 feet
DEPTH: 20 feet
HISTORY: Sold in 1914. Renamed *Glenlivet*. Renamed *Saskatchewan* in 1926. Scrapped in 1937 in Midland, Ontario.

VESSEL: *W. F. White*
REGISTRY: US 213555
YEAR BUILT: 1915
SHIPYARD: American Ship Building Co., Lorain. Hull No. 712
ENGINE: Triple expansion
GROSS TONNAGE: 7,180
LENGTH—KEEL: 530 feet
LENGTH—OVERALL: 550 feet
BEAM: 60 feet
DEPTH: 31 feet
HISTORY: Part of the Bradley Transportation Co. Sent to the East Coast in 1962 and operated on Chesapeake Bay. Returned to the lakes in 1965. Sold Canadian in 1976 and renamed *Erindale*. Scrapped in 1985 at Port Colborne, Ontario.

VESSEL: *Sir Joseph Whitworth*
REGISTRY: US 53255
YEAR BUILT: 1889
SHIPYARD: American Steel Barge Co., Superior. Hull No. 102
ENGINE: Barge
GROSS TONNAGE: 1,192
LENGTH—KEEL: 253 feet
BEAM: 36 feet
DEPTH: 19 feet
HISTORY: Sold in 1905. Renamed *Bath*. Sank off Virginia in 1905.

NAME: *Peter A. B. Widener*
REGISTRY: US 203677
YEAR BUILT: 1906
SHIPYARD: Chicago Ship Building Co., Chicago. Hull No. 71
ENGINE: Triple expansion
GROSS TONNAGE: 7,718
LENGTH—KEEL: 580 feet
LENGTH—OVERALL: 601 feet
BEAM: 58 feet
DEPTH: 32 feet
HISTORY: Sold in 1980 for use as storage barge. Scrapped in 1986 in Portugal.

VESSEL: *Homer D. Williams*
REGISTRY: US 215159
YEAR BUILT: 1917
SHIPYARD: American Ship Building Co., Lorain. Hull No. 720
ENGINE: Triple expansion
GROSS TONNAGE: 7,742
LENGTH—KEEL: 580 feet
LENGTH—OVERALL: 600 feet
BEAM: 60 feet
DEPTH: 32 feet
HISTORY: Repowered with diesel engine in 1951. Gross tonnage increased to 8,369. Scrapped in 1985 in Thunder Bay, Ontario.

VESSEL: *Thomas Wilson*
REGISTRY: US 145616
YEAR BUILT: 1892
SHIPYARD: American Steel Barge Co., Superior. Hull No. 119
ENGINE: Triple expansion
GROSS TONNAGE: 1,713
LENGTH—KEEL: 308 feet
BEAM: 38 feet
DEPTH: 24 feet
HISTORY: Sank on June 7, 1902, off the Duluth ship canal after colliding with steamer *George G. Hadley*.

VESSEL: *A. B. Wolvin*
REGISTRY: US 107563
YEAR BUILT: 1900
SHIPYARD: American Ship Building Co., Cleveland. Hull No. 401
ENGINE: Triple expansion
GROSS TONNAGE: 2,286
LENGTH—KEEL: 242 feet
BEAM: 42 feet
DEPTH: 26 feet

HISTORY: Sold in 1909 and renamed *Portland*. Sent to the ocean in 1911 and lost on Dec. 23, 1916, in the North Atlantic.

NAME: *Zenith City*
REGISTRY: US 28129
YEAR BUILT: 1895
SHIPYARD: Chicago Shipbuilding Co., Chicago. Hull No. 15
ENGINE: Triple expansion
GROSS TONNAGE: 3,850
LENGTH—KEEL: 387 feet
LENGTH—OVERALL: 400 feet
BEAM: 48 feet
DEPTH: 28 feet
HISTORY: Sold in 1944. Scrapped in 1947 in Hamilton, Ontario.

VESSEL: *August Ziesing*
REGISTRY: US 215870
YEAR BUILT: 1917
SHIPYARD: Great Lakes Engineering Works, Ecorse. Hull No. 170
ENGINE: Triple expansion
GROSS TONNAGE: 7,969
LENGTH—KEEL: 580 feet
LENGTH—OVERALL: 600 feet
BEAM: 60 feet
DEPTH: 32 feet
HISTORY: Laid up after 1974. Sold for scrap in 1983 and scrapped in 1986 in Port Colborne, Ontario.

VESSEL: *105*
REGISTRY: US 53258
YEAR BUILT: 1890
SHIPYARD: American Steel Barge Co., Superior. Hull No. 105
ENGINE: Barge
GROSS TONNAGE: 1,295
LENGTH—KEEL: 276 feet
BEAM: 36 feet
DEPTH: 19 feet
HISTORY: Sold in 1905. Renamed *Baroness*. Sank on the Atlantic Ocean in 1910.

VESSEL: *107*
REGISTRY: US 53260
YEAR BUILT: 1890
SHIPYARD: American Steel Barge Co., Superior. Hull No. 107
ENGINE: Barge
GROSS TONNAGE: 1,295
LENGTH—KEEL: 276 feet

BEAM: 36 feet
DEPTH: 19 feet
HISTORY: Sold in 1905. Renamed *Bay City* and *Thurmond.* Sank off Nantucket Shoals, Mass., in 1914.

VESSEL: *109*
REGISTRY: US 53265
YEAR BUILT: 1890
SHIPYARD: American Steel Barge Co., Superior. Hull No. 109
ENGINE: Barge
GROSS TONNAGE: 1,227
LENGTH—KEEL: 265 feet
BEAM: 36 feet
DEPTH: 22 feet
HISTORY: Sold in 1907. Renamed *Baravia.* Sank off Long Island, N.Y., in 1924.

VESSEL: *110*
REGISTRY: US 53266
YEAR BUILT: 1891
SHIPYARD: American Steel Barge Co., Superior. Hull No. 110
ENGINE: Barge
GROSS TONNAGE: 1,227
LENGTH—KEEL: 265 feet
BEAM: 36 feet
DEPTH: 22 feet
HISTORY: Sold in 1905. Later renamed *Badger* and *Pure Lubewell.* Burned in New Orleans, La., in 1932.

VESSEL: *111*
REGISTRY: US 53267
YEAR BUILT: 1891
SHIPYARD: American Steel Barge Co., Superior. Hull No. 111
ENGINE: Barge
GROSS TONNAGE: 1,227
LENGTH—KEEL: 265 feet
BEAM: 36 feet
DEPTH: 22 feet
HISTORY: Sold in 1905. Later renamed *Ivie.* Sank off Hampton Roads, Va., in 1916.

VESSEL: *116*
REGISTRY: US 53269
YEAR BUILT: 1891
SHIPYARD: American Steel Barge Co., Superior. Hull No. 116
ENGINE: Barge
GROSS TONNAGE: 1,169

LENGTH—KEEL: 256 feet
BEAM: 36 feet
DEPTH: 20 feet
HISTORY: Sold in 1905. Later renamed *Britannia* and *Pure Tioleene*. Scrapped in Texas in 1946.

VESSEL: *117*
REGISTRY: US 53271
YEAR BUILT: 1891
SHIPYARD: American Steel Barge Co., Superior. Hull No. 117
ENGINE: Barge
GROSS TONNAGE: 1,310
LENGTH—KEEL: 285 feet
BEAM: 36 feet
DEPTH: 22 feet
HISTORY: Sold in 1909. Renamed *Providence*. Sold British in 1929.

VESSEL: *118*
REGISTRY: US 53272
YEAR BUILT: 1891
SHIPYARD: American Steel Barge Co., Superior. Hull No. 118
ENGINE: Barge
GROSS TONNAGE: 1,310
LENGTH—KEEL: 285 feet
BEAM: 36 feet
DEPTH: 22 feet
HISTORY: Sold in 1909. Over the years renamed *Boston, Freeport Sulphur No. 3, Pure Oil S.S. Co. Barge No. 9,* and *Pure Detonox*. Scrapped in Texas in 1946.

VESSEL: *126*
REGISTRY: US 53273
YEAR BUILT: 1892
SHIPYARD: American Steel Barge Co., Superior. Hull No. 126
ENGINE: Barge
GROSS TONNAGE: 1,128
LENGTH—KEEL: 264 feet
BEAM: 36 feet
DEPTH: 22 feet
HISTORY: Sold in 1905. Renamed *Baden*. Wrecked off Massachusetts in 1906.

VESSEL: *127*
REGISTRY: US 53274
YEAR BUILT: 1892
SHIPYARD: American Steel Barge Co., Superior. Hull No. 127
ENGINE: Barge

GROSS TONNAGE: 1,128
LENGTH—KEEL: 264 feet
BEAM: 36 feet
DEPTH: 22 feet
HISTORY: Sold in 1905. Later renamed *Jeanie* and *Dallas*. Scrapped in Louisiana in 1936.

VESSEL: *129*
REGISTRY: US 53276
YEAR BUILT: 1893
SHIPYARD: American Steel Barge Co., Superior. Hull No. 129
ENGINE: Barge
GROSS TONNAGE: 1,310
LENGTH—KEEL: 292 feet
BEAM: 36 feet
DEPTH: 22 feet
HISTORY: Collided in autumn 1902 with steamer *Maunaloa* as the steamer attempted to take the barge under tow during a storm. Sank without loss of life thirty miles northwest of Vermilion Point on Lake Superior.

VESSEL: *130*
REGISTRY: US 53277
YEAR BUILT: 1893
SHIPYARD: American Steel Barge Co., Superior. Hull No. 130
ENGINE: Barge
GROSS TONNAGE: 1,311
LENGTH—KEEL: 292 feet
BEAM: 36 feet
DEPTH: 22 feet
HISTORY: Sold in 1909. Later renamed *Lynn*. Scrapped in 1924.

VESSEL: *131*
REGISTRY: US 53278
YEAR BUILT: 1893
SHIPYARD: American Steel Barge Co., Superior. Hull No. 131
ENGINE: Barge
GROSS TONNAGE: 1,311
LENGTH—KEEL: 292 feet
BEAM: 36 feet
DEPTH: 22 feet
HISTORY: Sold in 1909. Over the years renamed *Salem, Freeport Sulphur No. 4, Pure Oil No. 10,* and *Pure Nulube*. Scrapped in 1946.

VESSEL: *132*
REGISTRY: US 53279
YEAR BUILT: 1893

SHIPYARD: American Steel Barge Co., Superior. Hull No. 132
ENGINE: Barge
GROSS TONNAGE: 1,311
LENGTH—KEEL: 292 feet
BEAM: 36 feet
DEPTH: 22 feet
HISTORY: Sold in 1909. Renamed *Portsmouth*. Sank off Texas in 1927.

VESSEL: *133*
REGISTRY: US 53280
YEAR BUILT: 1893
SHIPYARD: American Steel Barge Co., Superior. Hull No. 133
ENGINE: Barge
GROSS TONNAGE: 1,311
LENGTH—KEEL: 292 feet
BEAM: 36 feet
DEPTH: 22 feet
HISTORY: Sold in 1909. Renamed *Searsport*. Sank off New York in 1911.

VESSEL: *134*
REGISTRY: US 53281
YEAR BUILT: 1893
SHIPYARD: American Steel Barge Co., Superior. Hull No. 134
ENGINE: Barge
GROSS TONNAGE: 1,311
LENGTH—KEEL: 292 feet
BEAM: 36 feet
DEPTH: 22 feet
HISTORY: Sold in 1909. Renamed *Bangor*. Stranded off Virginia in 1912.

VESSEL: *137*
REGISTRY: US 53284
YEAR BUILT: 1896
SHIPYARD: American Steel Barge Co., Superior. Hull No. 137
ENGINE: Barge
GROSS TONNAGE: 2,481
LENGTH—KEEL: 345 feet
BEAM: 45 feet
DEPTH: 26 feet
HISTORY: Sold in 1926. Scrapped in Hamilton, Ontario, 1965.

VESSEL: *201*
REGISTRY: US 59439
YEAR BUILT: 1890
SHIPYARD: Handren and Robins Atlantic Dock Co., Brooklyn, N.Y. Hull No. 201
ENGINE: Barge

GROSS TONNAGE: 948
LENGTH—KEEL: 182 feet
LENGTH—OVERALL: 192
BEAM: 42 feet
DEPTH: 17 feet
HISTORY: Lengthened to 244 feet in 1897. Sold in 1905 to Baltimore and Boston Barge Co. Renamed *Cassie* in 1907. Lost off New Jersey in 1917.

VESSEL: *202*
REGISTRY: US 59443
YEAR BUILT: 1890
SHIPYARD: Handren and Robins Atlantic Dock Co., Brooklyn, N.Y. Hull No. 202
ENGINE: Barge
GROSS TONNAGE: 948
LENGTH—KEEL: 182 feet
LENGTH—OVERALL: 192
BEAM: 42 feet
DEPTH: 17 feet
HISTORY: Lengthened to 244 feet in 1897. Sold in 1905 to Baltimore and Boston Barge Co. and renamed *Fannie*. Sank off New Jersey on Jan. 24, 1908.

Sources: *USS Vessel Registry*, 1968, 1984; *Great Lakes Bulk Carriers, 1869-1985*; *List of Merchant Vessels of the United States, 1900*.

Appendix 4

Original Vessels of the Pittsburgh Steamship Company

PITTSBURGH STEAMSHIP CO.

Clarence A. Black
Cornell
Griffin
Harvard
Joliet
Lafayette
LaSalle
William R. Linn
Princeton
Rensselaer
Wawatam
Bryn Mawr (barge)
Carrington (barge)

MINNESOTA STEAMSHIP CO.

Malietoa
Manola
Maricopa
Marina
Mariposa
Mariska
Maritana
Maruba
Masaba
Mataafa
Matoa
Maunaloa
Madeira (barge)
Magna (barge)

Minnesota Steamship Co. (con't)

Maia (barge)
Maida (barge)
Malta (barge)
Manda (barge)
Manila (barge)
Marcia (barge)
Martha (barge)
Marsala (barge)

MENOMINEE TRANSIT CO.

Briton
German
Grecian
Roman
Saxon

MUTUAL STEAMSHIP CO.

Cambria
Coralia
Corona
Corsica

AMERICAN STEAMSHIP CO.

Crescent City
William Edenborn
Isaac L. Ellwood
Empire City
John W. Gates
W. H. Gilbert

American Steamship Co. (con't)

James J. Hill
William P. Palmer
Queen City
Superior City
A. B. Wolvin
Zenith City

BESSEMER STEAMSHIP CO.

E. B. Bartlett
Sir Henry Bessemer
Robert W. E. Bunsen
Joseph L. Colby
James B. Colgate
Henry Cort
James B. Eads
John Ericsson
Sir William Fairbairn
Robert Fulton
Douglass Houghton
Colgate Hoyt
Samuel Mather
Alexander McDougall
Samuel F. B. Morse
James B. Neilson
General Orlando M. Poe
Frank Rockefeller
Sir William Siemens
George Stephenson
A. D. Thompson
John B. Trevor
Charles R. Van Hise
James Watt

Bessemer Steamship Co. (con't)

Thomas Wilson
Sir Isaac Lothian Bell (barge)
George H. Corliss (barge)
John Fritz (barge)
Alexander Holley (barge)
Alfred Krupp (barge)
William LeBaron Jenney (barge)
James Nasmyth (barge)
John A. Roebling (barge)
John Scott Russell (barge)
John Smeaton (barge)
Sidney G. Thomas (barge)
Sir Joseph Whitworth (barge)
105 (barge)
107 (barge)
109 (barge)
110 (barge)
111 (barge)
116 (barge)
117 (barge)
118 (barge)
126 (barge)
127 (barge)
129 (barge)
130 (barge)
131 (barge)
132 (barge)
133 (barge)
134 (barge)
137 (barge)
201 (barge)
202 (barge)

Sources: *USS Vessel Registry—1968; Blue Book of American Shipping, 1901.*

Fleet	Steamers	Barges	Total
Bessemer	25	31	56
Minnesota	12	10	22
American	12	0	12
Pittsburgh	11	2	13
Menominee	5	0	5
Mutual	4	0	4
Total	69	43	112

NOTES

CHAPTER 1

1. A. O. Backert, ed. *The ABC of Iron and Steel* (Cleveland: Penton Publishing Company, 1925), 33-48.
2. The Rev. Edward J. Dowling, "The 'Tin Stackers': The Story of the Ships of the Pittsburgh Steamship Company," *Inland Seas* 9 (Summer 1953): 80.
3. Ibid., 81-84.
4. Joseph Frazier Wall, *Andrew Carnegie* (New York: Oxford University Press, 1970), 259-61, 530-31; John K. Winkler, *Morgan the Magnificent: The Life of J. Pierpont Morgan, 1837-1913* (Garden City, N.Y.: Garden City Publishing Company, 1930), 205.
5. Douglas A. Fisher, *Steel Serves the Nation, 1901-1951: The Fifty-Year Story of United States Steel* (New York: United States Steel Corporation, 1951), 19; Wall, *Andrew Carnegie*, 533.
6. Al Miller, "A Tugman's Story: Life and Times of B. B. Inman," *American Neptune* 56 (Summer 1996): 201.
7. Winkler, *Morgan the Magnificent*, 125.
8. Dowling, "Tin Stackers," *Inland Seas* 9 (Fall 1953): 177; Wall, *Andrew Carnegie*, 598-99; Alexander McDougall, *The Autobiography of Captain Alexander McDougall* (1932; reprint, Cleveland: Great Lakes Historical Society, 1968), 40-49.
9. David Freeman Hawke, *John D.: The Founding Father of the Rockefellers* (New York: Harper & Row, 1980), 208-10.
10. Dowling, "Tin Stackers," *Inland Seas* 9 (Fall 1953): 175.
11. Wall, *Andrew Carnegie*, 623; Hawke, *John D.*, 210.
12. Henry Oliver Evans, *Iron Pioneer: Henry W. Oliver* (New York: Dutton, 1942), 252-3; *Marine Review*, 12 October 1899; McDougall, *Autobiography*, 48.
13. *Marine Review*, 12 October 1899.
14. Ibid.
15. Charter of the Pittsburgh Steamship Company, 6 November 1899, West Virginia Division of Culture and History, Charleston, W.V.; *Marine Review*, 18 January 1900; Bills of Sale of Enrolled Vessels, 1899-1900, Lake Supe-

rior Maritime Visitors Center, Duluth, Minn. [hereafter abbreviated LSM]; *Duluth Evening Herald,* 25 January 1900; Wall, *Andrew Carnegie,* 623.
16. *Duluth Evening Herald,* 22 June 1899.
17. Fisher, *Steel Serves the Nation,* 19.
18. Winkler, *Morgan the Magnificent,* 204.
19. Ibid., 207-8; Fisher, *Steel Serves the Nation,* 18-19; Wall, *Andrew Carnegie,* 784.
20. Fisher, *Steel Serves the Nation,* 20.
21. Ibid., 21.
22. Wall, *Andrew Carnegie,* 788-89.
23. *Marine Review,* 28 February, 3 March 1901.
24. Winkler, *Morgan the Magnificent,* 213.
25. *Marine Review,* 21 March 1901.
26. Dowling, "Tin Stackers," *Inland Seas* 9 (Winter 1953): 271-75.
27. Arundel Cotter, *The Authentic History of the United States Steel Corporation* (Moody Magazine & Book Co., 1916), 45.
28. USS Great Lakes Fleet, Vessel Register, 1968.
29. Harvey C. Beeson, *Inland Marine Directory* (Chicago, 1901), 172.

CHAPTER 2
1. *Marine Review,* 31 January 1901; *Duluth News-Tribune,* 28 April 1895; Beeson, *Inland Marine Directory* (1901), 170-72.
2. *Duluth News-Tribune,* 28 April 1895.
3. Bill of Sale for Enrolled Vessel, 2 March 1901, LSM.
4. Dowling, "Tin Stackers," *Inland Seas* 9 (Fall 1953): 175.
5. *Marine Review,* 14 May, 20 February 1896.
6. Miller, "Tugman's Story," 208-9.
7. *Marine Review,* 23 March 1899.
8. Ibid., 8 June 1899.
9. Ibid., 12 October 1899.
10. Bills of Sale for Enrolled Vessels, 1900-1901, LSM; Dowling, "Tin Stackers," *Inland Seas* 9 (Fall 1953): 176.
11. *Marine Review,* 30 March 1899.
12. Bill of Sale for Enrolled Vessel, 23 August 1900, LSM; *Marine Review,* 5 April, 14 June, 29 November 1900.
13. Andrew Edgar Collard, *Passage to the Sea: The Story of Canada Steamship Lines* (Toronto: Doubleday Canada Limited, 1991), 100-101.
14. *Marine Record,* 17 January 1901.
15. Ibid., 31 January 1901.
16. *Marine Review,* 17 January 1901.
17. Ibid., 18 April 1901; *Marine Record,* 18 April 1901.
18. *Duluth News-Tribune,* 23 April 1901.
19. Ibid., 2 May 1901.
20. William Buhrmann, interview with author, 15 January 1997.

21. *Duluth News-Tribune,* 22 April 1901.
22. *Marine Review,* 25 April 1901.
23. Ibid.
24. Ibid., 9 May 1901.
25. *Duluth News-Tribune,* 9 May 1901.
26. *Marine Record,* 16 May 1901; *Duluth Evening Herald,* 16 May 1901.
27. *Duluth Evening Herald,* 17 May 1901.
28. *Marine Review,* 8, 22 August 1901.
29. *Duluth Evening Herald,* 25 June 1901.
30. *Marine Record,* 27 June, 4, 11 July 1901.
31. Ibid., 26 September 1901.
32. Byron B. Inman to May R. Inman, 10 October 1901, Inman Collection, Northeast Minnesota Historical Center, Duluth.
33. Henry Elmer Hoagland, "Wage Bargaining on the Vessels of the Great Lakes," *University of Illinois Studies in the Social Sciences* 6 (September 1917): 41-42.
34. *Marine Record,* 10 October 1901.
35. Lake Carriers' Association [hereafter abbreviated LCA], *Annual Report,* 1901, p. 10.
36. Julius F. Wolff, Jr., *Lake Superior Shipwrecks,* ed. Thom Holden (Duluth, Minn.: Lake Superior Port Cities Inc., 1990), 97.
37. *Marine Record,* 3 July 1902.
38. *Labor World,* 19 July 1902.
39. Wolff, *Lake Superior Shipwrecks,* 97.
40. *Marine Review and Marine Record,* 10 October 1902.
41. Hoagland, "Wage Bargaining," 44.
42. Ibid., 55-57.
43. Ibid., 58.
44. *Labor World,* 19 December 1903.

CHAPTER 3

1. Walter Havighurst, *Vein of Iron* (Cleveland: World Publishing, 1958), 65-82; LCA, *Annual Report,* 1928, pp. 149-53.
2. *Marine Review and Marine Record,* 17 December 1903.
3. Havighurst, *Vein of Iron,* 65-82; LCA, *Annual Report,* 1928, pp. 149-53.
4. *Duluth Evening Herald,* 4, 6 January 1904.
5. Ibid., 12 January 1904.
6. Charles P. Larrowe, *Maritime Labor Relations on the Great Lakes* (East Lansing: Michigan State University, Labor and Industrial Relations Center, 1959), 12.
7. Ibid., 13.
8. Hoagland, "Wage Bargaining," 79-80.
9. Ibid., 81.
10. Ibid., 58.

11. Ibid., 44.
12. LCA, *Annual Report*, 1904, pp. 3-7.
13. Hoagland, "Wage Bargaining," 62; *Duluth Evening Herald*, 29 April, 20 May 1904.
14. LCA, *Annual Report*, 1904, pp. 3-7.
15. *Marine Review*, 5 January 1905.
16. Hoagland, "Wage Bargaining," 71.
17. Ibid., 74.
18. Backert, *ABC of Iron and Steel*, 47; Gary S. Dewar, "A Forgotten Class," *Telescope* 37 (April/May 1989): 31-39; *Marine Review*, 29 June 1904.
19. *Marine Review*, 29 June 1904.
20. Ibid.
21. Ibid., 25 June 1905.
22. "The Ship with Golden Rivets," *The Scanner* 6 (March 1974): 7; Dowling, "Tin Stackers," *Inland Seas* 9 (Winter 1953): 280.
23. *Marine Review*, 2 February 1905.
24. Ibid., 2 March 1905.
25. *Duluth News-Tribune*, 7 November 1905.
26. Ibid., 28 November 1905; Notes by H. W. Richardson, U.S. Weather Bureau, on photograph, Northeast Minnesota Historical Center, Duluth.
27. Notarized statement of Silas H. Hunter, 11 December 1905, LSM.
28. Notarized statement of Captain Frank Rice, 1 December 1905, LSM; *Duluth News-Tribune*, 24 December, 29 November 1905.
29. Notarized statement of Captain Frank Rice, 1 December 1905, LSM.
30. *Duluth Evening Herald*, 1 December 1905.
31. Notarized statement of Captain D. P. Wright, 2 December 1905, LSM; *Duluth Evening Herald*, 1 December 1905; Wolff, *Lake Superior Shipwrecks*, 114-15.
32. Notarized statement of Captain D. P. Wright, 2 December 1905. LSM.
33. *Duluth News-Tribune*, 1 December 1905.
34. Notarized statements of Captain George W. Balfour, 1 December 1905, and Captain F. D. Selee, 2 December 1905, both at LSM.
35. Notarized statement of Silas H. Hunter, 11 December 1905, LSM.
36. Notarized statement of William F. Hormig, 11 December 1905, LSM.
37. Wolff, *Lake Superior Shipwrecks*, 114-15; *Duluth Evening Herald*, 1 December 1905.
38. Notarized statement of Captain John Dissette, 1 December 1905, LSM.
39. *Duluth News-Tribune*, 30 November 1905.
40. Notarized statement of Captain F. A. Bailey, 13 December 1905, LSM.
41. Wolff, *Lake Superior Shipwrecks*, 116.
42. Notarized statements of Captain C. H. Cummings, 2 December 1905, and Alex S. Brown, 2 December 1905, both at LSM.
43. Notarized statement of Captain R. F. Humble, 1 December 1905, LSM; Wolff, *Lake Superior Shipwrecks*, 116-18; *Duluth News-Tribune*, 29

November-1 December 1905; *Duluth Evening Herald,* 29 November-1 December 1905.
44. *Duluth News-Tribune,* 1 December 1905.
45. Ibid., 2 December 1905.
46. Notarized statement of Captain C. H. Cummings, 2 December 1905, LSM.
47. *Duluth News-Tribune,* 3 December 1905.

CHAPTER 4

1. *Marine Review,* 19 January 1905.
2. Pittsburgh Steamship Company, Operating Statistics, 1907-1908, LSM.
3. Dowling, "Tin Stackers," *Inland Seas* 10 (Spring 1954): 44.
4. Paul F. Brissenden, *Employment System of the Lake Carriers' Association* (Washington: U.S. Department of Labor, Bureau of Labor Statistics, January 1918), 7-9.
5. Hoagland, "Wage Bargaining," 78.
6. Brissenden, *Employment System,* 46.
7. Ibid.
8. Ibid., 54-56.
9. *Marine Review,* 16 April 1908.
10. Pittsburgh Steamship Company, Operating Statistics, 1907-1908, LSM.
11. Hoagland, "Wage Bargaining," 87-88.
12. Brissenden, *Employment System,* 26.
13. Larrowe, *Maritime Labor,* 33-40.
14. Ibid., 33-40; LCA, *Annual Report,* 1909, p. 17; Beeson, *Inland Marine Directory* (1910), 101.
15. Brissenden, *Employment System,* 8-9.
16. Hoagland, "Wage Bargaining," 92-96.
17. Ibid.
18. Pittsburgh Steamship Company, Minutes of Third Annual Convention of Engineers and Officers, 26-28 January 1911, p. 14, Institute for Great Lakes Research, Bowling Green State University.
19. Ibid., 15.
20. Ibid., 25.
21. Ibid., 69-70.
22. Ibid., 41.
23. Dwight Boyer, *True Tales of the Great Lakes* (New York: Dodd, Mead, 1971), 275.
24. Ibid., 245.
25. Frank Barcus, *Freshwater Fury* (Detroit: Wayne State University Press, 1960), 76.
26. Boyer, *True Tales,* 244-46.
27. Ibid., 278-79.
28. Barcus, *Freshwater Fury,* 63.
29. Ibid., 13-15.

CHAPTER 5

1. LCA, *Annual Report*, 1916, pp. 21-23.
2. Dowling, "Tin Stackers," *Inland Seas* 10 (Spring 1954): 45-46.
3. LCA, *Annual Report*, 1917, p. 32; ibid., 1918, pp. 46-51.
4. Ibid., 1917, p. 32.
5. Lake Carriers' Association, *Whistle Signals* (Cleveland: Lake Carriers' Association, 1 May 1918), 36-51.
6. LCA, *Annual Report*, 1918, pp. 46-51.
7. Ibid., 1919, p. 24.
8. Ibid., 1920, pp. 19-30.
9. Wolff, *Lake Superior Shipwrecks*, 168-69.
10. Larrowe, *Maritime Labor*, 46-50.
11. LCA, *Annual Report*, 1921, pp. 180-82; *Skillings Mining Review*, 17 September 1921, pp. 4-5, and 12 September 1981, p. 27.
12. LCA *Annual Report*, 1921, pp. 180-82.
13. Gerald F. Micketti, *The Bradley Boats* (Traverse City, Mich.: Gerald F. Micketti, 1995), 5-9.
14. Ibid.
15. *Marine Review*, March 1932, p. 42.
16. LCA, *Annual Report*, 1923, p. 123; Fisher, *Steel Serves the Nation*, 208.
17. LCA, *Annual Report*, 1929, pp. 149-53.
18. Pittsburgh Steamship Company, "Confidential Letter to Masters, Sept. 8, 1926," Great Lakes Historical Society, Vermilion, Ohio [hereafter abbreviated GLHS].
19. Pittsburgh Steamship Company, "Confidential Letter, May 31, 1927," GLHS.
20. Pittsburgh Steamship Company, "Confidential Letter, Sept. 30, 1929," GLHS.
21. *U.S. Steel News*, September 1937, p. 19.

CHAPTER 6

1. *Skillings Mining Review*, 26 October 1929, p. 7, and 16 November 1929, p. 17.
2. LCA, *Annual Report*, 1931, pp. 13-14; Guido Gulder, interview with author, May 1996; *Skillings Mining Review*, 30 May 1931, p. 9.
3. LCA, *Annual Report*, 1932, p. 9; Gulder interview; *U.S. Steel News*, September 1937, p. 14.
4. LCA *Annual Report*, 1933, pp. 45-55.
5. Larrowe, *Maritime Labor*, 40-50.
6. Ibid., 52.
7. *U.S. Steel News*, September 1937, pp. 12-17.
8. Gulder interview; William L. Wallace, "Up the Grades on Ore Boats: From Coal Passer to Wheelsman to Third Mate," *Inland Seas* 53 (Fall 1997): 178-91.
9. William D. Wilson, interview with author, April 1996.
10. Gulder interview.
11. Ibid.

12. Ibid.
13. Ibid.
14. *U.S. Steel News,* June 1937, pp. 21-22, and September 1937, pp. 20-21.
15. Proposal by American Ship Building Company to Pittsburgh Steamship Company, 16 February 1937, LSM.
16. *U.S. Steel News,* April 1937, p. 7.
17. Gordon Bugbee, "In the Triple's Wake . . . ," *Telescope* 13 (July 1964): 150-54; Howard Varian, chief engineer of Great Lakes Engineering Works, "Observations in Development of Machinery in Great Lakes Vessels," speech presented 26 May 1956 to Michigan Quarterdeck Society, copy at LSM.
18. Dwight Boyer, *Strange Adventures of the Great Lakes* (New York: Dodd, Mead, 1974), 82.
19. *Duluth News-Tribune and Herald,* 10 November 1985.
20. George J. Joachim, *Iron Fleet: The Great Lakes in World War II* (Detroit: Wayne State University Press, 1994), 30-31.
21. Boyer, *Strange Adventures,* 96-97.
22. Pittsburgh Steamship Company, "Record of Casualty to Vessel," 18 November 1940, LSM.
23. Gulder interview.

CHAPTER 7

1. Gulder interview.
2. LCA, *Annual Report,* 1948, p. 128.
3. LCA, "Record Monthly Traffic through Sault Canals," *Bulletin,* November 1941, pp. 1-2; Joachim, *Iron Fleet,* 101.
4. Joachim, *Iron Fleet,* 38-39.
5. Gulder interview.
6. Pittsburgh Steamship Company, "Confidential Letter to Masters, April 21, 1942," GLHS.
7. Pittsburgh Steamship Company, "Confidential Letter to Masters, July 7, 1942," GLHS.
8. American Bureau of Shipping "Survey Report," 24 July 1942, LSM.
9. "Memorandum to Captain W. F. Meister," from J. S. Cottier, 21 July 1942, LSM.
10. Pittsburgh Steamship Company, "Confidential Letter to Masters, July 7, 1942," GLHS.
11. Pittsburgh Steamship Company, "Specifications for Hulls No. 285 & 286," and Pittsburgh Steamship Company, memo by Leland Haggan, 20 December 1940, both at LSM.
12. Pittsburgh Steamship Company, "Memo re Prospective New Boats," 27 December 1940, LSM.
13. LCA, *Annual Report,* 1942, pp. 9-12; ibid., 1943, pp. 9-12; LCA, "Biographies of Great Lakes Fleets . . . the Pittsburgh Steamship Company," *Bulletin,* September 1950, pp. 14-19.

14. Larrowe, *Maritime Labor,* 62-65.
15. Ibid., 63.
16. Ibid.
17. Ibid.
18. Ibid., 64.
19. Pittsburgh Steamship Company, "Circular Letter to Masters, Jan. 20, 1944," GLHS.
20. Edmund Siegrist, interview with author, 6 October, 1996.
21. Wendell Barrow, interview with author, July 1996.
22. Siegrist interview.
23. Gulder interview.
24. LCA, *Annual Report,* 1946, pp. 9-15.
25. Ibid., 1945, pp. 40-41; ibid., 1946, pp. 52-53.
26. LCA, "Biographies of Great Lakes Fleets . . . The Pittsburgh Steamship Company," *Bulletin,* September 1950, p. 18.
27. Gulder interview.
28. *Report to the Lake Carriers' Association on Great Lakes Radar Operational Research Project* (Washington, D.C.: Jansky & Bailey, 1947); *U.S. Steel News,* October 1947.
29. Wilson interview.
30. LCA, "Biographies of Great Lakes Fleets . . . the Pittsburgh Steamship Company," *Bulletin,* September 1950, pp. 14-19.

CHAPTER 8

1. Fisher, *Steel Serves the Nation,* 108.
2. Charles L. Horn, Jr., "The Iron Ore Industry of Minnesota and the Problem of Depleted Reserves" (Master's thesis, Princeton University, 1956), 61, 145.
3. Harry Benford and William A. Fox, eds., *A Half Century of Maritime Technology, 1943-1993* (Society of Naval Architects and Marine Engineers, 1993), 240-57.
4. Society of Naval Architects and Marine Engineers, *Transactions, Vol. 58, 1950* (Easton, Penn.: Mack Printing Co., 1950), 67-115.
5. USS Great Lakes Fleet, Vessel Register, 1968.
6. USS Great Lakes Fleet, Vessel Register, 1968; letter from United States Steel Corporation to Collector of Customs, Duluth, Minn., 14 February 1952, LSM.
7. Larrowe, *Maritime Labor,* 76-77.
8. Ibid.; "Agreement between Pittsburgh Steamship Division, United States Steel Corporation, and United Steelworkers of America," 16 July 1954, Historical Collections of the Great Lakes, Bowling Green State University.
9. "History of the Pittsburgh Steamship Company," *Pittsburgh Sidelights,* September-October 1954, pp. 3-13.
10. "A Mountain Every Minute," *Pittsburgh Sidelights,* August 1953, pp. 4-6.
11. "A Supermarket for Ships," *Pittsburgh Sidelights,* September 1953, pp. 3-10.

12. USS Great Lakes Fleet, correspondence with author, 16 December 1996.
13. "A Supermarket for Ships," 3-10.
14. "Bogie at 15,000 Feet," *Pittsburgh Sidelights,* June 1956, pp. 3-5.
15. *Duluth News-Tribune and Herald,* 18 September 1983.
16. "It'll Never Catch On, Mr. Ford," *Pittsburgh Sidelights,* May-June 1961, pp. 16-17.
17. Guest Log of Stmr. *William A. Irvin,* LSM.
18. "The Rescue Story," *Pittsburgh Sidelights,* July 1953, pp. 3, 16, 18-19; Wolff, *Lake Superior Shipwrecks,* 202-3.
19. LCA, *Annual Report,* 1954, pp. 9-13.
20. Ibid.
21. Horn, *The Iron Ore Industry,* 145-51.
22. E. W. Davis, *Pioneering with Taconite* (St. Paul: Minnesota Historical Society, 1964), 18-40.
23. Ibid., 41-61.
24. Ibid., 108-41.
25. LCA, *Annual Report,* 1955, pp. 64-65; ibid., 1956, p. 66; ibid., 1957, p. 68.
26. Davis, *Pioneering with Taconite,* 230; "A Look at the Log," *Pittsburgh Sidelights,* November 1955, p. 22.
27. Marine Engineers Benevolent Association-American Maritime Officers District 2, *Coulby to Potts to Beukema: U.S. Steel Pulls the Strings at Lake Carriers' Association* (n.p.: Marine Engineers Benevolent Association-American Maritime Officers District 2, n.d.), 21.
28. Larrowe, *Maritime Labor,* 78-79.
29. Donald Potts to Ralph Rovinsky, 30 July 1956, LSM.
30. Larrowe, *Maritime Labor,* 79-80; MEBA-AMO District 2, *Coulby to Potts to Beukema.*
31. Larrowe, *Maritime Labor,* 80.
32. Ibid., 80-81.
33. *Skillings Mining Review,* 15 September 1956, p. 22.
34. Larrowe, *Maritime Labor,* 81.
35. Ibid., 82.

CHAPTER 9

1. "A Visit with Our New President," *Pittsburgh Sidelights,* April 1957, pp. 4-5.
2. "Pittsburgh Fleet Anticipates New Ship Construction Program," *Great Lakes and Inland Waterways,* December 1957, p. 18.
3. Pittsburgh Steamship Division, Minutes of Annual Meeting, 1960, LSM.
4. Ibid.
5. Ibid.
6. Ibid.
7. Wilson interview.
8. "North to Salt Water: A Pittsburgh Ship Opens a New Trade Route," *Pittsburgh Sidelights,* Autumn 1962, pp. 3-12.

9. Wilson interview.
10. LCA, *Annual Report,* 1963, p. 59.
11. *Skillings Mining Review,* 4 January 1964, p. 10.
12. Ibid., 12 September 1964.
13. Bruce Liberty, interview with author, May 1996.
14. *Duluth News-Tribune,* 1 May 1956.
15. Liberty interview.
16. Liberty interview; Ralph Bertz, interview with author, 26 September 1996.
17. LCA, *Annual Report,* 1965, pp. 61-62; ibid., 1966, pp. 61-63; ibid., 1967, p. 56.
18. Buhrmann interview.
19. *Skillings Mining Review,* 8 July 1967, p. 26.
20. Ibid.
21. Wilson interview; USS Public Affairs Office, USS Great Lakes Fleet history, author's collection.
22. *Skillings Mining Review,* 4 August 1956, p. 28.
23. Micketti, *The Bradley Boats,* 43.
24. Ibid., 35-42.

CHAPTER 10

1. LCA, *Annual Report,* 1967, p. 38.
2. Buhrmann interview.
3. LCA, *Annual Report,* 1967, p. 9.
4. Buhrmann interview; Bertz interview.
5. *Skillings Mining Review,* 4 November 1967, p. 30.
6. Ibid., 4, 18 November 1967, 6 January 1968.
7. LCA, *Annual Report,* 1969, pp. 9-10.
8. Ibid., 10.
9. Buhrmann interview.
10. LCA, *Annual Report,* 1971, p. 36; Bertz interview; Marsh & McLennan Inc., "Statement of Claim under Builder's Risk Insurance. Case of the M.V. *Roger Blough,*" 16 October 1972, LSM.
11. *Skillings Mining Review,* 15 May 1971, p. 25.
12. Bertz interview; Jesse B. Cooper, speech to the Wisconsin Marine Historical Society in Milwaukee on 7 November 1986, reprinted 10 November 1997 in the *Minneapolis Star-Tribune.*
13. Buhrmann interview; *Skillings Mining Review,* 3 April 1971, p. 6, and 4 December 1971, p. 32; Alexander C. Meakin, "Four Long and One Short: A History of the Great Lakes Towing Company," *Inland Seas* 31 (Winter 1975): 289.
14. Christian Beukema, "The Demonstration: U.S. Steel, Winter 1970-71," *Seaway Review* 2 (Summer 1971): 11-15.
15. *Lorain Journal,* 29 March 1970.
16. Beukema, "The Demonstration," 11-15.

17. Ibid.
18. *Soo Evening News*, 30 January 1972.
19. "Second Maiden Voyage," *Pittsburgh Sidelights*, 1 July 1974, p. 1.
20. Robert J. Hemming, *Gales of November* (Chicago: Contemporary Books, 1981), 82.
21. *Duluth News-Tribune and Herald*, 10 November 1985.
22. Dwight Boyer, *Ships and Men of the Great Lakes* (New York: Dodd, Mead, 1977), 175.
23. *Duluth News-Tribune and Herald*, 10 November 1985.
24. Hemming, *Gales of November*, 177.
25. *Duluth News-Tribune and Herald*, 10 November 1985; Cooper speech.
26. *Duluth News-Tribune and Herald*, 10 November 1985.
27. Logbook of the Stmr. *Philip R. Clarke*, 1975-76, author's collection.
28. LCA, *Annual Report*, 1976, pp. 12-14; ibid., 1977, pp. 9-11; *Industry Week*, 5 February 1979, pp. 52-55.
29. Christian F. Beukema, oral history recorded 1980, Northeast Minnesota Historical Center, University of Minnesota-Duluth.
30. Buhrmann interview; Bertz interview.
31. Alexander C. Meakin, *Master of the Inland Seas* (Vermilion, Ohio: Great Lakes Historical Society, 1988), 355.
32. Buhrmann interview; Meakin, *Master of the Inland Seas*, 327, 355.
33. Bertz interview.
34. Ibid.
35. Ibid.
36. Ibid.
37. U.S. Steel Corporation press release, n.d.
38. *Milwaukee Journal*, 16 November 1980, Insight section; *U.S. Steel News*, 1980/2, p. 16.
39. Bertz interview.
40. Ibid.; LCA, *Annual Report*, 1979, p. 34.
41. Wilson interview.
42. USS Great Lakes Fleet, Vessel Register updates, 1984, LSM.
43. Benford and Fox, *Half Century of Maritime Technology*, 240-57.

CHAPTER 11

1. LCA, *Annual Report*, 1980, p. 65.
2. U.S. Steel Corporation news release, 5 June 1981.
3. Buhrmann interview.
4. Ibid.
5. U.S. Steel Corporation news release, 26 February 1981.
6. Liberty interview.
7. "Three Ships Converted to Self-Unloaders," *Missabe Iron Ranger* 67 (Spring-Summer 1982): 12-14.
8. Bertz interview.

9. *Milwaukee Journal*, 16 November 1980, Insight section.
10. Buhrmann interview.
11. LCA, *Annual Report*, 1982, pp. 17-20; USS Great Lakes Fleet, Vessel Register updates, 1984, LSM.
12. Buhrmann interview.
13. Liberty interview; Benford and Fox, *Half Century of Maritime Technology*, 240-57.
14. Bertz interview; Benford and Fox, *Half Century of Maritime Technology*, 240-57.
15. *Duluth News-Tribune*, 1 August, 3 October 1986.
16. Ibid., 21 June 1988.
17. Buhrmann interview; *Duluth News-Tribune*, 21 June 1988; USS Great Lakes Fleet Inc., correspondence with author, 16 December 1996.
18. Buhrmann interview.
19. "Home Port: Duluth. The Historic USS Great Lakes Fleet, Reinvented for the 1990s, Sails from Duluth," *Minnesota's World Port* 25 (Spring 1993): 8-9.
20. Bertz interview; Thom Holden, "A Second Maiden Voyage," *Nor'Easter* 6 (May-June 1981): 1.
21. Buhrmann interview.
22. "On the Move with Transtar," *Minnesota's World Port* 29 (Summer 1997): 8-9.
23. USS Great Lakes Fleet Inc., correspondence with author.
24. Thom Holden, "Observations," *Nor'Easter* 20 (July-August 1995): 11.
25. LCA, *Annual Report*, 1994, pp. 15, 36; Thom Holden, "Observations," *Nor'Easter* 21 (March-April 1996): 9.
26. USS Great Lakes Fleet brochure.
27. USS Great Lakes Fleet Inc., correspondence with author.
28. Ibid.
29. *Skillings Mining Review*, 12 July 1997, p. 16.
30. Ibid.
31. Thom Holden, "Observations," *Nor'Easter* 22 (November-December 1997): 10.
32. Buhrmann interview.
33. LCA, *Annual Report* 1994, p. 4.
34. USS Great Lakes Fleet Inc., correspondence with author.
35. Ibid.

BIBLIOGRAPHY

PRIMARY SOURCES

American Bureau of Shipping Survey Reports. Various ships and incidents. Lake Superior Maritime Visitors Center, Duluth, Minnesota.

Bertz, Ralph H., Bruce E. Liberty, and Ian D. Sharp. "The Design and Operational History of the Caterpillar Model 3612 Diesel Engine in a Marine Environment on the MV *George A. Sloan.*" Paper presented in Lafayette, Indiana, 15 October 1987, to Great Lakes and Great Rivers Section of the Society of Naval Architects and Marine Engineers.

Beukema, Christian F. Oral history recorded 1980. Northeast Minnesota Historical Center, University of Minnesota-Duluth.

Bills of Sale of Enrolled Vessels, 1891-1901. Lake Superior Maritime Visitors Center, Duluth, Minn.

Guest Log of Stmr. *William A. Irvin,* Lake Superior Maritime Visitors Center, Duluth, Minn.

Marsh & McLennan Inc. "Statement of Claim under Builder's Risk Insurance. Case of the M.V. *Roger Blough.*" 16 October 1972. In the collection of the Lake Superior Maritime Visitors Center, Duluth, Minn.

Pittsburgh Steamship Company. "Agreement between Pittsburgh Steamship Division, United States Steel Corporation and United Steelworkers of America." 16 July 1954. Historical Collections of the Great Lakes, Bowling Green State University.

———. Charter. West Virginia Division of Culture and History, Charleston, W.V.

———. Circular letters to masters and various correspondence for various years. Great Lakes Historical Society, Vermilion, Ohio.

———. Minutes of Third Annual Convention of Engineers and Officers, 26-28 January 1911. Institute for Great Lakes Research, Bowling Green State University.

———. Minutes of the Tenth Annual Convention of Masters and Mates, 23-25 March 1914, and Minutes of the Sixth Annual Convention of Engineers and Officers, 24 March 1914. Institute for Great Lakes Research, Bowling Green State University.

———. Operating Statistics, 1907-1908. Photocopy on file at Lake Superior Maritime Visitors Center.

———. Specifications for Hulls No. 285 & 286 (GLEW) Corrected to 4 March 1936 and 24 January 1941. Lake Superior Maritime Visitors Center, Duluth, Minnesota.

———. Various notes, minutes, statements, and correspondence. Lake Superior Maritime Visitors Center, Duluth, Minnesota.

Pittsburgh Steamship Division. Minutes of Annual Meeting, 1960. Lake Superior Maritime Visitors Center, Duluth, Minnesota.

True, Dwight. "Sixty Years of Shipbuilding." Ed. Harry Benford. Paper presented to Great Lakes Section of the Society of Naval Architects and Marine Engineers, 5 October 1956.

USS Great Lakes Fleet. Logbook of the Stmr. *Philip R. Clarke*, 1975-1976. Author's collection.

———. Vessel Register 1968. Author's collection.

———. Vessel Register Updates, 1983, 1984, Lake Superior Maritime Visitors Center, Duluth, Minnesota.

USS Public Affairs Office. USS Great Lakes Fleet history. Author's collection.

Varian, Howard. "Observations in Development of Machinery in Great Lakes Vessels." Speech by Howard Varian, chief engineer of Great Lakes Engineering Works, presented 26 May 1956, to Michigan Quarterdeck Society. Copy in files of Lake Superior Maritime Visitors Center, Duluth, Minn.

BOOKS AND ARTICLES

Aho, Jody. "The *William A. Irvin*, Beneath the Silver Stack." *Nor'Easter* 20 (May-June 1995): 1-5.

Backert, A. O., ed. *The ABC of Iron and Steel.* Cleveland: Penton Publishing Company, 1925.

Barcus, Frank. *Freshwater Fury.* Detroit: Wayne State University Press, 1960.

Beeson, Harvey C. *Inland Marine Directory.* Chicago, 1901-1909.

Benford, Harry, and William A. Fox, eds. *A Half Century of Maritime Technology, 1943-1993.* Society of Naval Architects and Marine Engineers, 1993.

Beukema, Christian. "The Demonstration: U.S. Steel, Winter 1970-71." *Seaway Review* 2 (Summer 1971): 11-15.

Boyer, Dwight. *Ships and Men of the Great Lakes.* New York: Dodd, Mead, 1977.

———. *Strange Adventures of the Great Lakes.* New York: Dodd, Mead, 1974.

———. *True Tales of the Great Lakes.* New York: Dodd, Mead, 1971.

Brissenden, Paul F. *Employment System of the Lake Carriers' Association.* Washington: U.S. Department of Labor, Bureau of Labor Statistics, January 1918.

Bugbee, Gordon. "In the Triple's Wake . . ." *Telescope* 13 (July 1964): 147-62.

Collard, Andrew Edgar. *Passage to the Sea: The Story of Canada Steamship Lines.* Toronto: Doubleday Canada Limited, 1991.

Cotter, Arundel. *The Authentic History of the United States Steel Corporation.* Moody Magazine & Book Co., 1916.
Davis, E. W. *Pioneering with Taconite.* St. Paul: Minnesota Historical Society, 1964.
Devendorf, John F. *Great Lakes Bulk Carriers, 1869-1985.* Niles, Mich.: John Devendorf, 1996.
Dewar, Gary S. "A Forgotten Class." *Telescope,* April/May 1989, pp. 31-39.
Dowling, The Rev. Edward J. "The Tin Stackers: The Story of the Ships of the Pittsburgh Steamship Company." *Inland Seas* 9 (Summer 1953): 79-85; 9 (Fall 1953): 175-80; 9 (Winter 1953): 271-80; 10 (Spring 1954): 43-50.
"Duluth Shiploader Project on Schedule." *Missabe Iron Ranger* 67 (Spring-Summer 1982): 8-11.
Evans, Henry Oliver. *Iron Pioneer: Henry W. Oliver.* New York: Dutton, 1942.
Fisher, Douglas A. *Steel Serves the Nation, 1901-1951: The Fifty-Year Story of United States Steel.* New York: United States Steel Corporation, 1951.
"Fraser Completes Triple Conversion." *Nor'Easter* 7 (May-June 1982): 1.
Hatcher, Harlan. *A Century of Iron and Men.* Indianapolis: Bobbs-Merrill, 1950.
Havighurst, Walter. *Vein of Iron.* Cleveland: World Publishing, 1958.
Hawke, David Freeman. *John D.: The Founding Father of the Rockefellers.* New York: Harper & Row, 1980.
Hemming, Robert J. *Gales of November.* Chicago: Contemporary Books, 1981.
Hoagland, Henry Elmer. "Wage Bargaining on the Vessels of the Great Lakes." *University of Illinois Studies in the Social Sciences* 6 (September 1917): 7-9, 31-33, 43-49, 55-57, 91.
Holden, Thom. "A Second Maiden Voyage." *Nor'Easter* 6 (May-June 1981): 1.
Horn, Charles L., Jr. "The Iron Ore Industry of Minnesota and the Problem of Depleted Reserves." Master's thesis, Princeton University, 1956.
Joachim, George J. *Iron Fleet: The Great Lakes in World War II.* Detroit: Wayne State University Press, 1994.
King, Franklin A. "Two Harbors: Minnesota's First Iron Ore Port." *Nor'Easter* 9 (Sept.-Oct. 1984): 1-4.
Lake Carriers' Association. *Annual Report of Lake Carriers' Association.* Cleveland: Lake Carriers' Association, 1901-1997.
Lake Carriers' Association. *Whistle Signals.* Cleveland: Lake Carriers' Association, May 1, 1918.
Lake Superior Iron Ore Association. *Lake Superior Iron Ores.* Cleveland: Lake Superior Iron Ore Association, 1938.
Larrowe, Charles P. *Maritime Labor Relations on the Great Lakes.* East Lansing: Michigan State University, Labor and Industrial Relations Center, 1959.
Law, W. H. *Heroes of the Great Lakes.* Detroit: Pohl, 1905.
List of Merchant Vessels of the United States—1900. Washington, D.C.: Government Printing Office, 1900.
McDougall, Alexander. *The Autobiography of Captain Alexander McDougall.* Cleveland: Reprinted by the Great Lakes Historical Society, 1968.

Marine Engineers Benevolent Association–American Maritime Officers District 2. *Coulby to Potts to Beukema: U.S. Steel Pulls the Strings at Lake Carriers' Association.* N.p.: Marine Engineers Benevolent Association–American Maritime Officers District 2, n.d.

Meakin, Alexander C. "Four Long and One Short: A History of the Great Lakes Towing Company." *Inland Seas* 31 (1975): 289.

———. *Master of the Inland Seas.* Vermilion, Ohio: Great Lakes Historical Society, 1988.

———. *The Story of the Great Lakes Towing Company.* Vermilion, Ohio: Great Lakes Historical Society, 1984.

Micketti, Gerald F. *The Bradley Boats.* Traverse City, Mich.: Gerald F. Micketti, 1995.

Miller, Al. "The Record Load." *Nor'Easter* 18 (July–August 1993): 1-3.

———. "A Tugman's Story: Life and Times of B. B. Inman." *American Neptune* 56 (Summer 1996): 191-213.

Report to the Lake Carriers' Association on Great Lakes Radar Operational Research Project. Washington: Jansky & Bailey, 1947.

Society of Naval Architects and Marine Engineers. *Transactions, Vol. 58, 1950.* Easton, Penn.: Mack Printing Co., 1950.

Sykes, Alan. "Clifford F. Hood in Retrospect." *Nor'Easter* 6 (Jan.–Feb. 1981): 1-2.

"Three Ships Converted to Self-Unloaders." *Missabe Iron Ranger* 67 (Spring–Summer 1982): 12-14.

Van Brunt, Walter. *Duluth and St. Louis County, Minnesota: Their Story and People.* Chicago: American Historical Society, 1921.

Wall, Joseph Frazier. *Andrew Carnegie.* New York: Oxford University Press, 1970.

Wallace, William L. "Up the Grades on Ore Boats: From Coal Passer to Wheelsman to Third Mate." *Inland Seas* 53 (Fall 1997): 178-91.

Winkler, John K. *Incredible Carnegie: The Life of Andrew Carnegie 1835-1919.* New York: Vanguard Press, 1931.

———. *Morgan the Magnificent: The Life of J. Pierpont Morgan, 1837-1913.* Garden City, N.Y.: Garden City Publishing Company, 1930.

Wolff, Julius F., Jr. *Lake Superior Shipwrecks.* Ed. Thom Holden. Duluth, Minn.: Lake Superior Port Cities Inc., 1990.

Wright, Richard J., *Freshwater Whales.* Kent, Ohio: Kent State University Press, 1969.

INDEX

105, 62, 263, 315, 322
107, 62, 263, 315, 322
109, 62, 263, 316, 322
110, 62, 263, 316, 322
111, 263, 316, 322
116, 62, 263, 316, 322
117, 62, 263, 317, 322
118, 263, 317, 322
126, 62, 263, 317, 322
127, 263, 317-18, 322
129, 46, 263, 318, 322
130, 263, 318, 322
131, 263, 318, 322
132, 263, 318-19, 322
133, 263, 319, 322
134, 263, 319, 322
137, 264, 319, 322
201, 62, 263, 319-20, 322
202, 62, 263, 320, 322

AAA class vessels: construction, 165-67; lengthening, 224; self-unloader, conversion to, 245-47
Abram, Charles, 65
Affleck, B. F., 249, 261, 265, 267
American Sheet Steel Co., 28
American Steamship Co., 21, 29, 40
American Steel and Wire Co., 17, 28, 137
American Steel Barge Co., 19, 23, 29, 34
American Steel Hoop Co., 28
American Tin Plate Co., 28

Anderson, Arthur M.: construction, 164-67; Edmund Fitzgerald, 226-29; extended season, 221, 224; Labrador ore, 199-200; lengthening, hull, 224; lengthening, unloading boom, 256; self-unloader, conversion to, 245-47; 262, 267
Arnold, Lon, 97
Ashby, Capt. George, 216, 221
Avery, Sewell, 153, 261, 265, 268

Baganz, Capt. Ed, 142
Bailey, Capt. Fred, 72-73, 177-78
Baker, George F., 85-86, 261, 264, 268
Balfour, Capt. George, 68-69
Banker, Capt. George, 122-24
Barrow, Wendell, 157-58
Bartlett, E. B., 62, 263, 268, 322
Bell, Sir Isaac Lothian, 264, 268-69, 322
Benson, Fred, 72
Bertz, Ralph, 217, 234, 237, 250
Bessemer, Sir Henry, 264, 269, 322
Bessemer Steamship Co., 20-21, 23, 29, 40
Bessemer steel-making process, 18
Beukema, Christian F., 207, 219, 244
Black, Clarence A., 23, 150
Blackstone Capital Partners, 251
Blough, Roger, 213-15, 217-19, 262, 269-70

Bourlier, William, 139
Bradley, Carl D., (I), 115, 262, 265, 273-74
Bradley, Carl D., (II), 209, 262, 265, 270
Bradley Transportation Co., 115, 168, 206-7
Brinker, Capt. C. J., 148
Briton, 263, 270, 321
Brown, Walter, 79
Bryn Mawr, 24, 264, 270, 321
Buffington, Eugene, 17
Buffington, Eugene J., 86, 148-52, 240, 261, 265, 270-71
Buhrmann, William, 234, 244-45
Bunsen, Robert W. E., 264, 271, 322
Bush, Walter, 78
Byrne, Charles, 78

Calcite, 115, 262, 265, 271
Calcite II, 209, 261, 271
Callaway, Cason J.: construction, 164-67; extended season, 221, 224; lengthening, hull, 224; lengthening, unloading boom, 256; self-unloader, conversion to, 245-47, 262, 272
Cambria, 17, 263, 272, 321
Cameron, Capt. M. C., 45
Campau, Capt. W. H., 42
Carnegie, Andrew: and Carnegie Steel, 17-18; and Henry Oliver, 21-23; forms Pittsburgh Steamship Co., 23-24; sells mills, 26
"Carnegie fleet," 116
Carnegie Steel Co., 17-18
Carnegie-Illinois Steel Co., 137
Carrington, 23, 263, 272, 321
Cedarville, 209-10, 262, 265, 272-73
Chambers, Capt. G. C., 148, 151
Chicago Ship Building Co., 59, 85
Clark, Morgan, 227
Clarke, Philip R.: construction of, 164-67; extended season, 221-22; lengthening, hull, 224; lengthening, unloading boom, 256; and new Poe Lock, 216; self-unloader, conversion to, 245-47; 262, 273
Clemson, D. M., 108, 181, 240, 261, 265, 273
Clemson, Daniel M., 39, 49
Cleveland, Ohio, 137
Cleveland Vessel Owners' Association, 55
Clyde, William G., 209, 261, 271
Clymer, Irvin L., 252-53, 262, 265, 273
Colby, Joseph L., 62, 263, 274, 322
Cole, Thomas F., 86, 141, 146, 240, 261, 265, 274
Colgate, James B., 263, 274, 322
Collins, E. C., 104, 261, 264, 269
Comins, Capt. G. T., 121-22
Conneaut, Ohio, 23, 137-38
Cooper, Capt. Jesse B., 141, 226-29
Coralia, 17, 72, 264, 275, 321
Corey, William E.: launching of, 61-62; 72-73, 83; in *Steinbrenner* rescue, 180-82, 261, 264, 275
Corliss, George H., 99, 264, 275, 322
Cornell, 24, 98-99, 264, 275, 321
Corona, 17, 263, 276, 321
Corsica, 17, 263, 276, 321
Cort, Henry, 97-98, 264, 276-77, 322
Coulby, Harry: annual meetings, 92-97; and labor unions, 54-59, 87-92; Pittsburgh Steamship Co., 53-54; retires, 119; ships, construction, 59-62; storm, 1905, 80-83
Crawford, Capt. R. M., 156
Crawford, George G., 86, 100, 261, 265, 277
Crescent City, 34, 44, 64-66, 264, 277, 321
Cummings, Capt. C. H., 73-74

Daggett, Arthur, 65
Dana, Capt. Arthur, 98, 147

Davenport, Capt. Frank, 160
Dickson, William B., 86, 261, 265, 277
Dinkey, Alva C., 86, 98, 240, 261, 265, 278
Duluth, 42, 219
Duluth and Iron Range Railroad, 25
Duluth, Missabe & Iron Range Railway, 136, 247, 251

Eads, James B, 264, 278, 322
Edenborn, William, 36, 70-71, 264, 278, 321
Edna G., 71, 81
Ellwood, Isaac, 17
Ellwood, Isaac L., 36, 73-74, 264, 279, 321
Emigh, Herbert, 78-79
Empire City, 34, 264, 279, 321
Empire Transportation Co., 33
Ericsson, John, 264, 279, 322
Everett, Capt. A. M., 180
Extended season, 220-24, 229-33

Fairbairn, Sir William, 264, 279-80, 322
Fairless, Benjamin, 139
Fairless, Benjamin F.: extended season, 221; fuel oil, converted to, 217; sold, 249; straight-decker, last, 248; Super Dupers, 152, 261, 265, 280
Fairport, Ohio, 137
Farrell, James A., 86, 177, 240, 261, 265, 280
Farrell, James A., 126
Federal Steel Co., 24-26, 28
Ferbert, A. H. (I), 104, 261, 264, 309
Ferbert, A. H. (II), 152, 160, 265, 280
Ferbert, A. H., 139, 146, 154, 161
Field, Marshall, 16
Filbert, William J., 261, 265, 280-81
Fitzgerald, Edmund, 226-29
Fraser, Leon, 152, 200, 249, 261, 265, 281
Fraser Shipyards, 224, 246

Frick, Henry C., 261, 264, 281
Fritz, John A., 99, 264, 281-82, 322
Frontier, 171, 173, 261, 264, 282
Fulton, Robert, 264, 282, 322

Gary, Elbert H., 59-60, 261, 264, 282
Gary, Elbert H., 17, 24-26, 28-29
Gary, Indiana, 137
Gary class vessels, 59-62
Gates, John W., 36, 264, 282-83, 321
Gates, John W., 17
Gayley, James, 38-40
German, 321, 263, 283
Gilbert, W. H., 33, 36, 263, 283, 322
Gogebic Range, 22, 206
Gott, Edwin H., 235-40, 245, 256-57, 262, 283
Governor Miller, 139-40, 177, 240, 261, 265, 284
Great Depression, 125-29
Grecian, 97, 263, 284, 321
Griffin, 23, 263, 284, 321
Gulder, Capt. Guido, 128, 133, 145, 158

Hammett, Loran, 207
Hamonic, 122-24
Harbottle, Capt. Thomas F., 195-96
Harvard, 24, 264, 284-85, 321
Harvey, A. F., 141, 207, 261, 264, 311, 322
Harvey, A. F., 38, 54, 120-22, 139, 145
Hatfield, Joshua A., 118, 249, 261, 265, 285
Hayes, Joseph, 38
Hemingway, Walter C., 161, 168, 182
Higgs, Seymour, 89
Hill, James J., 36, 264, 285, 322
Holdridge, Capt. George, 98
Holley, Alexander, 99, 264, 285, 322
Hood, Clifford F., 202, 262, 265, 285-86
Hormig, William, 70
Houghton, Douglass, 264, 286, 322
House, Francis E., 86, 121, 261, 265, 286
Hoyt, Colgate, 62, 263, 286-87, 322

Hoyt, James H., 33, 43
Hulett unloading machine, 137-38
Hulst, John, 139, 249, 261, 265, 287
Humble, Capt. Richard F., 74-82
Hunter, Silas, 70

Iler, Capt. Walter, 100
Illinois Steel Co., 17, 24-25
Inman, Byron B., 34, 44
International Seamen's Union, 129. See also Lake Seamen's Union
Iron ore: loading and unloading, 135-38; reserves, 163-64; transportation of, 36
Irvin, William A., 139, 177-79, 249, 261, 265, 287
Irvin, William A., 139
Isthmian Steamship Lines, 116

Jenney, W. LeBaron, 264, 287, 322
Johnson, Horace, 125, 249, 261, 265, 288
Johnson, James, 71
Joliet, 23, 97, 263, 288, 321

Kerr, D. G.: 104; record load, 111-15; Steinbrenner rescue, 181, 240, 261, 265, 288
Khoury, Charles R., 191, 193, 207, 219
Krupp, Alfred, 264, 286, 322

Labor. See Unions, labor
Lafayette, 24, 66-70, 263, 289, 321
Lake Carriers' Association: anti-union efforts, 57, 87-92; extended season, 220; and Great Depression, 128
Lake Seamen's Union, 48, 111. See also International Seamen's Union
Lake Superior Iron Co., 23
Lamont, Thomas W., 126, 202-3, 261, 265, 289
LaSalle, 23, 263, 289, 321
Lawrence, Charles, 99
Lehne, Capt. G. A., 181

Liberty, Bruce, 204-6
Lindabury, Richard V., 118, 240, 261, 265, 289-90
Linn, William R., 23, 264, 290, 321
Litton Great Lakes Corp., 233-34, 258
Lorain, Ohio, 137
Lucas, Robert H., 194,
Lynch, Thomas, 85, 261, 264, 290

MacBeth, Capt. M. H., 147
Madeira, 70-72, 263, 290, 321
Magna, 264, 291, 321
Maia, 72, 264, 291, 321
Maida, 99, 263, 291, 321
Male, Capt. E. K., 142
Malietoa, 36, 142, 264, 291-92, 321
Malta, 263, 292, 321
Manda, 263, 292, 321
Manila, 67-69, 99, 264, 292, 321
Manola, 263, 293, 321
Marcia, 99, 263, 321
Maricopa, 121, 293, 321
Marina, 263, 293-94, 321
Marine Engineers Beneficial Association, 58, 88, 91-92, 184-89. See also Unions, labor
Mariposa, 72, 264, 294, 321
Mariska, 263, 294, 321
Maritana, 264, 294, 321
Maritime class vessels, 153
Marsala, 264, 294-95, 321
Martha, 99, 264, 295, 321
Maruba, 263, 295, 321
Masaba, 263, 295, 321
Masters, Mates and Pilots Union, 185, 188. See also Unions, labor
Mataafa, 36, 74-82, 264, 296, 321
Mather, Samuel, 263, 296, 322
Mather, Samuel, 20
Matoa, 100-101, 263, 296, 321
Maunaloa, 46, 264, 296-97, 321
McDonald, J. J., 148, 152
McDougall, Alexander, 98, 264, 297, 322
McElroy, Capt. William P., 111-15
McGonagle, William A., 104-7, 147-48, 240, 261, 265, 297

INDEX === 343

McLean, John H. See *Hood, Clifford F.*
Meister, Capt. William, 152
Menominee Range, 16
Menominee Transit Co., 16, 29, 40
Merchant Marine Act of 1970, 216
Mesabi Range, 19, 163-64
Michigan Limestone and Chemical Co., 115-16, 207
Mills, Edwin S., 23, 39
Minnesota Iron Co., 25
Minnesota Steamship Co., 25, 29, 321
Minntac, 215-16
Mitchell, Pentecost, 104, 261, 264, 297
Morgan, J. Pierpont, 85, 261, 264, 298
Morgan, J. P., Jr., 86, 240, 261, 265, 298
Morgan, J. Pierpont, 25, 27, 29
Morgan class vessels, 85-87
Morse, Samuel F. B., 264, 298, 322
Munson, John G.: construction, 165; extended season, 225; lengthened, 224, 262, 298-99
Murphy, Simon J., 264, 299
Mutual Steamship Co., 17, 29, 40

National Maritime Union, 154-56, 169
National Steel Co., 28
National Tube Co., 25, 28, 137
Navigation, aids to, 159-61, 257
Naysmith, James, 74-75, 264, 299, 322
Neilson, James B., 264, 299, 322
Noble, Capt. John, 98

Ojard, Adolph, 257-58
Ojibway, 173-75, 261, 300
Olcott, William J., 86, 261, 265, 300
Olds, Irving S., 152; to burn fuel oil, 217, 221, 249, 261, 265, 300
Oliver, Henry W., 22-24
Olsen, Capt. G. J., 199

Palagyi, Capt. F. L., 176
Palmer, William P., (I) 36, 263, 300-301, 322
Palmer, William P., (II) 240, 261, 265, 301

Panic of 1893, 19-20
Pargny, Eugene W., 108; repowering, 167-68, 249, 261, 265, 301
Parilla, Joseph, 207, 219
Parke, Capt. J. F., 121
Parsons, Capt. Robert, 142
Pennsylvania and Lake Erie Dock Co., 137
Penzenhagen, C. A., 148, 151
Perkins, George W., 59, 141, 261, 264, 301
Phipps, Henry, 86, 97, 261, 265, 302
Pickands Mather and Co., 52-53
Pittsburgh and Conneaut Dock Co., 137
Pittsburgh Steamship Co.: annual meetings, 92-97; formation of, 23; and Great Depression, 125-29, 140; iron ore, loading, 135-37; iron ore, unloading, 137-38; and labor unions, 42, 57-59, 87-92, 154-55; original fleet, size of, 29; renamed, 168; signals, whistle, 109; U.S. Steel, acquired by, 28-29, 168; vessel operations, 135-38. See also Pittsburgh Steamship Division; USS Great Lakes Fleet, Inc.
Pittsburgh Steamship Division: and cost controls, 193-96; reorganized and renamed, 200; and labor unions, 168-70; scheduling vessels for, 170-71. See also Pittsburgh Steamship Co.; USS Great Lakes Fleet, Inc.
Pittsburgh Fleet: combined with Bradley fleet and renamed, 207
Platt, William, 68
Poe, General Orlando M., 264, 302, 322
Poe class vessels, 233-40
Pollution, 217
Porter, H. H., 16
Potts, Donald C., 147, 160, 161, 185-88, 191
Powell, L. W., 54

Pratt, Capt. Forrest, 141-42
Presque Isle, 233-34, 254, 258, 262, 302
Princeton, 24, 264, 302-3, 321

Quebec-Cartier Mining Co., 194, 199-200
Queen City, 34, 264, 303, 322

Radar, 160-61
Radiotelephone, 159-60
Ransom, William, 219
Ream, Norman B., 85, 261, 264, 303
Rensselaer, 24, 264, 303, 321
Rice, Capt. Frank, 64-66
Richardson, R. R., 104, 261, 264, 303-4
Roberts, Percival, Jr., 86, 261, 265, 304
Robinson, T. W., 217, 262, 265, 304,
Rockefeller, Frank, 264, 304-5, 322
Rockefeller, John D., 19-23, 29
Roebling, John A., 99, 264, 305, 322
Rogers, Henry H., 85, 261, 265, 305
Rogers City, 265, 305
Rogers City, Mich., 207
Rolfson, J. N., 139
Rolfson, Joseph Neil, Jr., 218
Roman, 263, 305-6, 321
Rosati, Robert S., 257
Russell, John Scott, 62, 263, 306, 322

Sandy, 261, 264, 306
Saunders, Fred, 78
Saxon, 263, 306, 321
Schelb, William, 72
Schiller, William B., 86, 118-19, 147-48, 184, 261, 265, 306-7
Schlesinger, Ferdinand E., 16
Schreiber, Rudolf, 89
Schwab, Charles M., 18-19, 27-28,
Seeley, Capt. F. D., 83, 98
Shaw, Howard L., 264, 307
Ships: AAA class, 165-67, 224, 245-47; bow thrusters, 203; construction incentives, 153, 216;

environmental laws, 217; Gary class, 59-62; Maritime class, 153; Morgan class, 85-87; Poe class, 233-40; repowering, 167-68, 202-3, 249-50; self-unloaders, 115-16, 245-47; "Super Dupers," 152
Shiras, MacGilvray, 104, 147, 261, 264, 307
Siegrist, Edmund, 156-57
Siemens, Sir William, 264, 307, 322
Simonds, Capt. William, 238
Sloan, George A., 153, 210; repowered, 249, 256, 261, 308
Small, Capt. Thomas, 216
Smeaton, John, 99, 264, 308, 322
Smith, W. W., 38
Soo Warehouse, 171-75
South Chicago, Ill., 137
Speer, Edgar B., 237-40, 245, 262, 308
Stalker, R. Neil, 253, 257
Stanbrook, R. C., 160
Stanley, Robert C., 153, 261, 265, 309
Steel Chemist, 116
Steel Electrician, 116
Steelmotor, 116
Steelvendor, 116
Stephenson, George, 264, 309, 322
Stevens, Hattie, 66
Storms: Nov. 27-29, 1905, 63-83; Nov. 7-10, 1913, 97-101; Nov. 11, 1940, 140-42; Nov. 10-11, 1975, 226-29
Strikes. *See* Unions, labor
Strom, Herman C., 264, 309
Stuller, Frank, 176
"Super Dupers," 152
Superior, 171, 309-10
Superior City, 34, 110-11, 263, 310, 322
Superior Ship Building Co., 59

Taconite: commercial production, 182-84; flux pellets, 254; Minntac, 215-16

Talbot, Capt. A. J., 70
Taylor, Myron C., 125, 207-9, 261, 310
Thayer, Nathanial, 16
Thomas, Eugene P., 126, 202, 249, 261, 265, 310
Thomas, Sydney G., 99, 264, 311, 322
Thompson, A. D., 62, 263, 311, 322
Thomson, Thomas, 66
"Tin Stackers," 62
Transtar Inc., 251
Trevor, John B., 97, 263, 311, 322
Trimble, Richard, 86, 99, 261, 265, 311-21
Turbine engines, 139-40
Two Harbors, Minn., 25, 136, 200, 206

USS Great Lakes Fleet, Inc.: designated common carrier, 244; marketing, 255-56; sold, 250-51; renamed, 243
Unions, labor: American Association of Masters and Pilots, 57; International Seamen's Union, 129; and the Lake Carriers' Association, 56-58, 87-92; Lake Seamen's Union, 48, 111; Marine Cooks and Stewards Union, 57; Marine Engineers Beneficial Association, 57; Marine Firemen, Oilers and Water Tenders Union, 57; Masters, Mates and Pilots Union, 185, 188; National Maritime Union, 154-56, 169; United Steelworkers of America Local 5000, 168-70
United States Steel Corp.: formation, 25-29; buys Michigan Limestone Co., 115-16; and extended season, 220-24, 229-33; downsizing, 243-44
United States Steel Corporation Great Lakes Fleet, 207, 243
U.S. Steel Export Co., 116

United Steelworkers of America, 168-70. *See also* Unions, labor

Van Hise, Charles R., 42, 263, 312, 322
Vermilion Range, 16
Voorhees, Enders M., 152, 221, 261, 265, 312

Wade, Patrick, 69
Water Quality Improvement Act, 217
Watson, Ralph H., 139, 261, 265, 312
Watt, James, 264, 312-13, 322
Wawatam, 23, 263, 313, 321
West Bay City Ship Building Co., 59
Whalebacks, 19, 41, 62
White, W. F., 115, 262, 265, 313
Whitworth, Sir Joseph, 62, 263, 313, 322
Widener, Peter A. B., 85, 141, 240, 261, 265, 314
Williams, Homer D., 108, 167-68, 249, 261, 265, 314
Wilson, Capt. Bill, 160-61, 197-99, 238, 240
Wilson, John, 238
Wilson, Thomas, 45-46, 263, 314, 322
Winter navigation. *See* Extended season
Wissbeck, Capt. Herold, 203
Wolvin, A. B., 36, 263, 314, 322
Wolvin, Augustus B., 21, 31-41, 46-49
Wolvin, John W., 37
Wolvin, Roy M., 37
Woodgate, Thomas, 78
World War I, 103, 109-10
World War II, 143-47, 153-58
Wright, Capt. Dell, 66-69
Wright, Henry, 78
Zenith City, 34, 264, 315, 322
Zenith Transit Co., 21, 34
Ziesing, August, 108, 216, 249, 261, 265, 315

Titles in the Great Lakes Books Series

Freshwater Fury: Yarns and Reminiscences of the Greatest Storm in Inland Navigation, by Frank Barcus, 1986 (reprint)

Call It North Country: The Story of Upper Michigan, by John Bartlow Martin, 1986 (reprint)

The Land of the Crooked Tree, by U. P. Hedrick, 1986 (reprint)

Michigan Place Names, by Walter Romig, 1986 (reprint)

Luke Karamazov, by Conrad Hilberry, 1987

The Late, Great Lakes: An Environmental History, by William Ashworth, 1987 (reprint)

Great Pages of Michigan History from the Detroit Free Press, 1987

Waiting for the Morning Train: An American Boyhood, by Bruce Catton, 1987 (reprint)

Michigan Voices: Our State's History in the Words of the People Who Lived It, compiled and edited by Joe Grimm, 1987

Danny and the Boys, Being Some Legends of Hungry Hollow, by Robert Traver, 1987 (reprint)

Hanging On, or How to Get through a Depression and Enjoy Life, by Edmund G. Love, 1987 (reprint)

The Situation in Flushing, by Edmund G. Love, 1987 (reprint)

A Small Bequest, by Edmund G. Love, 1987 (reprint)

The Saginaw Paul Bunyan, by James Stevens, 1987 (reprint)

The Ambassador Bridge: A Monument to Progress, by Philip P. Mason, 1988

Let the Drum Beat: A History of the Detroit Light Guard, by Stanley D. Solvick, 1988

An Afternoon in Waterloo Park, by Gerald Dumas, 1988 (reprint)

Contemporary Michigan Poetry: Poems from the Third Coast, edited by Michael Delp, Conrad Hilberry and Herbert Scott, 1988

Over the Graves of Horses, by Michael Delp, 1988

Wolf in Sheep's Clothing: The Search for a Child Killer, by Tommy McIntyre, 1988

Copper-Toed Boots, by Marguerite de Angeli, 1989 (reprint)

Detroit Images: Photographs of the Renaissance City, edited by John J. Bukowczyk and Douglas Aikenhead, with Peter Slavcheff, 1989

Hangdog Reef: Poems Sailing the Great Lakes, by Stephen Tudor, 1989

Detroit: City of Race and Class Violence, revised edition, by B. J. Widick, 1989

Deep Woods Frontier: A History of Logging in Northern Michigan, by Theodore J. Karamanski, 1989

Orvie, The Dictator of Dearborn, by David L. Good, 1989

Seasons of Grace: A History of the Catholic Archdiocese of Detroit, by Leslie Woodcock Tentler, 1990

The Pottery of John Foster: Form and Meaning, by Gordon and Elizabeth Orear, 1990

The Diary of Bishop Frederic Baraga: First Bishop of Marquette, Michigan, edited by Regis M. Walling and Rev. N. Daniel Rupp, 1990

Walnut Pickles and Watermelon Cake: A Century of Michigan Cooking, by Larry B. Massie and Priscilla Massie, 1990

The Making of Michigan, 1820-1860: A Pioneer Anthology, edited by Justin L. Kestenbaum, 1990

America's Favorite Homes: A Guide to Popular Early Twentieth-Century Homes, by Robert Schweitzer and Michael W. R. Davis, 1990

Beyond the Model T: The Other Ventures of Henry Ford, by Ford R. Bryan, 1990

Life after the Line, by Josie Kearns, 1990

Michigan Lumbertowns: Lumbermen and Laborers in Saginaw, Bay City, and Muskegon, 1870-1905, by Jeremy W. Kilar, 1990

Detroit Kids Catalog: The Hometown Tourist by Ellyce Field, 1990

Waiting for the News, by Leo Litwak, 1990 (reprint)

Detroit Perspectives, edited by Wilma Wood Henrickson, 1991

Life on the Great Lakes: A Wheelsman's Story, by Fred W. Dutton, edited by William Donohue Ellis, 1991

Copper Country Journal: The Diary of Schoolmaster Henry Hobart, 1863-1864, by Henry Hobart, edited by Philip P. Mason, 1991

John Jacob Astor: Business and Finance in the Early Republic, by John Denis Haeger, 1991

Survival and Regeneration: Detroit's American Indian Community, by Edmund J. Danziger, Jr., 1991

Steamboats and Sailors of the Great Lakes, by Mark L. Thompson, 1991

Cobb Would Have Caught It: The Golden Age of Baseball in Detroit, by Richard Bak, 1991

Michigan in Literature, by Clarence Andrews, 1992

Under the Influence of Water: Poems, Essays, and Stories, by Michael Delp, 1992

The Country Kitchen, by Della T. Lutes, 1992 (reprint)

The Making of a Mining District: Keweenaw Native Copper 1500-1870, by David J. Krause, 1992

Kids Catalog of Michigan Adventures, by Ellyce Field, 1993

Henry's Lieutenants, by Ford R. Bryan, 1993

Historic Highway Bridges of Michigan, by Charles K. Hyde, 1993

Lake Erie and Lake St. Clair Handbook, by Stanley J. Bolsenga and Charles E. Herndendorf, 1993

Queen of the Lakes, by Mark Thompson, 1994

Iron Fleet: The Great Lakes in World War II, by George J. Joachim, 1994

Turkey Stearnes and the Detroit Stars: The Negro Leagues in Detroit, 1919-1933, by Richard Bak, 1994

Pontiac and the Indian Uprising, by Howard H. Peckham, 1994 (reprint)

Charting the Inland Seas: A History of the U.S. Lake Survey, by Arthur M. Woodford, 1994 (reprint)

Ojibwa Narratives of Charles and Charlotte Kawbawgam and Jacques LePique, 1893-1895. Recorded with Notes by Homer H. Kidder, edited by Arthur P. Bourgeois, 1994, co-published with the Marquette County Historical Society

Strangers and Sojourners: A History of Michigan's Keweenaw Peninsula, by Arthur W. Thurner, 1994

Win Some, Lose Some: G. Mennen Williams and the New Democrats, by Helen Washburn Berthelot, 1995

Sarkis, by Gordon and Elizabeth Orear, 1995

The Northern Lights: Lighthouses of the Upper Great Lakes, by Charles K. Hyde, 1995 (reprint)

Kids Catalog of Michigan Adventures, second edition, by Ellyce Field, 1995

Rumrunning and the Roaring Twenties: Prohibition on the Michigan-Ontario Waterway, by Philip P. Mason, 1995

In the Wilderness with the Red Indians, by E. R. Baierlein, translated by Anita Z. Boldt, edited by Harold W. Moll, 1996

Elmwood Endures: History of a Detroit Cemetery, by Michael Franck, 1996

Master of Precision: Henry M. Leland, by Mrs. Wilfred C. Leland with Minnie Dubbs Millbrook, 1996 (reprint)

Haul-Out: New and Selected Poems, by Stephen Tudor, 1996

Kids Catalog of Michigan Adventures, third edition, by Ellyce Field, 1997

Beyond the Model T: The Other Ventures of Henry Ford, revised edition, by Ford R. Bryan, 1997

Young Henry Ford: A Picture History of the First Forty Years, by Sidney Olson, 1997 (reprint)

The Coast of Nowhere: Meditations on Rivers, Lakes and Streams, by Michael Delp, 1997

From Saginaw Valley to Tin Pan Alley: Saginaw's Contribution to American Popular Music, 1890-1955, by R. Grant Smith, 1998

The Long Winter Ends, by Newton G. Thomas, 1998 (reprint)

Bridging the River of Hatred: The Pioneering Efforts of Detroit Police Commissioner George Edwards, 1962-1963, by Mary M. Stolberg, 1998

Toast of the Town: The Life and Times of Sunnie Wilson, by Sunnie Wilson with John Cohassey, 1998

These Men Have Seen Hard Service: The First Michigan Sharpshooters in the Civil War, by Raymond J. Herek, 1998

A Place for Summer: One Hundred Years at Michigan and Trumbull, by Richard Bak, 1998

Early Midwestern Travel Narratives: An Annotated Bibliography, 1634-1850, by Robert R. Hubach, 1998 (reprint)

All-American Anarchist: Joseph A. Labadie and the Labor Movement, by Carlotta R. Anderson, 1998

Michigan in the Novel, 1816-1996: An Annotated Bibliography, by Robert Beasecker, 1998

"Time by Moments Steals Away": The 1848 Journal of Ruth Douglass, by Robert L. Root, Jr., 1998

The Detroit Tigers: A Pictorial Celebration of the Greatest Players and Moments in Tigers' History, updated edition, by William M. Anderson, 1999

Letter from Washington, 1863-1865, by Lois Bryan Adams, edited and with an introduction by Evelyn Leasher, 1999

Father Abraham's Children: Michigan Episodes in the Civil War, by Frank B. Woodford, 1999 (reprint)

A Sailor's Logbook: A Season aboard Great Lakes Freighters, by Mark L. Thompson, 1999

Huron: The Seasons of a Great Lake, by Napier Shelton, 1999

Wonderful Power: The Story of Ancient Copper Working in the Lake Superior Basin, by Susan R. Martin, 1999

Tin Stackers: The History of the Pittsburgh Steamship Company, by Al Miller, 1999

www.ingramcontent.com/pod-product-compliance
Lightning Source LLC
Chambersburg PA
CBHW070402100426
42812CB00005B/1610